Basics in Dermatological Laser Applications

Current Problems in Dermatology

Vol. 42

Series Editors

Peter Itin Basel
Gregor Jemec Roskilde

Basics in Dermatological Laser Applications

Volume Editors

Inja Bogdan Allemann Zurich
David J. Goldberg New York, N.Y.

56 figures, 53 in color, and 19 tables, 2011

KARGER

Basel · Freiburg · Paris · London · New York · New Delhi · Bangkok · Beijing · Tokyo · Kuala Lumpur · Singapore · Sydney

Current Problems in Dermatology

Inja Bogdan Allemann, MD
Klinik für Plastische Chirurgie und Handchirurgie
Universitätsspital Zürich
Rämistrasse 100
CH-8091 Zürich
Switzerland

David J. Goldberg, MD
Skin Laser & Surgery Specialists of New York & New Jersey
115 E. 57th Street, Suite 710
New York, NY 10022
USA

Library of Congress Cataloging-in-Publication Data

Basics in dermatological laser applications / volume editors, Inja Bogdan Allemann, David J. Goldberg.
p. ; cm. -- (Current problems in dermatology, ISSN 1421-5721 ; vol. 42)
Includes bibliographical references and indexes.
ISBN 978-3-8055-9788-3 (hard cover : alk. paper) -- ISBN 978-3-8055-9789-0 (electronic version)
1. Skin--Laser surgery. 2. Skin--Diseases--Treatment. I. Allemann, Inja Bogdan. II. Goldberg, David J., M.D. III. Series: Current problems in dermatology ; v. 42. 1421-5721
[DNLM: 1. Skin Diseases--therapy. 2. Cosmetic Techniques. 3. Laser Therapy--methods. W1 CU804L v.42 2011 / WR 650]
RL120.L37B37 2011
617.4'770598--dc23

2011023285

Bibliographic Indices. This publication is listed in bibliographic services, including MEDLINE/Pubmed.

Disclaimer. The statements, opinions and data contained in this publication are solely those of the individual authors and contributors and not of the publisher and the editor(s). The appearance of advertisements in the book is not a warranty, endorsement, or approval of the products or services advertised or of their effectiveness, quality or safety. The publisher and the editor(s) disclaim responsibility for any injury to persons or property resulting from any ideas, methods, instructions or products referred to in the content or advertisements.

Drug Dosage. The authors and the publisher have exerted every effort to ensure that drug selection and dosage set forth in this text are in accord with current recommendations and practice at the time of publication. However, in view of ongoing research, changes in government regulations, and the constant flow of information relating to drug therapy and drug reactions, the reader is urged to check the package insert for each drug for any change in indications and dosage and for added warnings and precautions. This is particularly important when the recommended agent is a new and/or infrequently employed drug.

www.karger.com
Printed in Switzerland on acid-free and non-aging paper (ISO 9706) by Reinhardt Druck, Basel
ISSN 1421–5721
ISBN 978–3–8055–9788–3
e–ISBN 978–3–8055–9789–0

Contents

Indications for Laser Applications

Non-Laser Technologies/Light-Based Systems

Preface

In the last two decades, there has been a virtual explosion in the use of lasers in medicine. This is most obvious in the field of cosmetic dermatology, where both the uses and indications for laser treatment are continually expanding. In fact, many of the clinical conditions presented today are solely treated by lasers.

When discussing the term 'lasers', one must be aware that we are talking about many different types of lasers and other similar energy-based devices.

This relatively recent and dramatic increase in the use of lasers for skin-related conditions has helped shape a small group of experts who have greatly influenced almost the entire sphere of technological development seen in this area. However, younger physicians at the beginning of their careers who look upon this vast field often find themselves facing an extremely complex physics-based area of medicine with a veritable jungle of different devices on offer.

Although there are already many excellent books and reviews on laser surgery, many of these assume the reader has certain pre-existing knowledge and experience. In contrast, this book is aimed at the novice laser physician. It is intended to provide a structured and comprehensive overview of the physical knowledge relevant to understanding laser medicine and surgery, without delving into too much unnecessary detail. Moreover, the various clinical indications and treatments performed with lasers are clearly laid out and discussed. Because the field of laser surgery is changing rapidly, we have purposely tried to be generic when talking about the various lasers. We did however focus on specific brand name devices that are currently on the market. This approach will hopefully increase the longevity of this book in what is after all a rapidly evolving field.

We, the editors, have intentionally combined the talents of a younger physician and accomplished laser specialist (Inja Bodgan Allemann) with the expertise of one of the most internationally well-known cutaneous laser physicians (David J. Goldberg) in order to bring forth this unique book *Basics in Dermatological Laser Applications*. Many different authors have kindly agreed to contribute to this book. All of them are experts in their field and have provided concise and topical chapters, and we would like to take this opportunity to thank each of them for their time and valuable contributions.

Inja Bogdan Allemann, Zurich
David J. Goldberg, New York, N.Y.

Bogdan Allemann I, Goldberg DJ (eds): Basics in Dermatological Laser Applications.
Curr Probl Dermatol. Basel, Karger, 2011, vol 42, pp 1–6

History of Lasers in Dermatology

Michael L. Geiges

Institute and Museum of Medical History, University of Zurich and Department of Dermatology, University Hospital Zurich, Zurich, Switzerland

Abstract

In the 1950s, based on the theory of stimulating radiant energy published by Albert Einstein in 1916, the collaboration of physicists and electrical engineers, searching for monochromatic radiation to study the spectra of molecules, led to the invention of the first laser in 1960. Ophthalmologists and dermatologists were the first to study the biological effects and therapeutic possibilities of laser beams. The construction of new laser systems emitting energy at different wavelengths or with different durations, as well as the development of new concepts of the biomedical effects, led to its broad use in surgery in the treatment of vascular and pigmented lesions as well as cosmetic applications.

It was not predictable that an analytical tool for physicists to examine the molecular structure of molecules would become one of the most important inventions of the 20th century. Theodore Maiman was enthusiastic about the first laser he constructed, but regarded it as 'a solution looking for a problem' [1]. The expectations during the Cold War that lasers could be used as powerful weapons remained unfulfilled. Nonetheless, today laser systems are present in almost every household, at least in CD players. The acceptance of lasers in medicine by doctors as well as by patients has always been very high. With a positive attitude towards such a powerful source of harmless-looking light, lasers became regarded as magic bullets for the treatment of cancer, replacing the bloody knife of the surgeons.

The history of laser is very young and tells of an invention of great economic value. However, this makes it very difficult to interpret the different references, which are predominantly scientific papers and autobiographical retrospections of the people involved. Depending on the intention and the background of the authors, selection and emphasis vary.

Inventing the Laser

In 1916, Albert Einstein discussed the possibility of stimulating radiant energy based on Niels Bohr's theory that atoms emitted energy in quanta when transitioning from excited states back to resting states [2]. The first experimental proof of his theory was published by the German physicists Rudolf Ladenburg and Hans Kopfermann in 1928 [3]. However, stimulated emission received little attention from experimentalists during the 1920s and 1930s when atomic and molecular spectroscopy were of central interest to many physicists [4]. In 1939, Valentin A. Fabrikant defended his doctoral thesis, 'The emission mechanism of a gas discharge', at the P.N. Lebedev Physical Institute in Moscow. It discussed experimental evidence for the existence of negative absorption (what was later called stimulated emission) and suggested experiments on light amplification [5].

Although the essential ideas for constructing a laser were known around 1930, it was not before the early 1950s that physicists and electrical engineers began to collaborate with the research on monochromatic radiation of constant amplitude at very small wavelengths studying the microwave and radio frequency spectra of molecules. In this context, in 1953 and 1954, several physicists independently suggested the use of stimulated emission for microwave amplification, creating the acronym MASER to stand for 'microwave amplification by stimulated emission of radiation' [6].

In 1953, the American physicist Joseph Weber at the University of Maryland published a proposal for a microwave amplifier that was based on stimulated emission in a paramagnetic solid [7]. In 1954, Nikola G. Basov and Alexander M. Prokhorov of the Lebedev Institute in Moscow and J.P. Gordon, H.J. Zeiger, and Charles H. Townes of Columbia University in New York reported on two molecular devices for generating microwave radiation, both using the ammonia molecule as the active species [8, 9].

Charles H. Townes, Nikolay G. Basov, and Alexander M. Prokhorov received the Nobel Prize in Physics 1964 for their 'fundamental work in the field of quantum electronics which has led to the construction of oscillators and amplifiers based on the maser-laser principle' [10].

The ammonia beam maser itself was not particularly useful as its operation was limited to the resonant frequency of the ammonia molecule and could only be used at barely detectable power levels [1]. In 1958, C.H. Townes and his brother-in-law Arthur Leonard Schawlow, professor at Stanford University, showed that masers could theoretically be made to operate in the optical and infrared regions [11]. The same year, G. Makov, C. Kikuchi, J. Lambe and R.W. Terhune at the University of Michigan developed and built a solid-state maser [12]. They used crystalline corundum (ruby) in a large magnetic field and a strategy similar to that known as optical pumping, suggested by Nicolas Bloembergen at Harvard University in 1956 [13]. Theodore H. Maiman at the Hughes Corporation Research Laboratories took over the Kikuchi ruby maser. In 1960, he presented the first functional optical ruby maser excited by a xenon flash lamp to produce a bright pulse of 693.7 nm, deep red light of about a 1-ms duration and a power output of about a billion watt per pulse [14]. His invention rapidly led to the development of multiple other optical masers, now called laser (light amplification by stimulated emission of radiation). In 1961, Fred J. McClung and Robert W. Hellwarth introduced the quality-switching (Q-switching) technique to shorten the pulse length to nanoseconds with the use of an electro-optical shutter that permitted the storage and subsequent release of a peak power up to gigawatts of energy [15, 16].

Medical Use of Lasers

The medical specialists who were already treating diseases with sunlight and technical light sources were also the first to carry out biomedical research with lasers.

One year after Maiman had presented the first ruby laser, ophthalmologists using xenon lamps for retinal photocoagulation published on ocular lesions experimentally produced in a rabbit by an optical maser [17]. These studies were soon followed by clinical experience on patients treated for retinal tears, flat detachments, angiomas, and tumors [18].

In dermatology, the treatment of skin diseases with light has a long tradition – e.g. lupus vulgaris with the Finsen lamp in 1899, wound healing and rickets with artificial UV light sources after

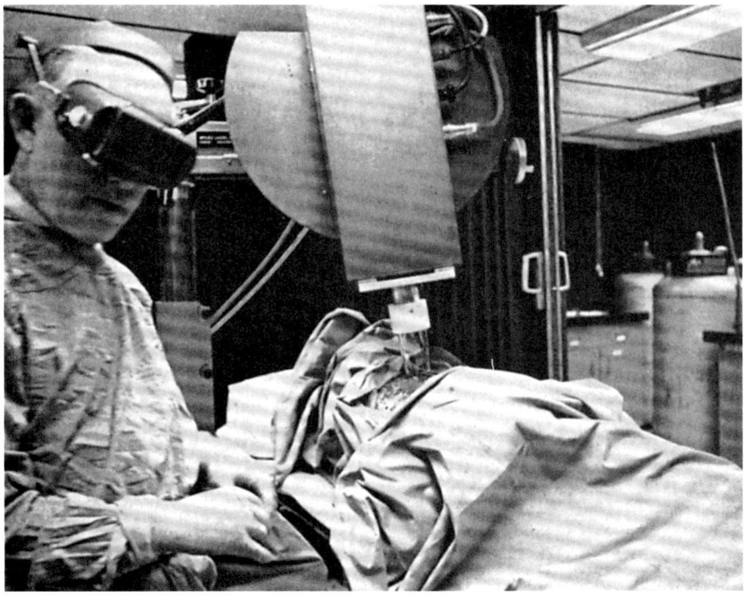

Fig. 1. Experimental treatment of metastatic cutaneous melanoma with a pulsed neodymium laser by Leon Goldman [23].

1901, and psoriasis with the combination of light and tar in 1925. Hence, it is no surprise that the American Society for Laser Medicine and Surgery honors a dermatologist, Leon Goldman, as the 'father of lasers in medicine in the United States' [19]. Leon Goldman did his dermatological training in Zürich, London, and Cincinnati. He was Chairman of the Department of Dermatology at the University of Cincinnati when he learned of the invention of Theodore Maiman's ruby laser in 1960. He was convinced of the great potential of lasers in medicine. In 1961, he founded the first biomedical laser laboratory at the University of Cincinnati [20]. In 1963, Goldman and his coworkers published the first study on the effects of lasers on skin describing the selective destruction of pigmented structures of the skin including hair follicles with the beam of the ruby laser. They noted highly selective injury of pigmented structures (black hair) with no evident change in the white skin underneath [21, 22].

Goldman published on the possible treatment of nevi, melanomas, and tattoos using the pulsed ruby laser: 'The most striking results have been obtained with the removal of tattoos, especially with the Q switched laser'. He expected the laser to bring substantial benefits to the treatment of skin cancer: 'Because of the accessibility and color, laser surgery can be used extensively in the field of skin cancer. The most significant treatments have been given for that black cancer of man, melanoma. Here, our laboratory has done laser operations even in delicate areas such as melanoma near the brain. It is too early to tell how permanent the effects will be. . .' [23] (fig. 1).

He preformed clinical and histopathological studies on vascular malformations with the argon laser. In 1973, Goldman published promising effects on angiomas with the continuous-wave neodymium:yttrium-aluminium-garnet (Nd:YAG) laser. His book *Biomedical Aspects of the Laser* published in 1967 is a comprehensive overview over the possibilities, problems, and ideas of the use of the laser in medicine at that time, also emphasizing the need for protection from laser energy. In addition, he discussed ideas of using the laser as a diagnostic tool (transillumination) to detect foreign bodies, hard tumors, or bone defects, and presented data on the use of laser in dentistry.

Photoexcision (the optical scalpel) was possible with continuous-wave lasers all invented in 1964; first the CO_2 laser, followed by the Nd:YAG-laser and then the argon laser. For the CO_2 laser, the color of the target area was not of any great significance and with an out-of-focus beam and larger spot size, hemostasis was also possible making it a helpful tool for surgery on vasculated organs (liver, oral mucosa, gynecology). Developments in fiber optics made it possible to transmit far-infrared laser beams, increasing the flexibility of CO_2 lasers for endoscopic surgery. The argon laser showed superior absorption by hemoglobin and was used for treating port wine stains and teleangiectasia of the face and early rhinophyma [24].

The early continuous-wave lasers emitted an uninterrupted beam of light that was effective in destroying the desired target, but also exposed the surrounding healthy tissue to laser energy for prolonged periods. The result of this collateral damage was unacceptably high rates of hypertrophic scarring and pigment alteration. The first attempt to minimize this nonspecific tissue injury involved making the continuous-wave lasers discontinuous or quasi-continuous by using a mechanical shutter to interrupt the beam of light. In the treatment of vascular lesions, the development of the tunable yellow light dye laser with the absorption peak closer to oxyhemoglobin than the early argon lasers reduced the risk of side effects. In 1996, the erbium (Er):YAG laser with a very short wavelength of 2,940 nm allowed a more superficial vaporization of tissue and was used together with CO_2 lasers for skin resurfacing. Very recently, the new technical concept of fractional photothermolysis was introduced. It received FDA approval in 2004 for skin resurfacing and in 2005 for the treatment of melasma [25].

Selective Photothermolysis

These developments support the common assumption that progress in medicine (and laser dermatology) depends mainly on new technology; however, technical progress reflects only one part of the history of the laser. Both acceptance of and interest in observations depends very much on the attitude of the involved persons. Even within the short history of lasers in medicine, the example of the pulsed laser systems shows how new ideas foster new use of old technology [26].

Leon Goldman wrote in 1967: 'There is every indication that Q-switched lasers will remain an important tool in the physicists' laboratories' [23]. Despite the various treatment possibilities that Goldman proposed and the encouraging results he published, there was an initial lack of interest in the development and support of the use of pulsed lasers by the government, industry, and the armed forces. Surgeons focused their interest on continuous rather than pulsed lasers [24]. The ruby laser was ineffective when used as an optical scalpel for cutting or coagulation, and when using high-energy pulses the effect became unpredictable because of cavitations (vapor bubbles). The attempts to use the pulsed Nd:YAG laser were not more successful as tissue fragments were spattered all over the operating room [26]. In the 1980s, the pulsed ruby laser was commercialized in Japan for the treatment of tattoos and pigmented lesions, while being abandoned in Europe and the USA where tattoo removal was performed by CO_2 laser vaporization [27].

With the flashlamp-pumped pulsed dye laser in the early 1980s, R. Rox Anderson and John A. Parrish from the Department of Dermatology at the Harvard Medical School in Boston developed the theory of selective photothermolysis that revolutionized the practice of cutaneous laser surgery [28]. The authors recognized that the collateral thermal damage in the surrounding tissue of the target chromophore resulted from prolonged exposure to the laser's energy. By the appropriate manipulation of wavelength and pulse duration, and in dependence upon the chromophore's relaxation time, therapeutic destruction could be maximized while minimizing thermal damage to the surrounding tissue [29].

The new theoretical understanding of the advantages of pulsed lasers led to a revival of the Q-switched lasers (ruby [30], alexandrite [31], Nd:YAG [32]) for the treatment of benign pigmented cutaneous lesions, especially for tattoo and hair removal [33]. More than 3 decades later, a nearly identical ruby laser to the one used by Goldman in 1963 became the first device approved by the FDA in 1989 for permanent removal of pigmented hair, and the Q-Switched Nd:YAG received FDA approval as a treatment modality for tattoos in 1991 [25, 26].

References

1 Maiman TH: The Laser Odyssey. Blane, Laser Press, 2000
2 Einstein A: Zur Quantentheorie der Strahlung. Physikalische Gesellschaft Zürich 1916;18:47–62 (the same paper was published on 15 March 1917, Physikalische Zeitschrift 1917;18:121–128).
3 Kopfermann H, Ladenburg R: Untersuchungen über die anomale Dispersion angeregter Gase II Teil. Anomale Dispersion in angeregtem Neon Einfluß von Strom und Druck, Bildung und Vernichtung angeregter Atome. Zeitschrift für Physik 1928;48:26–50
4 Townes CH: Production of coherent radiation by atoms and molecules. Nobel Lecture, December 11, 1964. http://nobelprize.org/nobel_prizes/physics/laureates/1964/townes-lecture.pdf (accessed 30 October 2010).
5 Lukishova S: Valentin A. Fabrikant: negative absorption, his 1951 patent application for amplification of electromagnetic radiation (ultraviolet, visible, infrared and radio spectral regions) and his experiments. J Eur Optical Soc 2010(rapid publications 5):10045s.
6 Townes CH: Early history of quantum electronics. J Modern Optics 2005;52:1637–1645.
7 Weber J: Amplification of microwave radiation by substances not in thermal equilibrium. Trans Inst Radio Eng PGED-3 1953;1.
8 Basov NG, Prokhorov AM: Application of molecular beams to the radio spectroscopic study of the rotation spectra of molecules. Zh Eksp Theo Fiz 1954;27:431
9 Gordon JP, Zeiger HJ, Townes CH: The Maser – new type of microwave amplifier, frequency standard, and spectrometer. Phys Rev 1955;99:1264–1274.
10 Edlén B: Nobel prize award ceremony speech. 1964. http://nobelprize.org/nobel_prizes/physics/laureates/1964/press.html (accessed 30 Oct 2010).
11 Schawlaow AL, Townes CH: Infrared and optical masers. Phys Rev 1958;112:1940–1949.
12 Makov G, Kikuchi C, Lambe J, Terhune RW: Maser action in ruby. Phys Rev 1958;109:1399–1400.
13 Bloembergen N: Proposal for a new type solid state maser. Phys Rev 1956;104:324–327.
14 Maiman TH: Stimulated optical radiation in ruby. Nature 1960;187:493.
15 Hellwarth RW: Theory of the pulsation of fluorescent light from ruby. Phys Rev Lett 1961;6:9–11.
16 Hellwarth RW, McClung FJ: Giant pulsations from ruby. Appl Phys 1962;33:838–841 and Bull Am Phys Soc 1962;6:414.
17 Zaret MM, Breinin GM, Schmidt H, Ripps H, Siegel IM, Jolon LR: Ocular lesions produced by an optical maser (laser). Science 1961;134:1525.
18 Campbell CJ, Noyori KS, Rittler MC, Koester C: Retinal coagulation clinical studies. Ann N Y Acad Sci 1965;122:780.
19 American Society for Laser Medicine and Surgery: History of ASLMS. www.aslms.org/aslms/history.shtml (accessed 30 Oct 2010).
20 Coras B, Landthaler M: Leon Goldman (1905–1997); in Loeser C, Plewig G (Eds): Pantheon der Dermatologie – Herausragende historische Persönlichkeiten. Heidelberg, Springer, 2008, pp 357–361.
21 Goldman L, Blaney DJ, Kindel DJ, Franke EK: Effect of the laser beam on the skin. J Invest Dermatol 1963;40:121–122.
22 Goldman L, Blaney DJ, Kindel DJ Jr, Richfield D, Franke EK: Pathology of the effect of the laser beam on the skin. Nature 1963;197:912–914.
23 Goldmann L: Biomedical Aspects of the Laser: The Introduction of Laser Applications into Biology and Medicine. Springer, Berlin, 1967.
24 Goldman L: Historical perspective: personal reflections; in Arndt KA, Noe JM, Rosen S: Cutaneous Laser Therapy: Principles and Methods. New York, Wiley, 1983, p 7.
25 Houk LD, Humphreys T: Masers to magic bullets: an updated history of lasers in dermatology. Clin Dermatol 2007:25;434–442.
26 Anderson RR: Dermatologic history of the ruby laser: the long story of short pulses. Arch Dermatol 2003:139:70–74.
27 Bailin PL, Ratz JL, Levine HL: Removal of tattoos by CO_2 laser. J Dermatol Surg Oncol 1980;6:997–1001.
28 Anderson RR, Parrish JA: Selective photothermolysis: precise microsurgery by selective absorption of pulsed radiation. Science 1983:220:524–527.

29 Parrish JA, Anderson RR, Harrist T, Paul B, Murphy GF: Selective thermal effects with pulsed irradiation from lasers: from organ to organelle. J Invest Dermatol 1983;80(suppl):75s–80s.
30 Taylor CR, Gange RW, Dover JS, Flotte TJ, Gonzalez E, Michaud N, Anderson RR: Treatment of tattoos by Q-switched ruby laser: a dose-response study. Arch Dermatol 1990;126:893–899.
31 Fitzpatrick RD, Goldman MP, Ruiz-Esparza J: Use of the alexandrite laser (755 nm, 100 nsec) for tattoo pigment removal in an animal model. J Am Acad Dermatol 1993;28:745–750.
32 Kilmer SL, Anderson RR: Clinical use of the Q-switched ruby and the Q-switched Nd:YAG (1064 nm and 532 nm) lasers for treatment of tattoos. J Dermatol Surg Oncol 1993;19:330–338.
33 Sherwood KA, Murray S, Kurban AK, Tan OT: Effect of wavelength on cutaneous pigment using pulsed irradiation. J Invest Dermatol 1989;92:717–720.

Michael L. Geiges, MD
Institute and Museum of Medical History, University of Zurich
Hirschengraben 82
CH–8001 Zurich (Switzerland)
Tel. +41 634 20 19, Fax +41 634 23 49, E-Mail michael.geiges@mhiz.uzh.ch

Bogdan Allemann I, Goldberg DJ (eds): Basics in Dermatological Laser Applications.
Curr Probl Dermatol. Basel, Karger, 2011, vol 42, pp 7–23

Laser Principles

Inja Bogdan Allemann[a] · Joely Kaufman[b]

[a]Klinik für Plastische Chirurgie und Handchirurgie, Universitätsspital Zürich, Zürich, Switzerland;
[b]University of Miami Miller School of Medicine and Private Practice Coral Gables, Miami, Fla., USA

Abstract

Since the construction of the first laser in the 1960s, the role that lasers play in various medical specialities, including dermatology, has steadily increased. However, within the last 2 decades, the technological advances and the use of lasers in the field of dermatology have virtually exploded. Many treatments have only become possible with the use of lasers. Especially in aesthetic medicine, lasers are an essential tool in the treatment armamentarium. Due to better research and understanding of the physics of light and skin, there is now a wide and increasing array of different lasers and devices to choose from. The proper laser selection for each indication and treatment requires a profound understanding of laser physics and the basic laser principles. Understanding these principles will allow the laser operator to obtain better results and help avoid complications. This chapter will give an in-depth overview of the physical principles relevant in cutaneous laser surgery.

It was in 1959 that the first laser was invented and developed by the physicist T.H. Maiman at the Hughes Research Laboratory in Malibu, California [1]. He used a flashlamp-pumped ruby crystal, which produced red light with a wavelength of 694 nm. The clinical use of lasers in dermatology started in 1963, when Dr. Leon Goldman used the ruby laser for various dermatological conditions [2, 3]. In the following years, the continuous-wave argon, CO_2, and Nd:YAG lasers followed. In 1983, the theory of selective photothermolysis, proposed by R. Anderson and J.A. Parrish revolutionized the use of lasers in dermatology [4]. Based on this theory, pulsed lasers were able to specifically and selectively target and destroy structures within the skin, without damage to surrounding tissues. This theory can be regarded as a milestone in cutaneous laser treatments. Finally, a recent technological advance was the introduction of the concept of fractional photothermolysis in 2003, which led to further development and expansion of laser resurfacing in dermatology and will also be discussed below [5].

Spontaneous and Stimulated Emission of Radiation

The word laser is an acronym for 'light amplification by the stimulated emission of radiation'. In contrast to sunlight for example, which is emitted spontaneously, laser light is emitted by stimulated emission. The concept of stimulated light emission, which the generation of laser light is essentially based on, was originally conceived by Albert

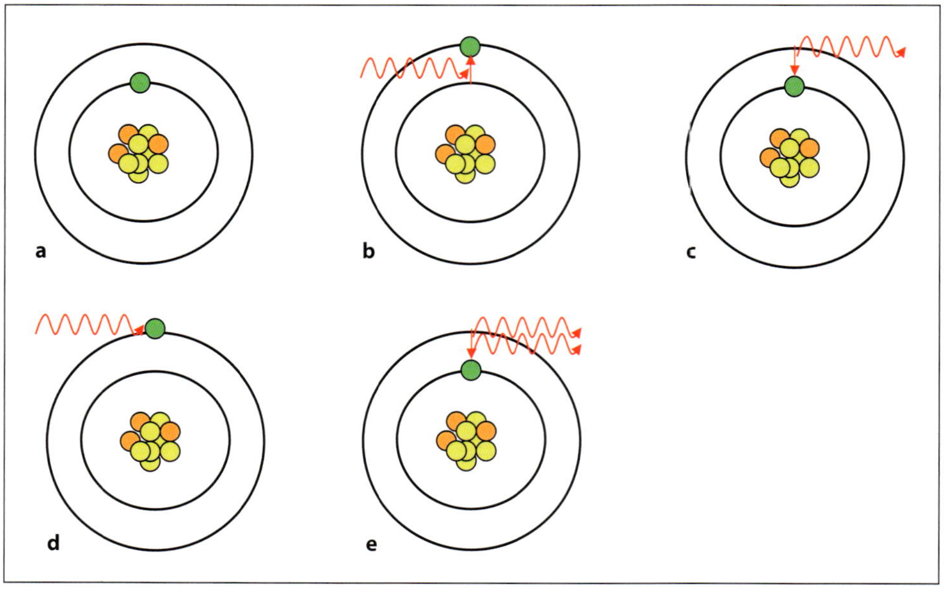

Fig. 1. Spontaneous and stimulated emission. **a** Resting state of the atom, with the electron in a low-energy orbiting position. **b** Absorption of a photon elevates the electron into an excited state. **c** The unstable excited electron falls back into the lower-energy resting state and releases the photon of energy. **d** If an excited electron absorbs another photon of energy, (**e**) it releases two photons with the same energy, direction, and wavelength when falling back into its resting state.

Einstein [6] in 1917, clearly long before the first laser was invented.

Spontaneous Emission of Radiation
Einstein based his theory on the original model of the atom, set forth by Neils Bohr in 1913.

Bohr's model explained that atoms are composed of a nucleus, containing positively charged protons and neutral neutrons. Negatively charged electrons orbit the nucleus, much like planets orbit the sun. The electrical attraction between the positive and the negative holds the entire complex together. Electrons that surround a molecule or atom orbit in a stable resting state. This state is the state of lowest energy and is close to the nucleus. If an electron absorbs energy, in the form of a photon (particle of light), it will jump to a higher energy level at a position further from the nucleus and will then be in an excited state. This state is a far less stable position. Atoms can only stay in an excited state for a short period of time and as the excited electrons return to their resting state, they release the energy previously absorbed in form of photons. This emission of photons is random and occurs in all directions. When it is random and occurs on a regular basis, it is referred to as spontaneous emission. In nature, spontaneous emission is dominant and the majority of electrons are in the resting state. Spontaneous emission is the source of sunlight and virtually any light observed in nature. However, this process of excitation followed by de-excitation and release of photons can also be induced, as proposed by Einstein in the early 1900s.

Stimulated Emission of Radiation
For this to happen, an already excited electron has to collide with yet another photon with the

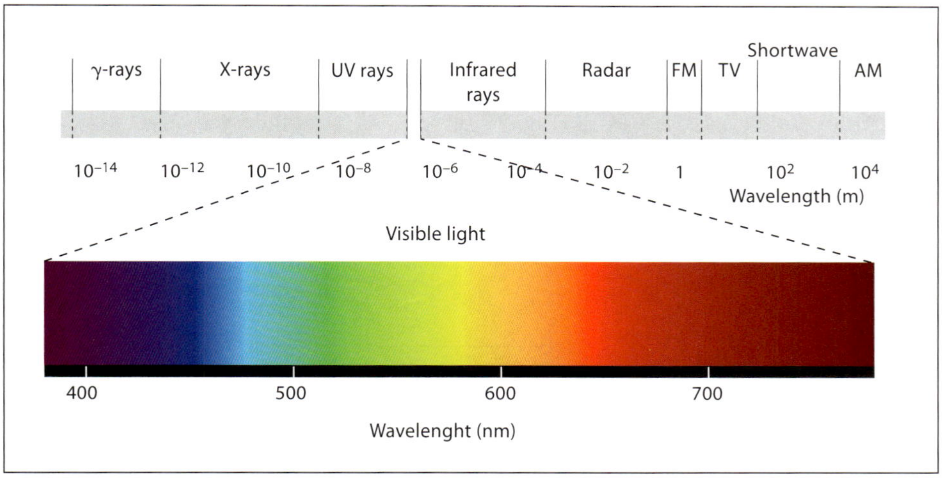

Fig. 2. The electromagnetic spectrum.

proper energy, which will lead to the emission of two photons when the electron returns to resting state. Importantly, these two photons will contain the same energy, frequency, and direction, which sets the basis for the unique properties of laser light discussed below (fig. 1). Finally, such emitted photons can again stimulate the emission of further photons. Stimulated emission needs a source of energy to induce the change. In the case of lasers, we have several sources of stimulation, referred to as the 'pump' portion of the device. This increasing stimulation of photon emission will ultimately lead to an environment where there are proportionately higher numbers of atoms in the excited state as compared to the resting state. This situation is called 'population inversion' and is a crucial prerequisite for the generation of laser light within an optical chamber.

Properties of Laser Light

Light is a form of energy, all forms of light are represented in the electromagnetic spectrum. Thus, laser light represents a part of the electromagnetic spectrum of energy. The electromagnetic spectrum encompasses all energy sources that travel through space in the form of waves, and ranges from the short wavelengths of X-rays and γ-rays to the long wavelengths of microwaves and radio waves. Most lasers that are used in medicine and dermatology generate light within the visible (400–760 nm), near-infrared (760–1,400 nm), mid-infrared (1.4–3 μm), infrared (>3 μm), and rarely the ultraviolet range (200–400 nm) of the electromagnetic spectrum (fig. 2). As the name states, laser light within the visible spectrum of the electromagnetic spectrum creates visible beams of various colors, depending on the respective wavelength. For example, the ruby laser at 694 nm generates red light, whereas the KTP laser at 532 nm generates green light. Lasers within the infrared section of the electromagnetic spectrum, such as YAG and CO_2, generate invisible beams of laser light.

Laser light has three main and unique characteristics that differentiate it from other light sources and represent the basis for its therapeutic use and effectiveness: monochromaticity, coherence, and collimation (fig. 3) [7].

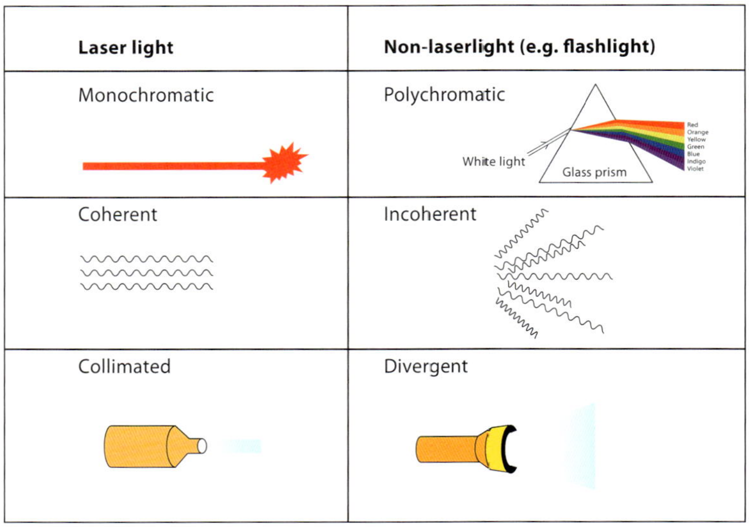

Fig. 3. Laser properties.

Monochromaticity
In contrast to sunlight, which encompasses a wide spectrum of wavelengths, laser light is monochromatic, which means it emits light of only one clearly defined single wavelength or a very narrow band of wavelengths. Laser light that is shone through a prism produces the same color. White light however that is shone through a prism produces the entire spectrum of colors. In the visible spectrum, every wavelength has its own color. The respective wavelength is defined by the lasing medium used in the laser, which will be discussed below.

Coherence
Laser light is also required to be coherent, which means that the light waves are in phase with respect to space and time. The coherence of laser light is based on the process of stimulated emission, where only light in the same phase and direction can be emitted.

Collimation
Finally, laser light is collimated. As the light waves are parallel, the laser beam has a defined narrow beam diameter with no divergence, which virtually does not increase even with increasing distance. This is best illustrated by a laser pointer, whose beam stays practically the same no matter at which distance. This fact allows laser light to not loose any of its energy even after travelling over longer distances. This is different from non-collimated light emitted by a flashlight, which with increasing distance, has an increasing divergent beam diameter and a decreasing intensity.

High Intensity
Finally, another crucial feature of laser light is that it can emit light with very high intensity.

The wavelength of lasers is usually measured in nanometers. One nanometer is 10^{-6} mm, which is 0.000001 mm.

Non-Laser Light Sources

Non-laser light sources, such as for example intense pulsed light (IPL) devices, which nowadays are used as frequently as lasers, have different features than true lasers. IPL, or broadband light devices use flashlamps and computer-controlled capacitor banks to generate pulsed polychromatic high-intensity light. In contrary to lasers, IPL devices emit light of many wavelengths, encompassing a wide range usually from 500 to 1,300 nm within the electromagnetic spectrum. Many times, cutoff filters are built in, in order to tailor the wavelength range more precisely to the target structure, such as vascular or pigmented ones [8]. Thus, for example vascular lesions can be targeted with an IPL device with a wavelength range of 500–670 and 870–1,400 nm, whereas pigmented lesions would be targeted with another device of 525–1,200 nm. The clinical effect of IPL on the target tissue is photothermal, which means generation of heat within the target tissue with subsequent destruction the target structure.

For such photothermolysis to occur, monochromaticity (one single wavelength or a very narrow range of wavelengths) is not necessarily required, as the main chromophores in the skin (melanin, hemoglobin, water) show broad absorption spectra.

As IPL devices employ multiple wavelengths at the same time, various chromophores, such as for example melanin and hemoglobin can be targeted with the same light exposure. This can be a desirable effect, when for example performing the treatment of photoaged skin, which is composed of pigmentary and vascular alterations. Clearly, in other clinical indications, such as for example the treatment of facial redness, where only one chromophore is targeted, the other one (melanin) might be competing, actually hindering the treatment.

IPL devices have pulse durations within the millisecond range which can be set according to the clinical indication. There are indications, such as photoepilation, where pulse durations within the millisecond range or longer are needed. However, as with lasers, ideally the pulse duration should be lower than the thermal relaxation time (TRT) of the target structure to prevent unselective damage to the surrounding tissue [9]. The usually large spot sizes represent another advantage of IPL devices, again especially useful in photoepilation.

There are numerous clinical indications that can be treated with IPL devices such as vascular lesions, pigmented lesions, photoaged skin, excess hair growth, and acne vulgaris.

A detailed description of IPL technology can be found in a dedicated chapter [Schoenewolf et al., pp. 166–172].

Fractional Technology

The novel concept of fractional photothermolysis was introduced to the market by Dieter Manstein and Rox Anderson in the year 2003 [10]. It was an answer to the need for effective yet low-risk resurfacing methods. Unlike conventional ablative and non-ablative lasers, fractional ablative and non-ablative lasers treat only a fraction of the skin, leaving up to a maximum of 95% of the skin uninvolved – thus the name 'fractional'. This is achieved by inducing microscopic small three-dimensional zones of thermal damage or ablation, surrounded by undamaged tissue allowing for rapid epidermal repair.

The target chromophore for fractional photothermolysis devices is water. Hence, various water-containing structures at differing depths throughout the skin, such as, epidermal keratinocytes, collagen, and blood vessels can be targeted resulting in a wide array of clinical indications [11].

Based on the wavelength's affinity for water, fractional technologies can be divided into two main categories. Those with wavelengths that are highly absorbed by water are termed 'ablative', while those wavelengths that are less avidly absorbed by water are termed non-ablative. This novel technology will be described in detail in chapter on fractional photothermolysis [Bogdan Allemann and Kaufman, pp. 56–66].

Construction of Lasers

Extremely simplified, a laser consists of a laser chamber containing the lasing medium, two mirrors, and an excitation source, i.e. the external pump [12]. The chamber is a highly reflective optical cavity, with two parallel mirrors on each side, one of which is fully reflective and the other partially reflective. Within this chamber is the lasing medium, which can be a solid, liquid, or gas (table 1). It is the lasing medium which defines the wavelength emitted by a laser. Examples of solid lasing media are ruby, alexandrite, Er:YAG, Nd:YAG, and diode. A liquid lasing medium used is dye, and gaseous lasing media are CO_2, argon, and excimer. The lasing medium represents the supply of electrons needed to produce a laser beam by stimulated emission of radiation and defines the wavelength of the laser light emitted. Many lasers are named after the lasing medium contained within the laser. The photons within the lasing medium in the chamber need to be externally stimulated, in order to be amplified and generate the laser light. This external energy source is referred to as the pump or pumping system. Pumps can be any form of energy source, including a high-powered flashlamp, electrical current or another laser. Within the lasing medium, the energy gets absorbed by the atoms, which with stimulation get excited and then release their energy in the form of photons. The photons within the chamber are reflected back and forth from one mirror to the other one, colliding with other atoms and hence stimulating further emission within the same axis. The more excited atoms in the medium (= population of atoms) there are, the more photons get produced. Eventually, there will be more excited atoms than resting ones, which is called 'population inversion'. Remember, in a normal population of atoms, the majority resides in a resting state. One mirror within the chamber then becomes partially reflective, which allows for some of those stored photons within the chamber to get emitted as a beam of light. In fact, only a fraction of light is emitted from the laser cavity, while the rest remains in the cavity to maintain the lasing process, which continues, as long as the pump keeps exciting the atoms in the lasing medium [13]. Of the light emitted, only parallel waves of the same wavelength can be emitted, which sets the basis for the collimated, monochromatic nature of laser light. Depending on whether the pump is continuous or pumped, a continuous wave of light or a pulsed wave of light gets emitted.

Once the light is emitted from the chamber through the partially reflective mirror, it enters the delivery system which will ultimately transmit the light to the handpiece. Some delivery systems are fragile articulated arms that contain mirrors reflecting the light, other delivery systems are build from fiberoptic cables. The handpiece finally focuses the light onto the skin (fig. 4).

Table 1. Lasing media and lasers

Laser type	Lasing media
Liquid	dye
Gas	CO_2
	argon
	excimer
Solid	ruby
	alexandrite
	Er:YAG
	Nd:YAG
	diode

The lasing medium represents the supply of electrons needed to produce a laser beam by stimulated emission of radiation and defines the wavelength of the laser light emitted

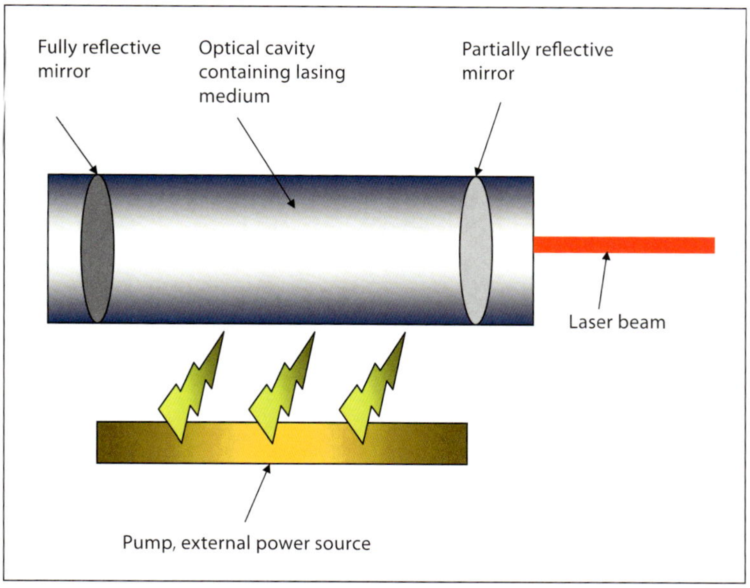

Fig. 4. Construction of lasers.

Tissue Optics

When laser light reaches the skin, it can interact with the tissue in four different ways: it can be absorbed, reflected, scattered, or transmitted (fig. 5). Most times, a combination of various interactions, each at different degrees, takes place at the same time [14].

Absorption

When using lasers for therapeutic purposes, the goal is for the laser light to be absorbed by a specific target. The Grothus-Draper law of photobiology states that in order to have a biological effect in the tissue, light must be absorbed by the target tissue. It is only this light which is absorbed that does real work on the tissue. When a target molecule absorbs a photon, the entire energy of the photon is transferred to the target molecule. The specific light-absorbing targets are known as chromophores. The three main chromophores in the skin are melanin, (oxy)hemoglobin, and water. Tattoo ink is the main external chromophore of importance in laser dermatology. The amount of light that gets absorbed by the specific chromophore depends on the wavelength used and whether it corresponds to the specific absorption spectrum of the respective chromophore.

Light that is not absorbed can either be scattered, transmitted, or reflected – all of which have no true biological effects [15]

Reflection

Roughly 4–6% of the light usually gets reflected when it hits the skin at a 90 degree angle. Reflection of the light mostly takes place at the stratum corneum and is the reason why protective eyeshields should be worn at all times when working with lasers. Reflection of light can be minimized by applying the incident laser beam perpendicularly to the tissue surface. With an increased angle of light incidence, the rate of reflection increases above the aforementioned 4–6% [16]. Furthermore, dry or scaly skin reflects even more light, which again can be minimized by applying a thin layer of clear

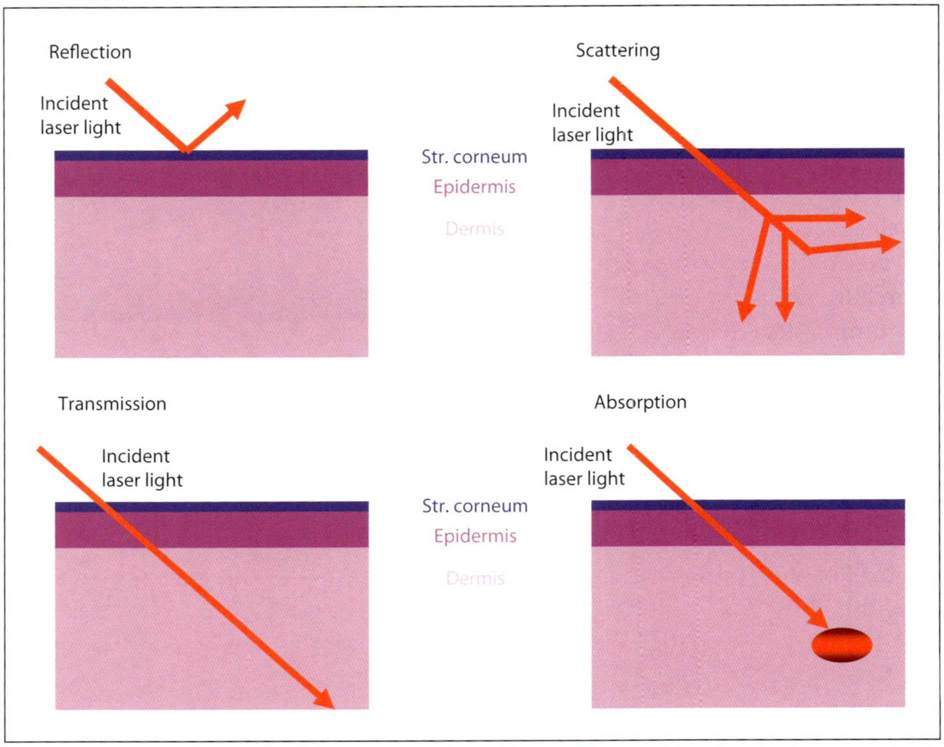

Fig. 5. Laser tissue interactions.

oil or gel onto the skin. It is crucial to try to keep the amount of reflection at the surface at its minimum as increased reflection means decreased fluence absorbed by the tissue.

Scattering
Once the light has passed the stratum corneum and into matter, it can then be scattered within the tissue. Scattering mainly occurs with shorter wavelengths – largely in the dermis, due to the collagen fibers – and predominantly in the forward direction. The amount of scattering of laser energy is inversely proportional to the wavelength of incident light, with shorter wavelengths scattered more and longer wavelengths less. With some exceptions, this also results in deeper penetration of longer wavelengths. The scattered light shows a different direction from the original direction of the incident light. This plays an important role in the spatial distribution of absorbed light energy. Hence, scattering defocuses the spot of light by spreading out the beam, which actually leads to the irradiation of a larger area. With larger laser beam diameters (spot size), less scattering occurs, while at the same time there is deeper penetration and less loss of energy with depth of penetration [17].

Transmission
Finally, residual light – which has not been reflected, absorbed, or scattered – will be transmitted into deeper structures, such as the subcutaneous tissue. The transmission of light is important for longer wavelengths to reach and target deeper structures

within the tissue. Shorter wavelengths of 300–400 nm will already have been scattered and only penetrate superficially, while scattering at 1,000–1,200 is minimal and hence penetration is great.

From 300 nm and higher, the depth of penetration of the laser light increases with increasing wavelengths, e.g. a 755 nm alexandrite laser beam will penetrate deeper than a 532 nm KTP laser beam [18]. At high water absorption wavelengths (2,940 nm), penetration is once again decreased by the trapping of light in the upper layers of the skin where water is the most abundant. Clinically, when choosing a laser for a certain indication, not only the absorption maximum of the target chromophore needs to be taken into account, but also the depth of penetration of the chosen laser wavelength in order to actually be able to reach the target within the tissue.

Light-Tissue Effects

As described before, a biological effect of laser light in the tissue can only be achieved if the light is absorbed and converted into thermal energy. The biological effect depends on the actual temperature achieved in the target chromophore or tissue, as well as on the period of time the target is at the respective temperature. Finally, tissue effects also depend on the conduction of the heat from the target to the surrounding tissue. This means the damage achieved in the tissue depends on the energy density, the pulse width/pulse duration, and the heat conduction [19].

Laser light that is absorbed by a chromophore can result in three biological reactions: photothermal, photomechanical, or photochemical.

Photothermal

Photothermal effects occur when the absorbed light energy within the chromophore is converted into thermal energy. This is the primary mechanism by which lasers function in skin. Depending on the actual temperature achieved within the target, various effects, such as coagulation or vaporization, can occur. A temperature increase in the tissue of only 5°C can lead to tissue injury with subsequent inflammation and repair. At temperatures above 60°C, a denaturation of proteins and DNA and a coagulation of the tissue occur. Finally, at temperatures above 100°C, the intracellular water exceeds the boiling point and vaporization occurs, which can be seen clinically as ablation of the tissue [16].

Photomechanical

Photomechanical reactions within the chromophore occur when extremely high energies get absorbed at short pulse durations, which lead to extremely rapid thermal expansion of the target and subsequent photomechanical destruction [20]. Such reactions play an important role in selective photothermolysis of melanin or tattoo ink particles treated with the nanosecond Q-switched lasers. Another example of photomechanical destruction of tissue is with the purpura induced by the high-fluence short-pulse-width pulsed dye laser.

Photochemical

Finally, photochemical reactions occur with endogenous or exogenous photosensitizers, such as those used in photodynamic therapy, where light-absorbing chromophores are introduced into the tissue and then elicit selective photochemical reactions by light absorption. Photosensitivity associated with porphyria can also be regarded as a photochemical reaction.

A more detailed description on light and tissue interactions can be found in a dedicated chapter [Weber et al., 24–34].

Selective Photothermolysis

The theory of selective photothermolysis, proposed by Anderson and Parrish in 1983, is one of the most important concepts to explain laser-tissue interactions and why laser light can be used for targeted therapeutic purposes [4]. It states that laser energy can be absorbed by a defined target chromophore, leading to its controlled destruction without significant damage to the surrounding tissue. This means that we can selectively destroy tissue, for example melanosomes or tattoo ink particles within the skin, without damaging the surrounding tissue, such as vessels or collagen. In order for this concept to hold true, a number of principles need to be applied:

Wavelength
First, the wavelength of the laser light needs to correspond to the absorption maximum or lie within the absorption spectrum of the respective target chromophore.

Pulse Duration
Second, the pulse duration, also called 'pulse width' of the laser beam must be equal to or shorter than the thermal relaxation time (TRT) of the target chromophore. The TRT is defined as the time needed for the target chromophore to dissipate 63% of its peak temperature. This time is directly proportional to the square size of the chromophore. Hence, small objects cool faster than large ones, while larger chromophores have a longer TRT than smaller chromophores. The pulse width is thus determined by the size of the target chromophore.

Energy Density
Third, the energy density delivered by the laser beam, also referred to as fluence, must be high enough to actually destroy the target chromophore within the defined pulse duration. Energy densities are measured in joules per centimeter squared (J/cm^2).

Based on these concepts, the wavelength, pulse duration, and fluence must all be tailored to the properties of the target chromophore and clinical indication in order to produce a desirable clinical outcome and avoid complications.

In most cases, the clinical target of the treatment is also the chromophore used in the treatment. However, in many cases, the chromophore and the target are not the same. In these cases, the pulse width selected may actually differ significantly from the TRT of the chromophore. Consider, for example, in laser hair removal, where the target and the chromophore are not the same. The target chromophore in hair removal is melanin; however, the clinical targets seem to be the hair matrix and the stem cells. The hair matrix contains melanin and hence can be destroyed by direct heating. However, the stem cells do not contain pigment and are found at a distance from the pigment-containing target chromophore, namely the melanosomes within the hair shaft. If we heat only long enough to destroy the hair itself (nanosecond pulse width), the hair itself will fragment, but no permanent removal will result. In order to destroy the non-pigmented stem cells, and achieve (semi)permanent hair removal, the heat must diffuse from the pigmented area (hair) to the target. Thus, the clinical target will be destroyed by heat diffusion rather than by direct heating. This can only be achieved if the TRT is longer than the TRT of the chromophore. This mechanism has been proposed as the concept of thermal damage time as an extended theory of selective photothermolysis for non-uniformly pigmented targets [21]. It basically states that thermal destruction of a clinical target at a distance from the chromophore can be achieved by heat diffusion. Importantly, the pulse width chosen must be longer than the TRT of the chromophore, in order to allow heat diffusion. Using an alexandrite laser with a nanosecond pulse width – shorter that the TRT of the melanosomes – and one with a several millisecond pulse width results in very different treatments (table 2). The short nanosecond pulse width (high, short peak heat) will result in destruction of melanin (or tattoo pigment), whereas the millisecond pulse width (with cooling) will result in hair removal. Much of this can also be attributed to epidermal cooling, which will be discussed below.

Summary

Principles of selective photothermolysis:
- Wavelength: preferentially absorbed by target chromophore
- Pulse width: shorter than TRT of target chromophore
- Fluence: high enough to destroy target chromophore

Parameters to adjust according to target:
- Wavelength
- Pulse width
- Fluence

Table 2. Thermal relaxation times of important laser targets

	Size, µm	Thermal relaxation times (approx.)
Tattoo ink particle	0.5–4	10 ns
Melanosome	0.5–1	1 µs
Erythrocyte	7	2 µs
Blood vessel	50	1 ms
Blood vessel	100	5 ms
Blood vessel	200	20 ms
Hair follicle	200	10–100 ms

Chromophores in the Skin

There are various chromophores in the skin that are able to absorb light; however, the three main endogenous chromophores in the skin are melanin, hemoglobin, and water. Each of these chromophores has its own absorption spectrum and absorption peaks, detailing their relative absorption for each wavelength (fig. 6). This figure is of major importance in laser dermatology and should be studied and understood in detail, as from this curve the corresponding wavelengths and hence laser devices to target the respective chromophore can be identified.

Melanin shows a decreasing absorption spectrum ranging from 400 to 750 nm. The absorption spectrum of hemoglobin reaches from 400 to 600 nm, with absorption peaks that can be targeted preferentially. Oxyhemoglobin shows its maximum absorption peak at 418 nm, followed by smaller peaks at 548 and 577 nm. These absorption peaks can be targeted specifically, in order to minimize absorption by competing chromophores.

Water shows increasing absorption, starting at mid-infrared and increasing towards the infrared portion of the electromagnetic spectrum.

Thus, visible and UV light are mainly absorbed by melanin and hemoglobin, while infrared light is mainly absorbed by water.

Tattoo ink is the main exogenous chromophore of importance in laser dermatology. Its corresponding wavelengths are based on the color of the ink particle and will be discussed in detail in the chapter 'Tattoo removal' [Adatto et al., 97–110].

Choosing the optimum wavelength for the absorption of the respective target chromophore is one important thing. However, for clinical results, one has to bear in mind that within the visible range, penetration depth increases with increasing wavelengths. Hence, although the highest melanin absorption is at short wavelengths within the visible range, those wavelengths will not penetrate deeply enough to, for example, reach melanin located in the dermis and consequently treat a dermal pigmented lesion. Thus, wavelength has to be considered with regards to absorption maximum as well as depth of penetration. For example, a Q-switched 532 nm laser may easily remove superficial pigmented lesions such as a lentigo, but will not be able to penetrate far enough to treat a nevus of Ito due to the deep location of the pigment.

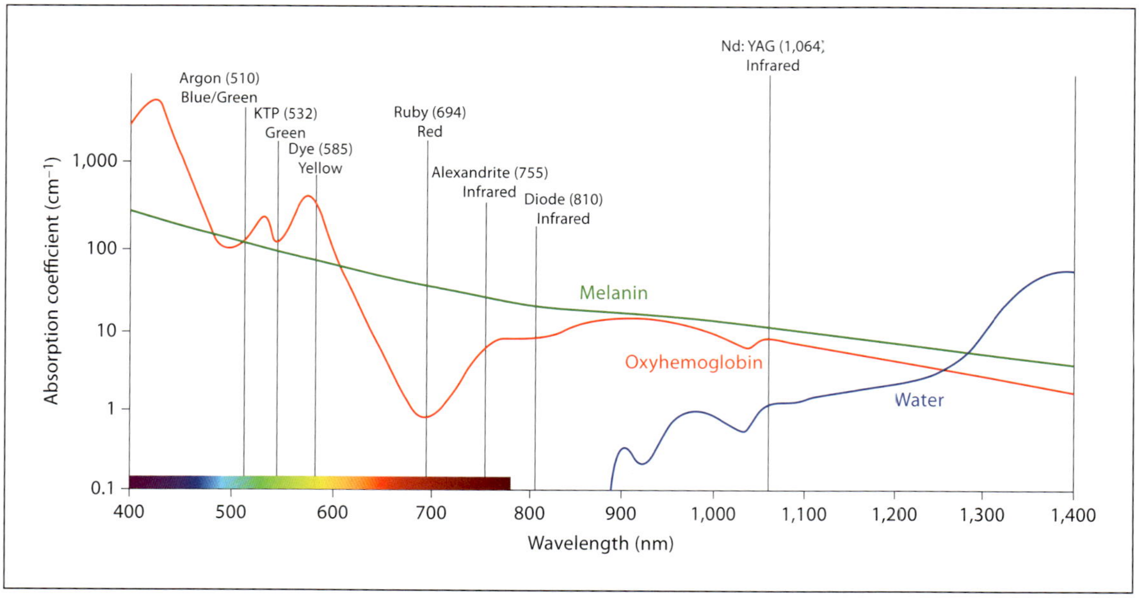

Fig. 6. Absorption spectra of the three main skin chromophores.

Summary

- The specific light-absorbing targets are known as chromophores.
- The three main chromophores in the skin are melanin, (oxy)hemoglobin, and water.
- Tattoo ink is the main external chromophore of importance in laser dermatology.
- The peak absorptions of chromophores are: melanin = 400–750 nm; hemoglobin = 400–600 nm; water = mid- and far-infrared.

Laser Properties and Parameters

When working with lasers, there are a few definitions that need to be clearly understood (table 3).

For continuous-wave lasers, time, power, and spot size are essential. For pulsed lasers, energy per pulse, pulse duration, fluence, and spot size are the most important parameters.

The energy of laser light refers to the number of photons delivered in a single pulse and is measured in joules. Joules are thus suitable to describe the energy of pulsed lasers. The energy per area is the fluence or energy density, which is expressed in joules per cm^2. The power of a laser is measured in watts and expresses the amount of energy the laser releases per unit time, i.e. how many joules are delivered per second (W = J/s). Watts are mainly used for continuous-wave lasers. Irradiance describes the power density, i.e. watts per square centimeter. It describes the intensity of a continuous-wave laser beam.

The time over which energy is delivered – i.e. the time of actual lasing, which is especially important in pulsed lasers – is the pulse duration or the pulse width. It can reach anywhere from nanoseconds to seconds. The frequency (pulse repetition rate) at which the single pulses are delivered is measured in hertz and 1 Hz equals 1 pulse per second. The wavelength characterizes the type of laser light and is measured in nanometers. It refers to the distance between two peaks of the light waves.

Finally, the spot size is the diameter of the laser beam, and is measured in millimeters.

Table 3. Laser properties and parameters

Energy	joules = watts × seconds
Fluence	energy density = joules/cm^2 = watts × seconds/cm^2
Power	watts = joules/second
Irradiance	power density = watts/cm^2
Pulse duration	seconds, milliseconds, nanoseconds
Frequency	hertz = pulses per second
Wavelength	nanometers
Spot size	millimeters

Fluence describes the energy density of a pulsed laser beam:

$$\text{Fluence} = \frac{\text{intensity} \times \text{time}}{\text{area}} = \frac{\text{watts} \times \text{seconds}}{\text{cm}^2} = \frac{\text{joules}}{\text{cm}^2}$$

Irradiance describes the power density of a continuous-wave laser beam:

$$\text{Irradiance} = \frac{\text{intensity}}{\text{area}} = \frac{\text{watts}}{\text{cm}^2}$$

Spot Size

The spot size of a laser is an important parameter, which in many lasers can be chosen within a range of possibilities. Spot size is not simply based on the size of the target area. In fact, the spot size can have a great deal of influence on the depth of penetration of the laser, regardless of wavelength. Clearly, it is easy to understand that the larger the area to be treated, i.e. when performing photoepilation of hairs on an entire male back, a larger spot size is more convenient and faster. The opposite would be true when removing unwanted hair from an upper lip. However, one needs to be aware that the spot size directly influences the fluence and the irradiance of the laser beam as well as the depth of penetration and the scattering of the laser light. The fluence and the irradiance of the laser beam, which measure energy and power density, respectively, are inversely proportional to the square of the radius of the spot size. Thus, decreasing the spot size by 50% will increase the fluence or irradiance by a power of 4. Hence, to maintain the same energy density with a spot size only half the diameter, one would have to reduce the fluence or irradiance by a factor of 4. Further, small spot sizes allow for greater scattering of the laser beam, thus limiting its penetration, while larger spot sizes penetrate more deeply into the tissue due to less scattering [18]. This increase in penetration with larger spot sizes does not hold true when considering microscopic spot sizes, as is seen with fractional resurfacing. Spot sizes of one hundred to several hundred micrometers somehow penetrate very deeply into the skin. Very high energies can be safely delivered without significant epidermal damage via micrometer spot sizes, allowing for deep penetration and rapid recovery.

Beam Profiles

The beam profile represents how the intensity of the light produced by a laser is distributed across the beam diameter. This special distribution of power is also referred to as transverse electromagnetic mode (TEM). Most commercial lasers used in dermatology produce a beam with a gaussian profile. This beam profile is called the fundamental mode or TEM_{00} of the laser. In this beam profile, the intensity is not the same across the beam diameter but its intensity peaks at the centre of the diameter. Roughly 86% of the power is contained at that impact spot. From the center of peak intensity, the intensity then falls off to both sides with a gaussian distribution (bell shaped). When focused through a lens, this mode of lasing gives the smallest focal point. However, such a distribution of light intensity clinically requires that the target be treated with some degree of overlap, in order to achieve more uniform energy throughout the tissue.

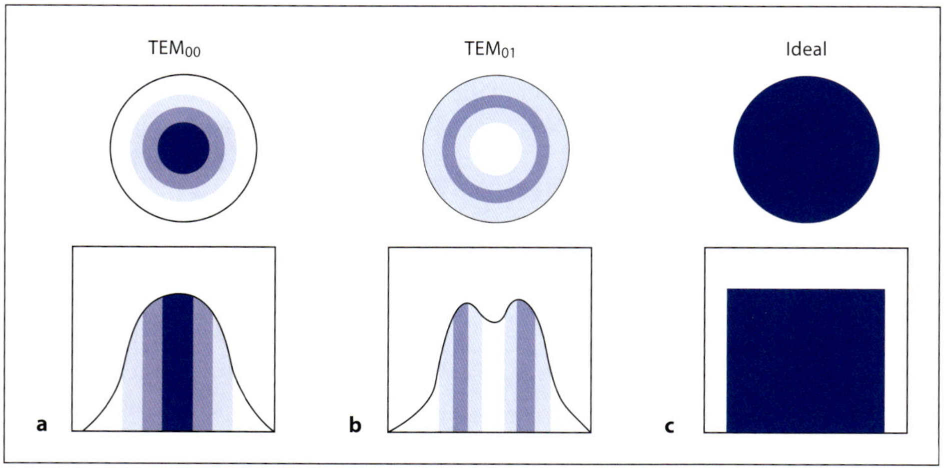

Fig. 7. Beam profiles. **a** Gaussian distribution with central hot spot which contains approximately 86% of the total energy. **b** Donut distribution with cold spot in the center. **c** Ideal energy distribution across the entire spot.

Other modes, so called higher order modes (TEM_{01}, TEM_{02}) of laser operation are for example doughnut-shaped or target-shaped modes. There, the intensity is greatest at the edge of the beam diameter (doughnut-shaped) or fluctuates across the beam diameter (target-shaped). The advantage of these modes is that they deliver a more constant intensity across the beam diameter and overlapping should not be performed [22, 14]. Ideally of course, the power would be distributed evenly over the entire surface area (fig. 7).

Pulse Duration

There are different modes of how laser light can be delivered. It can either be delivered in a continuous wave or a pulsed wave. The continuous-wave light is emitted over an uninterrupted period of time with a constant beam of light (fig. 8a). These continuous waves are of rather low power with limited peak energies. They are mostly gaseous lasers, such as the CO_2 laser. Continuous beams can be mechanically shuttered to produce beams with short emission times, which then are referred to as quasi-continuous-wave lasers. In fact these are just interrupted emissions of the usually continuous-wave laser energy (fig. 8b). Such mechanical shutters are able to generate pulses with durations between 1 ms and 1 s. However, those are not truly pulsed laser beams and the peak power is still low. Truly pulsed lasers can be either short pulse with nanosecond (10^{-9} s) pulse durations, or long pulse with pulse durations within the millisecond (10^{-3} s) range. There are also investigations into creating picosecond (10^{-12} s) pulsed lasers. Truly pulsed lasers are high-power lasers that emit ultra-short single pulses with extremely high energies (fig. 9). They are mostly solid state lasers, such as the Nd:YAG. Quality switching or Q-switching is the method of producing a short pulse of light with high peak powers by using an electro-optical switch with two polarizers within the laser chamber, suddenly creating a population inversion and emitting the stored energy in extremely short highly energetic pulses. Hence, Q-switched lasers can emit ultra-short pulses in the range of nanoseconds with extremely high peak power.

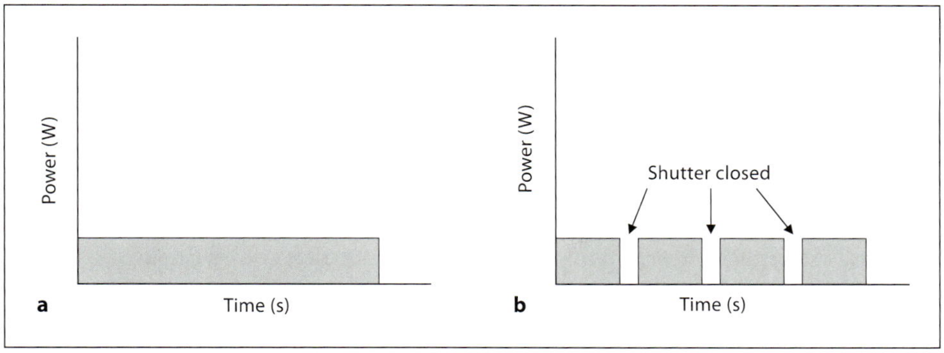

Fig. 8. **a** Continuous-wave lasers. **b** Quasi-continuous-wave lasers.

Continuous laser light leads to bulk heating and nonselective tissue damage. The concept of selective photothermolysis applies only to pulsed lasers which can emit light in times shorter than the TRT of the target chromophore and as such allow for selective damage. The pulsed mode further allows for the tissue to cool between pulses, reducing the spread of thermal damage.

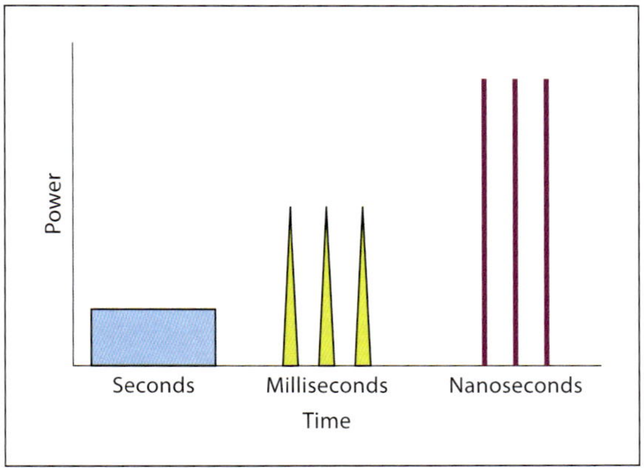

Fig. 9. Pulsed lasers.

Surface Cooling

When we aim at a target located in the dermis or subcutaneous tissue, the laser must first pass through the epidermis. If the target is not located in the epidermis, the epidermis needs to be protected from this transient passage of light. This can effectively be achieved by cooling. In many laser and light procedures, especially during photoepilation and the treatment of vascular lesions, cooling of the epidermis is the one crucial factor in avoiding adverse events.

Cooling can be performed before, during, and after treatment, and is hence referred to as pre-cooling, parallel cooling, or post-cooling. Pre-cooling is performed shortly before laser treatment in order to reduce or prevent excess heating of the epidermis during treatment. It is usually performed by a cryogen spray built into the laser handpiece, an air cooling device, contact cooling, or even ice. Parallel cooling is performed simultaneously with laser treatment, usually by cooled sapphire tips within the handpiece. Post-cooling helps to extracted excess heat from the tissue and prevent side effects such as post-inflammatory hyperpigmentation due to bulk heating and damage to surrounding tissue structures. Cooling at any stage of the treatment reduces pain and edema and patients appreciate its use.

However, care has to be employed when cooling during the targeting of epidermal structures. Too much cooling of the epidermis can result in ineffective treatments. Cryogen burns or post-inflammatory hyperpigmentation from over-aggressive cooling is also a danger. The proper balance between laser energy and cooling is always critical.

Summary

When using lasers in clinical applications, it is important to remember all aspects of the laser, including the wavelength, pulse duration, fluence, and cooling. These should always be viewed and adjusted according to the biology of the target structure and the tissue. Understanding the basic principles of lasers will allow you to expand your treatment indications and success.

References

1 Maiman T: Stimulated optical radiation in ruby. Nature 1960;187:493–494.
2 Goldman L, Blaney DJ, Kindel DJ, Kindel DJ Jr, Franke EK: Effect of the laser beam on the skin: preliminary report. J Invest Dermatol 1963;40:121–122.
3 Goldman L, Rockwell RJ, Meyer R, Otten R: Investigative studies with the laser in the treatment of basal cell epitheliomas. South Med J 1968;61:735–742.
4 Anderson RR, Parrish JA: Selective photothermolysis: precise microsurgery by selective absorption of pulsed radiation. Science 1983;220:524–527.
5 Manstein D, Herron GS, Sink RK, Tanner H, Anderson RR: Fractional photothermolysis: a new concept for cutaneous remodeling using microscopic patterns of thermal injury. Lasers Surg Med 2004;34:426–438.
6 Einstein A: Zur Quantentheorie der Strahlung. Physiol Z 1917;18:121–128.
7 Stratigos AJ, Alora MB, Urioste S, Dover JS: Cutaneous laser surgery. Curr Probl Dermatol 1998:10:127–174.
8 Raulin C, Greve B, Grema H: IPL technology: a review. Lasers Surg Med 2003;32:78–87.
9 Babilas P, Schreml S, Szeimies RM, Landthaler M: Intense pulsed light (IPL): a review. Lasers Surg Med 2010;42:93–104.
10 Huzaira M, Anderson RR, Sink K, Manstein D: Intradermal focusing of near-infrared optical pulses: a new approach for non-ablative laser therapy. Lasers Surg Med 2003;32(suppl 15):17–38.
11 Khan MH, Sink RK, Manstein D, Eimerl D, Anderson RR: Intradermally focused infrared laser pulses: thermal effects at defined tissue depths. Lasers Surg Med 2005;36:270–280.
12 Fuller TA: The physics of surgical lasers. Laser Surg Med 1980;1:5–14.
13 Reinisch L: Laser physics and tissue interactions. Otolaryngol Clin North Am 1996;29:893–914.
14 Ratz JL: Laser physics. Clin Dermatol 1995;13:11–20.
15 Anderson RR, Parrish JA: The optics of human skin. J Invest Dermatol 1981;77:13–19.
16 Herd RM, Dover JS, Arndt KA: Basic laser principles. Dermatol Clin 1997;15:355–372.
17 Goldberg DJ: Laser Dermatology. Berlin, Springer, 2005.
18 Parrish JA, Deutsch TF: Laser photomedicine. IEEE J Quantum Electron 1984;QE-20:1386–1396.
19 Herd RM, Dover JS, Arndt KA: Basic laser principles. Dermatol Clin 1997;15:355–372.
20 Watanabe S, Flotte TJ, McAuliffe DJ, Jacques SL: Putative photoacoustic damage in skin induced by pulsed ArF excimer laser. J Invest Dermatol 1988;90:761–766.
21 Altshuler GB, Anderson RR, Manstein D, Zenzie HH, Smirnov MZ: Extended theory of selective photothermolysis. Lasers Surg Med 2001;29:416–432.

Suggested Reading

Anderson RR: Laser-tissue interactions in dermatology; in Arndt KA, Dover JS, Olbricht SM (eds): Lasers in Cutaneous and Aesthetic Surgery. Philadelphia, Lippincott-Raven, 1997, pp 25–51.

Anderson RR, Parrish JA: Selective photothermolysis: precise microsurgery by selective absorption of pulsed radiation. Science 1983;220:524–527.

Bogdan Allemann I, Kaufman J: Fractional photothermolysis – an update. Lasers Med Sci 2010;25:137–144.

Herd RM, Dover JS, Arndt KA: Basic laser principles. Dermatol Clin 1997;15:355–372.

Houk LD, Humphreys T: Masers to magic bullets: an updated history of lasers in dermatology. Clin Dermatol 2007;25:434–442.

Hruza GJ, Geronemus RG, Dover JS, Arndt KA: Lasers in dermatology. Arch Dermatol 1993;129:1026–1035.

Manstein D, Herron GS, Sink RK, Tanner H, Anderson RR: Fractional photothermolysis: a new concept for cutaneous remodeling using microscopic patterns of thermal injury. Lasers Surg Med 2004;34:426–438.

Ratz JL: Laser physics. Clin Dermatol 1995;13:11–20.

Ross EV: Laser versus intense pulsed light: competing technologies in dermatology. Lasers Surg Med 2006;38:261–272.

Stratigos AJ, Dover JS: Overview of lasers and their properties. Dermatol Ther 2000;13:2–16.

Tanzi EL, Lupton JR, Alster TS: Lasers in dermatology: four decades of progress. J Am Acad Dermatol 2003;49:1–31, quiz 31–34.

I. Bogdan Allemann, MD
Klinik für Plastische Chirurgie und Handchirurgie, Universitätsspital Zürich
Rämistrasse 100
CH–8091 Zürich (Switzerland)
Tel. +41 44 255 1111, E-Mail Inja.bogdan@usz.ch

Bogdan Allemann I, Goldberg DJ (eds): Basics in Dermatological Laser Applications.
Curr Probl Dermatol. Basel, Karger, 2011, vol 42, pp 24–34

Laser-Induced Tissue Reactions and Dermatology

Rebecca J. Weber[a] · Brent R. Taylor[b] · Dendy E. Engelman[c]

Departments of [a]Cell and Molecular Pharmacology and Experimental Therapeutics and [b]Dermatology, Medical University of South Carolina, Charleston S.C., and [c]Skin Laser & Surgery Specialists of New York & New Jersey, New York, N.Y., USA

Abstract

Knowledge of laser tissue reactions and tissue properties allows the practitioner to tailor a treatment to an individual patient's need and goals. A laser's power, spot size and pulse duration may be manipulated to yield different tissue reactions. Five tissue reactions, each the result of varying laser pulse durations and energy densities, may be achieved. They are photochemical, photothermal, photoablation, plasma-induced ablation and photomechanical. Of these, photothermal reactions are most utilized in dermatology. When higher powered pulses are applied, tissue often undergoes multiple reactions simultaneously. An understanding of these reactions allows their effects to be predicted. In this chapter, the various reactions are reviewed, and the reactions caused by many of the most commonly used lasers in dermatology are discussed.

Laser-induced tissue reactions are dependent upon the parameters of the laser being utilized as well as the target tissue's properties. The laser's power, spot size and pulse duration may be manipulated to yield different tissue reactions [1]. Exposure time determines the ultimate temperature of the tissue, and is therefore the most crucial parameter determining which tissue reactions will occur [2]. The fluence of a laser is the amount of energy delivered in a single pulse. Fluence is measured in J/cm^2 and, as it is dependent upon area, changing the spot size will impact the laser's irradiance [3]. All medically relevant tissue effects occur between 1 and 1,000 J/cm^2 [2]. The net transmission of a given wavelength through tissue is determined by tissue coefficients of reflection, absorption and scattering [2]. The wavelength of a laser will predominantly determine penetration depth and location of energy absorption. The penetration depth of the tissue is defined as the depth where the collimated beam is attenuated by 37% and can be calculated as the reciprocal of the sum of the tissue absorption and scattering coefficients. Penetration will continue for visible light until water absorption becomes high, at which point the depth of penetration decreases rapidly. In general, blue light (e.g. 460 nm) has poor penetration, whereas red and near-infrared light (650–1,100 nm) exhibit deeper penetration [4]. Along with water, chromophores in the tissue influence where absorption occurs. A chromophore is a molecule that acts like a catalyst to allow a light-induced reaction to occur in an otherwise non-absorbing molecule. Of the multiple chromophores in the skin, the most relevant are melanin, hemoglobin,

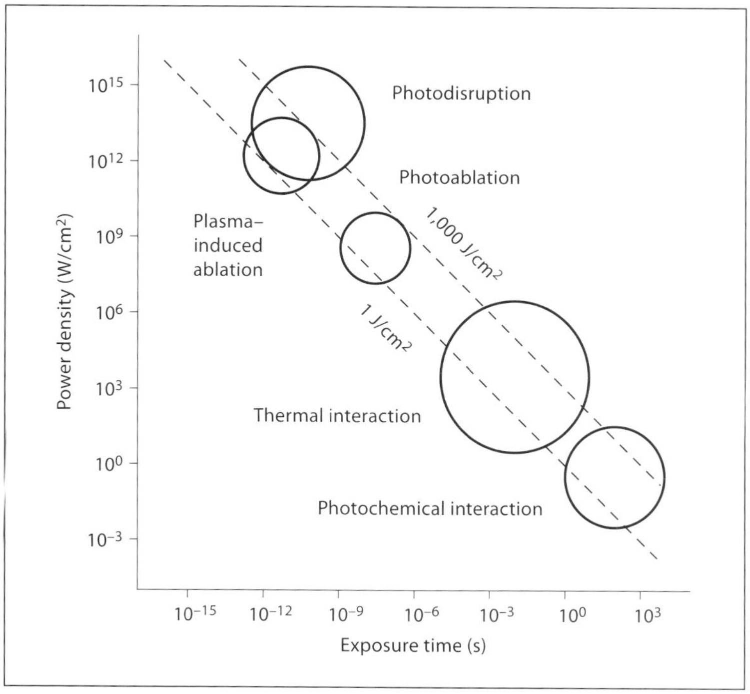

Fig. 1. Map of laser-tissue reactions. Circles indicate laser parameters for each tissue reaction. Reprinted from Niemz [fig. 3.1, 2].

oxyhemoglobin, DNA and protein and the various cytochromes, all of which maximally absorb different wavelengths. Ultraviolet light is absorbed by DNA, proteins and water; visible light is absorbed by melanin, hemoglobin, oxyhemoglobin, hemosiderin, bilirubin, carotenoids and cytochromes; near-infrared light is weakly absorbed by melanin, hemoglobin, oxyhemoglobin and cytochromes; mid-infrared light is absorbed by water and lipids; and far-infrared light is absorbed by water [5]. Most biological molecules exhibit a complex band structure of absorption between 400 and 600 nm. The 'therapeutic window' of tissue, between 600 and 1,200 nm, is the range of wavelengths that penetrate the tissue deeply due to low absorption [2]. The scattering of incident light occurs when the photon propagation is altered due to local changes in the refractive index. The majority of the scattering is done by melanin, cell nuclei and type I collagen in the dermis. Short wavelengths are scattered more than long; thus, allowing longer wavelengths to penetrate deeper into tissue [5]. Five tissue reactions that may be elicited in tissue by lasers are discussed below. They are photochemical, photothermal, photoablation, plasma-induced ablation and photomechanical. Each is the result of different pulse durations and energy density. Pulses less than 1 ns allow plasma-induced ablation and photomechanical effects to occur, 1-ns to 1-ms pulses allow photoablation to occur, 1-ms to 1-second pulses allow a thermal reaction, and greater than 1-second pulses allow photochemical effects [2]. The overlap in pulse durations is one factor allowing tissue to exhibit more than one reaction at a single time. The above tissue reactions are also separated in their required energy density (fig. 1). The major effect of lasers on the skin is photothermal and will be discussed in the most detail [6].

Tissue Reaction: Photomechanical/ Photodisruptive

A photomechanical reaction occurs when a high density of photons breaks intramolecular bonds such that ions are created (plasma generation). These ions rapidly expand, creating a shock wave that fragments surrounding tissue. A photomechanical reaction is not to be confused with plasma-induced ablation, a reaction wherein plasma generation remains localized (refer to next section).

Laser Parameters

To induce a photomechanical tissue reaction, high-pulse energies with durations of nanoseconds or shorter [7] are used to create high-energy plasma [2]. High irradiances of 10^8–10^{12} W/cm^2 can generate strong electric fields [8]. The high spatial density of photons must be coherent to allow their fields to add constructively [7]. The exposure rate used corresponds to 10^6–10^7 V/cm^2 [9] and is comparable to intermolecular Coulomb fields [7]. Thus, the electric fields are adequate to break intracellular structures and ionize tissue material [8].

Reaction Details: Sequence of Photomechanical Effects

Photomechanical effects rely on the generation of plasma, a process called optical breakdown, followed by shock wave generation, cavitation, and, if the reaction occurs near a solid boundary, jet formation [2]. These four different processes, separated by time yet reliant upon each other, explain the tissue effects observed in photomechanical reactions. Each process will be discussed individually as well as their relationships with the other processes.

Plasma Formation

Plasma formation begins during the pulse of the laser and only lasts a few nanoseconds before the free electrons generated diffuse into adjacent tissue. Plasma formation relies on optical breakdown, the process by which ionized molecules and free electrons are created and subsequently ejected from the surface to create a localized cloud called plasma [10]. Whether or not optical breakdown will occur is largely dependent upon the created local electric field strength, 'E'. When 'E' exceeds a certain threshold, the electric field required to induce the ionization of molecules and atoms, optical breakdown will occur. For plasma to be created, an electron avalanche must be generated in which multiple electrons are accelerated. This may occur via two distinct processes dependent upon the laser being used: thermionic emission as a result of using a Q-switched laser or via multiphoton ionization when using a mode-locked laser. Plasmas induced by Q-switched lasers are of a higher energy because of the associated increase in the threshold energy of plasma formation [2]. Thermionic emission involves a pulse that highly heats the target. A photon is absorbed and produces thermions, which are electrons released from their covalent bonds by thermal ionization [10]. Thermions can convert their energy into kinetic energy and free themselves from their surroundings. After the initial thermion is freed, it accelerates and can collide with another atom. This can lead to the ionization of the second atom and generate two free electrons each with less kinetic energy than the original thermion. These two free electrons can absorb additional incoming photons, accelerate and strike other atoms to generate an 'electron avalanche'. Tunneling ionization can cause plasma formation when a strong electric field can change the potential energy of an electron, defined as its Coulomb potential, so that a bound electron can escape and induce a cascade as described above [10]. In order to achieve optical breakdown, the irradiance utilized must be intense enough to cause rapid ionization to overcome losses such as inelastic collisions and the diffusion of electrons that threaten to quench the cascade. However, at very high irradiances, plasma can form a shield (discussed in photoablation) and this effect becomes more significant as the pulse energy increases [2].

Shock Wave Generation

Shock waves may be induced by phase explosion or the expansion and subsequent collapse of cavitation bubbles [10]. In a photomechanical tissue reaction, shock wave generation begins as the plasma expands [2]. The shock wave, associated with a high mechanical energy [10], quickly leaves the target and enters adjacent tissue. Thirty to fifty nanoseconds later, the shock wave slows down from supersonic speed to a normal acoustic wave. Once a shock wave is generated, it will precede both upward toward the air-surface interface as well as downward into tissue. The downward wave will propagate without interruption. These shock waves produce mechanical side effects, such as ablation and fracturing, which scale linearly with the absorbed energy. The spatial extent of the shock wave can reach 0.2 mm^2.

Cavitation

The next observed reaction is cavitation, a macroscopic effect that begins 50–150 ns after the advent of the pulse. Cavitation is observed when the laser is focused into the tissue as opposed to on its surface, and only if the plasma is generated in soft tissue. The basis of cavitation is the high plasma temperature generated which allows the vaporization of water near the target. The observed time delay is due to the requirement for the bubble's contents to be produced during the vaporization process.

As water vaporizes and gas is produced, work is done against the surrounding tissue. This allows the kinetic energy to be converted to potential energy that is stored in the expanded cavitation bubble. Within a millisecond, the bubble will implode due to the outer static pressure, which then causes the contents of the bubble, water vapor and CO_2 to undergo compression. This compression increases the pressure and temperature of the bubble, which allows the bubble to rebound. The bubble may expand and collapse multiple times within the span of a few microseconds with each cavitation having the potential to produce a shock wave, and, if near a solid boundary, form a jet. This process will repeat until all the bubble's energy has been dissipated and the gases are dispensed into adjacent tissue. The average amount of energy lost in the first cycle is nearly 84% and is mainly due to the emission of sound. Cavitation bubbles can reach a diameter of a few millimeters and thus induce macroscopic effects [2].

Jet Formation

Jet formation occurs when a high-speed ejection of fluid is created secondary to the collapse of a cavitation bubble in the vicinity of a solid boundary. Due to the necessity of location near a solid boundary, cavitation bubbles attached to solid structures have a greater chance of causing damage than those not formed near a solid boundary. The jet causes a high-impact pressure against the wall against which it is formed. The process of jet formation relies on the sequence of events where the fluid surrounding the bubble is accelerated toward the center of the collapsing bubble. An asymmetric collapse can occur at the side pointing toward the boundary, as less fluid is available in this area. The faster collapsing fluid on the opposite side will gain kinetic energy and gain in speed until the collapse has finished and a jet has formed in the opposite direction. The jet has the potential to damage and erode the boundary. A greater potential for damage exists if a gas bubble from an earlier pulse is hit with acoustic waves from a subsequent pulse. Gas bubbles with a small volume and strong surface tension are able to dissolve quickly, and thus present less of a risk for inducing damage than larger gas bubbles [2].

Applications
Photomechanical effects are utilized in tattoo removal with Q-switched nanosecond lasers. This is their primary application in dermatology. The Nd:YAG, Nd:YLF and Ti:sapphire are lasers often used in ophthalmology to elicit this reaction and have found a role in performing iridotomy and eye capsulotomies [7].

Tissue Reaction: Plasma-Induced Ablation

Laser Parameters
Plasma-induced ablation occurs secondary to optical breakdown and requires power densities exceeding 10^{11} W/cm^2 in solids and fluids. This can be achieved only with picosecond pulses as nanosecond pulses produce a higher energy density associated with optical breakdown and are accompanied by mechanical side effects [2].

Reaction Details
As this reaction relies on the induction of plasma, refer to details of plasma formation presented in the previous section.

Tissue Response
Overall, plasma-induced ablation is a very clean and well-defined reaction without evidence of thermal or mechanical damage. At higher pulse energies, and thus higher plasma energies, shock waves and other mechanical side effects might determine the tissue effect. Optical breakdown renders possible an energy deposition not only in pigmented tissue but also in weakly absorbing media due to the increased absorption of the induced plasma [2].

Applications
Plasma-induced ablation is not often utilized for dermatology. The same lasers used for photomechanical effects, Nd:YAG, Nd:YLF and Ti:sapphire, are utilized to elicit plasma-induced ablation for caries therapy or cornea surgery [2].

Tissue Reaction: Photoablation

As photoablation is associated with the chemical transition of bonds breaking, some sources refer to this reaction as a photochemical reaction. However, the term photochemical should be reserved for long-duration pulse interactions as opposed to the short pulses utilized to induce photoablation.

Laser Parameters
Photoablation is observed when nanosecond pulses are used with irradiances of 10^7–10^8 W/cm^2. The CO_2 and excimer lasers are typically utilized to produce this irradiance with pulse lengths of 10 ns. Only ablation caused by ultraviolet photons should be considered photoablation as only these lasers are able to produce adequate energy to dissociate bonds, whereas higher wavelengths are associated with an undesired thermal component. Generating higher harmonics of other lasers can also result in ultraviolet radiation.

Reaction Details
Photoablation is a combination of tissue evaporation followed by the expulsion of liquefied material by hydrodynamic forces. The laser energy absorbed is used for the phase transitions of vaporization and liquefication [8] with excess laser energy providing kinetic energy [7]. The high intensity of the laser radiation liquefies tissue, and the arising pressure can eject the molten material from the target [8]. A certain threshold of intensity must be applied to achieve photoablation, and is dependent on the bonds requiring dissociation [2] as well as the absorptive and elastic properties of tissue and its viscosity. The interaction tissue volume is determined by spot diameter and optical penetration depth [7].

Tissue Response
Photoablation is a very clean exact process with excellent predictability and little thermal damage [2]. As the reaction occurs rapidly and there is

little energy remaining [8], thermal effects do not have time to occur [2]. This is also due to the utilization of ultraviolet wavelengths that exhibit low penetration [7]. The ablation threshold is largely determined by the number of bonds which must be dissociated, which means that large fragments with a small surface-to-volume ratio require less energy than fragments with a higher ratio [2]. If energy below the evaporation threshold is used, thermal energy will increase [8] and lead to hyperthermal damage [7]. If the ablation threshold is surpassed, plasma will be produced which can lead to plasma shielding [8] and is considered the saturation limit. If the reached ablation depth is less than the optical absorption depth, subsequent pulses will enter the previously irradiated tissue and proceed into the deeper tissue. This translates into only the first few pulses being unique with subsequent pulses demonstrating a linear relationship between the number of pulses and ablation depth [2]. As discussed in photomechanical effects, the plasma produced in photoablation has a high power density and thus a high absorptive capacity that absorbs subsequent laser radiation to produce thermal effects and inhibit further ablation.

Applications
To induce photoablation, typically only photons from ultraviolet lasers, usually excimer (308 nm), provide enough energy to dissociating bonds [2]. ArF 1 (93 nm), KrF (249 nm), XeCl (308 nm) and XeF (351 nm) lasers all cause photoablation due to the high absorption of their wavelengths and can cut without heating, thus earning the name 'cold lasers' [7]. In dermatology, the excimer (308 nm) laser is most commonly used to induce photoablation. Some of its applications include treatment for psoriasis, vitiligo and hypopigmented striae. Some ophthalmologic applications of photoablation include correcting hyperopia, myopia and astigmatism. Other lasers used to create photoablation include the ErCr:YSGG (2,780 nm) and Er:YAG (2,940 nm), and are used for dental hard tissue due to their ability to cause ablation without thermal effects [7].

Tissue Reaction: Photothermal

Laser Parameters
Photothermal reactions are induced by irradiances of 1–10^6 W/cm^2 [8] for pulse durations of 1 ms to 100 s.

Reaction Details
Photothermal reactions are a two-step procedure whereby a molecule absorbs a photon, is excited, and then subsequently deactivates while kinetically activating its neighbor via an inelastic collision. The molecule's absorption of a photon is facilitated by its accessible vibrational states. The increase in kinetic energy translates into a rise in tissue temperature and the observed reaction which depends on the temporal evolution of temperature achieved, exposure time and placement of heat [2]. Different thermal tissue effects – such as coagulation, vaporization, carbonization and melting – may be distinguished and multiple thermal effects are usually induced in the tissue depending on the laser. There may be a range in the observed effects from carbonization at the tissue surface to only hyperthermia within the tissue [2]. The rate of heat generation depends on the rate of absorption of photons within tissue [4]. If a phase transition does not occur, a liner change in temperature is observed. The body's temperature is 37°C, and no observable tissue reaction will occur up to 42°C. Once 42–50°C is reached, bond destruction, conformational changes of molecules and membrane alteration can occur in a reaction called hyperthermia. If this stage lasts for several minutes without sufficient cooling, necrosis can occur. The probability of the cells staying alive at this point depends upon the duration and evolution of the temperature. Beyond 50°C, the cells begin to evidence a decrease in cell repair mechanisms, a reduction in enzyme activity and

a decreased chance of survival. At 60°C, protein and collagen will denature which leads to coagulation and necrosis [2]. Collagen melting may be a desirable reaction, as it causes the fibers to shrink with a noticeable skin tightening [5]. Membrane permeability is compromised at 80°C and the cell's chemical equilibrium is destroyed. Water will vaporize at 100°C and the resulting water loss causes the thermal properties of the tissue to change as both conductivity and diffusivity are reduced [4]. The large vaporization of water (2,253 kJ/kg) permits the vapor to carry away some heat and prevent further thermal damage. However, gas bubbles formed can cause mechanical damage and, as water expands in volume, an increase in pressure causes microexplosions [2]. Before removal of tissue, temperatures below the tissue surface exceed surface temperature. Afterwards, there is a sudden decrease in tissue temperature due to the loss of heated tissue [4]. Continued irradiation will occur in a repetitive cycle until all water vaporizes. If the laser energy continues, the tissue temperature will reach above 100°C where carbonization can occur. The tissue turns black and smokes which can reduce visibility during surgery. Beyond 300°C, melting can be achieved in certain targets. Although melting and carbonization require high temperatures, any laser can elicit these effects if the exposure time is sufficient. The temporal evolution and duration of temperature dictate whether or not the thermal reaction is irreversible or reversible. Carbonization and coagulation are always irreversible reactions. Hyperthermia may be irreversible or reversible dependent upon the tissue properties, the laser utilized and the duration and temporal evolution of temperature [2].

Tissue Response

In thermal reactions, absorption by water is important [2]. An absorption peak for water at 3 μm is due to its symmetric and asymmetric vibrational modes corresponding to 1.08×10^{11} and 1.13×10^{11} respectively [11]. As this tissue reaction is dependent upon heat generation, any mechanism that has the potential to carry away heat must be factored into estimating the tissue reaction. Heat transport is based on tissue properties such as heat conductivity and heat capacity. Losses of heat include heat conduction, heat convection due to blood flow and heat radiation. The latter two are usually only important for lasers with long pulse durations. Heat conduction is the primary means by which heat is transferred to adjacent tissue and is proportional to the temperature gradient [2]. To minimize thermal affects, the duration of pulse must be optimized [2]. This may be done by considering the thermal relaxation time (TRT), a measure of the thermal susceptibility of tissue [12]. The TRT is directly proportional to the area of the tissue and inversely proportional to absorptive coefficients of the tissue. Thus, small objects with a low TRT are spared while larger structures exhibit more heat dispersion during exposure [13]. Laser pulses less than the TRT do not allow heat diffusion to the distance given by optical penetration depth and thermal damage is negligible. Pulse durations greater than the TRT allow heat to diffuse to adjacent tissue and induce thermal damage [2]. The TRT of a target can be estimated through examination with a magnifying glass. A target with X millimeter vessels will respond best to $(X)^2$ ms [5]. The shortest TRT of 1 μs occurs at the absorption peak of water near 3 μm. This leads to the 'rule of one microsecond' that states laser pulses less than 1 μs are not usually associated with thermal damage and the tissue temperature will increase linearly with time. Pulses longer than 1 μs will allow heat diffusion to occur and an increased thermal damage zone where temperatures no longer linearly increase. Besides pulse duration, a high repetition rate may evoke temperature increase if the rate of heat transport is less than rate of heat generation.

Applications

In dermatology, a photothermal tissue reaction is most commonly utilized, especially coagulation and vaporization. The types of lasers used in practice are argon ion, pulsed dye lasers (PDL), CO_2, Nd:YAG, Er:YAG, ruby, argon, krypton and diode [2, 8]. Of these, the Nd:YAG or diode lasers are typically used for coagulation, while the CO_2 laser is effective for vaporization and thermal cutting of tissue [2]. The CO_2 laser is used to accomplish ablative tissue reactions that rely on skin appendages to supply new epidermal tissue [1]. It can be run on continuous wave or pulsed, and may be transmitted through fibers. Its penetration in water is 10 μm and its ablation depth in tissue is dependent upon the water content [7]. This laser targets extra- and intracellular water to allow for vaporization and coagulation in the dermis. Pulses less than 1 ms allow for minimal thermal damage and are achieved with ultra-short pulse or computer-controlled shutter systems. Five J/cm^2 of energy is needed to exceed the vaporization threshold [1]. The zone of cavitation extends laterally and vertically away from the dermal-epidermal junction 2–3 diameters outside of the target and 300–400 μm down the follicles [14]. The thermal zone extends 1–2 diameters outside the target and the complete zone of damage is about three times the diameter of contact [14]. The best result in treated tissue was at 350 mJ where vaporization was observed without thermal damage [14]. The argon and krypton ion lasers have a low efficiency due to a significant amount of energy lost via heat [7]. Argon's main wavelengths are at 488 nm (blue) and 514.5 nm (green), which are well absorbed by oxyhemoglobin and melanin. The absorption by oxyhemoglobin, even though not its most effective wavelength for absorption, will damage red blood cells in cutaneous vessels in the first few millimeters of skin. Minimal tissue blanching is the clinical endpoint [15]. The ruby laser emits red light with a wavelength of 694 nm and can be used in long-pulse or Q-switched modes [7]. It is mainly absorbed by melanin and weakly absorbed by non-pigmented tissue making it ideal for deep hair follicles and tattoo removal [8]. The Nd:YAG laser is a low efficacy (less than 2%) laser that can operate in a pulsed or continuous wave fashion. The most emitted frequency is 1,064 nm [7] and by utilizing optical crystals, this laser can generate second harmonics at 532 nm [8]. At 1,064 nm, it has a penetration depth of 3–5 mm and elicits hyperthermal reactions with good hemostasis. At 532 nm, it exhibits low penetration, but good coagulative properties [7]. The Er:YAG laser was created in response to a demand for a laser able to achieve skin rejuvenation with a less aggressive approach than the CO_2 laser. Its wavelength of 2,940 nm is effectively absorbed by water; thus, it exhibits low penetration in water-rich tissue. Its short-pulse option of 250 μs is absorbed 12–18 times more efficiently than the CO_2 laser. Each pass can ablate 5–20 μm at a fluence of 5 J/cm^2 with no thermal damage apparent past 15 μm. Three laser passes can result in total epidermal ablation with fast re-epithelization. The short-pulse Er:YAG exhibits worse hemostasis and less clinical improvements as compared to the CO_2 laser. The Er:YAG laser is better than the CO_2 laser for dark-skinned patients due to its lower risk of pigmentary changes [3].

One specific example of the utilization of photothermal effects in dermatology is for laser hair removal. This is accomplished by lasers selectively targeting melanin in the hair shaft with its absorption of wavelengths from 600 to 1,000 nm. Melanin in the epidermis acts as a competing absorber for laser energy; thus, high concentrations of melanin in the epidermis present a risk of thermal injury.

Another application in dermatology is the selective removal of vascular lesions via the selective photothermolysis of the endothelium and vessel wall. Lasers emitting green or yellow wavelengths (PDL, KTP and intense pulsed light) are often utilized [5]. PDL (585–600 nm) are the most selective for vascular lesions less than 2 mm deep and 0.05–1.0 mm in diameter. Variable PDL with pulse lengths between 0.45 and 40 ms can be selected based on vessel size and the patient's tendency for transient purpura development.

The KTP laser (532 nm) functions similarly to the PDL [5]. As the argon laser is strongly absorbed by both hemoglobin and melanin, it is used for the superficial treatment of lesion such as port wine stains (PWS) [2]. The treatment of deeper vessels is accomplished via long-pulse near-infrared lasers, such as the alexandrite (755 nm), diode (800–810 nm) or Nd:YAG (1,064 nm) with cooling [5]. These lasers' long wavelengths are less scattered and absorbed in skin versus shorter wavelength lasers to allow for greater optical penetration [2]. Skin resurfacing also relies on photothermal effects where the epidermis and superficial dermis is vaporized and the remaining damaged dermal layer shrinks to results in tighter skin. Lasers utilized to accomplish this reaction include scanned CO_2, pulsed CO_2, pulsed Er:YAG, fractional Er:YAG and combined CO_2 and Er:YAG [1]. These lasers are also used to reverse photoaging, remove benign tumors and reduce acne scarring [5].

The future of laser utilization for photothermal effects lies in tailoring each treatment to the patient. For example, port wine stains are currently being evaluated with noninvasive imaging and mathematical modeling to identify the most effective laser to be utilized. Lasers could be developed for tattoo removal that emit low-fluence sub-microsecond pulses to selectively destroy pigment with little associated mechanical damage. The utilization of adjuvants, such as macrophage-colony-stimulating factor, could stimulate a local macrophage reaction required for successful treatment. Other chemotactic factors, such as imiquimod, might similarly be utilized [15].

Tissue Reaction: Photochemical Effects (Photodynamic Therapy)

Laser Parameters

Photochemical tissue effects are utilized to accomplish photodynamic therapy (PDT). Long pulse durations of 10–1,000 s with low irradiances of 10^{-3} to 1 W/cm^2 [8] are utilized to achieve fluence rates below the hyperthermia threshold [7]. Most wavelengths used are in the visible range due to their efficacy and high penetration [2]. The optical window of tissue, between 620 and 1,100 nm, makes red light with its wavelength of 620 nm and penetration depth of 1–3 mm most effective [16]. Almost any laser with an output of 400–800 nm in the visible spectrum can be used to activate aminolevulinic acid (ALA) or methyl aminolevulinate (MAL), two commonly used photosensitizers. Noncoherent light sources are used for large skin lesion due to their large illumination field. Polychromatic light sources allow for the use of different photosensitizers with different absorption maximums.

Reaction Details

PDT relies on exogenous or endogenous photosensitizers [15] called chromophores, which remain inactive until irradiation [2]. Photochemical, and thus PDT, tissue reactions are dependent upon the generation of reactive oxygen species by these chromophores [17]. After irradiation, the photosensitizer performs several simultaneous or sequential decays that result in intramolecular transfer reactions. Light is absorbed, and an electron of the photosensitizer is transferred to its excited singlet state. There are then three potential ways for it to decay: nonradiative and radiative singlet decay to singlet ground state or intersystem crossing to excited triplet state. The latter may then proceed to a singlet ground state via nonradiative or radiative triplet decay. Radiative singlet decay lasts nanoseconds and is called 'fluorescence', whereas triplet decay lasts milliseconds to seconds and is called 'phosphorescence' [2]. The deactivation of the photosensitizer to its original state is accompanied by the release of cytotoxic reactants such as singlet oxygen species that cause the irreversible oxidation of cellular structures, which is called 'photosensitized oxidation.' Singlet oxygen has a lifetime of 10–40 ns [13] and once produced, only diffuses

10–20 nm. This localizes the oxidative reaction to where the majority of the photosensitizers are located [18].

Tissue Response
PDT may be dependent on thermolabile enzymes and a mild hyperthermia may be synergistic in producing the tissue reaction [8] or inducing the synthesis of beneficial heat shock proteins [19]. The heat capacity and conductivity of tissue can influence the reaction and are hard to estimate due to irregular tissue geometry and blood flow. As fluence rates below the threshold for hyperthermia are used [7], there is no primary heating of the tissue [8]. However, the level of hyperthermia must be approached as to avoid a long exposure [7] and if temperatures above 50°C are reached, the same reactions discussed in the photothermal section will be observed [8].

Applications
Almost any light between 400 and 800 nm (visible) can be used to activate ALA and MAL [20]. Various lasers and lights have been utilized for PDT including blue LED (405–420 nm), red LED (635 nm), PDL (585 nm), intense pulsed light (500–1,200 nm) [21], argon-pumped dye lasers (630 nm) or diode (633 nm) [8]. The light sources may also be coupled to fiber optics to allow organ-specific treatment [20]. Although PDT is only FDA-approved for nonhyperkeratotic actinic keratosis (AK) on the face and scalp, off-label uses include photorejuvenation, the improvement of facial telangiectasias, pigmentary dyschromias and skin texture, as well as the treatment of AK. ALA is currently undergoing clinical evaluations in the United States for non-hyperkeratotic AK on the face and scalp, basal cell carcinoma unsuitable for regular therapy (both red light), and sebaceous gland hyperplasia and hidradenitis suppurativa (both blue light) [15]. PDT can be used for the treatment of acne by exploiting *P. acnes*' production of the porphyrins, PpIX and coproporphryin III, which absorb blue light at a peak of 415 nm. A variety of lights such as halogen, xenon and tungsten have been used successfully. Alternatively, by using an 810 nm diode laser, the sebaceous gland can be destroyed [22].

Tissue Reaction: Biostimulation

Utilizing lasers to achieve biostimulation remains a controversial topic [7]. Biostimulation is believed to fall under the category of photochemical reactions and occurs at very low irradiances of 1–5 mW with an energy fluence of 1–10 J/cm^2. It has been proposed that wound healing and anti-inflammatory reactions may occur as a result of treatment with red or near- infrared wavelengths [2]. HeNe (1–5 mW at 633 nm) or GaAs lasers (5 mW at 850 nm) may promote local wound healing due to the low-intensity light's effect on cell proliferation [8]. The light may be a stimulus to increase cell proliferation in tissue with an environment that is inhibitory to proliferation, such as low oxygen or pH [2].

Future

The future of laser treatment in dermatology will lie in the ability to eventually tailor each laser treatment to the individual patient and treatment goals. This may be achieved with a better understanding of how to elicit a specific tissue reaction, how an individual's tissue will affect the outcome of the treatment and how to select a laser with the most appropriate parameters. Concurrent efforts to create more precise lasers with specific pulse durations and fluences will allow the applications of lasers in dermatology to continually increase.

References

1 Sandhu N, Goldberg D: Cutaneous laser resurfacing, carbon dioxide. http://emedicine.medscape.com/article/1120283-overview (updated 18 June 2010, accessed 27 December 2010).
2 Niemz M: Laser-Tissue Interactions: Fundamentals and Applications, ed 3. Berlin, Springer, 2007.
3 Tanzi E, Badler RS: Cutaneous laser resurfacing, Erbium-YAG. http://emedicine.medscape.com/article/1120936-overview (updated 15 July 2009, accessed 27 December 2010).
4 Welch AJ, Torres JH, Cheong WF: Laser physics and laser-tissue interaction. Tex Heart Inst J 1989;16:141–149.
5 Wolff K, Katz S, Goldsmith L: Fitzpatrick's Dermatology in General Medicine, ed 7. New York, McGraw-Hill, 2007.
6 Chen WR, Adams RL, Bartels KE, Nordquist RE: Chromophore-enhanced in vivo tumour cell destruction using an 808-nm diode laser. Cancer Lett 1996;88:125–131.
7 Peng Q, Juzeniene A, Chen J, Svaasand LO, Warloe T, Giercksky K-E, Moan J: Lasers in medicine. Rep Prog Phys 2008;71:1–28.
8 Knappe V, Frank F, Rhode E: Principles of lasers and biphotonic effects. Photomed Laser Surg 2004;22:411–417.
9 Boulnois J: Photophysical processes in recent medical laser developments: a review. Laser Med Sci 1986;1:47–66.
10 Lee H, Jeong Y, Chan K: The advent of laser therapies in dermatology and urology: underlying mechanisms, recent trends and future directions. J Opt Soc Korea 2009;13:312–329.
11 Pohl RW: Optik und Atomphysik. Berlin, Springer, 1976.
12 Hayes JR, Wolbarsht ML: Thermal model for retinal damage induced by pulsed lasers. Aerospace Med 1968;39:474–480.
13 Anderson RR, Parrirsh JA: Selective photothermolysis: precise microsurgery by selective absorption of pulsed radiation. Science 1983;220:524–527.
14 Fulton JW, Shitabat PK: CO_2 laser physics and tissue interactions in skin. Lasers Surg Med 1999;24:113–121.
15 Goldberg G: Laser Dermatology. Berlin, Springer, 2005.
16 Moan J, Peng Q, Sorensen R, Iani V, Nesland JM: The biophysical foundations of photodynamic therapy. Endoscopy 1998;30:387–391.
17 Halliwell B: Reactive oxygen species in living systems: source, biochemistry, and role in human disease. Am J Med 1991;91:14–22.
18 Moan J, Berg K: The photodegradation of porphyrins in cells can be used to eliminate the lifetime of singlet oxygen. Photochem Phtobiol 1991;53:549–553.
19 Hantash BM, Bedi VP, Sudireddy V, Struck SK, Herron GS, Chan KF: In vivo histological evaluation of a novel ablative fractional resurfacing device. Lasers Surg 2007;39:96–107.
20 Rao J: Photodynamic therapy for the dermatologist. http://emedicine.medscape.com/article/1121517-overview (updated 11 January 2010, accessed 27 December 2010).
21 Hongcharu W, Taylor CR, Chang Y, Aghassi D, Suthamjariya K, Anderson RR: Topical ALA – photodynamic therapy of the treatment of acne vulgaris. J Invest Dermatol 2000;115:183–192.
22 Pollock B, Turner D, Stringer RA, Bojar V, Goulden G, Stables I, Cunliffe WJ: Topical aminolevulinic acid-photodynamic therapy for the treatment of acne vulgaris: a study of clinical efficacy and mechanism of action. Br J Dermatol 2004;151:616–622.

Dendy Engelman, MD
Skin Laser & Surgery Specialists of New York & New Jersey
115 East 57th Street Suite 710
New York, NY 10022 (USA)
Tel. +1 212 750 8900, E-Mail dendyengelman@gmail.com

Bogdan Allemann I, Goldberg DJ (eds): Basics in Dermatological Laser Applications.
Curr Probl Dermatol. Basel, Karger, 2011, vol 42, pp 35–39

Laser Safety

Jacob Dudelzak[a] · David J. Goldberg[b,c]

[a]Accredited Dermatology, Laser & Cosmetic Surgery, Brooklyn, N.Y. and [b]Skin Laser & Surgery Specialists of New York & New Jersey, New York, N.Y., and [c]Mount Sinai School of Medicine, New York, N.Y., USA

Abstract

The use of lasers in medical practice has seen great expansion in the past decades. However, these devices may also pose a significant hazard. Laser hazards are generally divided into beam hazards and nonbeam hazards. Beam hazards inflict ocular and cutaneous injury, whereas nonbeam hazards stem from the laser device itself or its interaction with materials within the surgical environment. The latter include laser plume hazards, fire hazards, and electrical hazards inherent in a high-voltage system that is a laser device. Therefore, a thorough understanding of these hazards along with methods to reduce their risk is of paramount importance in order to ensure maximal safety for the surgeon, the staff, and the patient.

The use of lasers in medical practice has seen great expansion in the past decades. However, these devices may also pose a significant hazard. Laser hazards are generally divided into beam hazards and nonbeam hazards. Beam hazards inflict ocular and cutaneous injury, whereas nonbeam hazards stem from the laser device itself or its interaction with materials within the surgical environment. The latter include laser plume hazards and electrical hazards inherent in a high-voltage system that is a laser device.

Ocular Hazards

Ocular injury is a very serious complication arising from the use of a laser. Usually operating in the millisecond range of pulse duration, a laser beam cannot effectively be shielded by the eyelid blink reflex, which takes a tenth of a second to complete. Furthermore, lasers operating in the infrared spectrum do not emit a bright visible light necessary to elicit a blink reflex [1].

Ocular damage resulting from laser surgical procedures is tied directly to the laser wavelength utilized. As a general rule, optical radiation emitted by lasers operating in the ultraviolet (200–400 nm), mid-infrared (1,400–3,000 nm), and far-infrared (3,000–10,600 nm) range of wavelengths are absorbed by the anterior ocular segment, and inflict injury on the lens or the cornea. On the other hand, light emitted by lasers in the visible light (400–760 nm) and near-infrared (760–1,400 nm) spectra penetrate through, amplified by the above structures, and are absorbed in the posterior ocular segment by the retina and the vascular choroid [2].

These findings reflect the interaction between light of a given wavelength and the tissue. For example, the far-infrared 10,600 nm wavelength light emitted by the CO_2 laser is highly absorbed by water and does not penetrate beyond approximately 100 μm into the cornea, which has an aqueous composition of 78%. The resulting corneal injury is mediated by a thermal process, in which the optical radiation is absorbed by water, resulting in the generation of heat energy, the rate of which exceeds that of its dissipation by the surroundings, leading to an elevation in temperature [3].

The retina is particularly vulnerable to visible and near-infrared spectrum optical radiation, even that emitted by low-powered light devices, as a result of focusing by the ocular refractive media. Retinal injury from optical radiation is in large part related to the absorption by the melanin chromophore and, to a lesser degree, xanthophyll and hemoglobin. Concentrated densely within the retinal pigment epithelium, as well as focally within the choroid, melanin absorbs optical radiation of visible and near-infrared wavelengths [3, 4]. The injury sustained as a result of laser exposure may be insidious given the lack of nociceptors in the retina. Depending on the site of retinal injury, the insult may range from insignificant in the periphery, to a perceptible deterioration of visual acuity and color discrimination in the foveal region. The degree of visual loss is determined by the number of lost photoreceptor cells in the fovea. The inflammation and edema resulting from a laser injury in the parafoveal region may extend into the fovea, resulting in transient reduction in visual acuity that may recover over a period of days to weeks [2].

Certain lasers have the ability to inflict injury on multiple ocular structures. For instance, the 1,320 nm neodymium-doped yttrium aluminum garnet (Nd:YAG) laser may damage the lens and cornea, as well as the retina. The 755 nm alexandrite, 810 nm diode, and the 1,064 nm Nd:YAG lasers endanger the retina and the lens; the 2,940 nm erbium:YAG (Er:YAG) laser may damage the lens as well as the cornea [2]. Q-switched near-infrared lasers, such as the Q-switched alexandrite and Nd:YAG lasers present an even greater threat through a mechanism of photoacoustic waves in addition to their thermal effects [5]. Light emitted in very short pulse durations, on the order of pico- and nanoseconds, results in a photoacoustic laser-tissue interaction, a mechanical, rather than a thermal, process that is not pigment-dependent. The extremely high temperatures (exceeding 10,000°C) generated within the tissue lead to electron stripping, resulting in plasma formation, the rapid expansion of which produces tissue-damaging photoacoustic waves. Retinal injury resulting from photoacoustic waves may culminate in perforation [3, 4].

Ocular protection is, therefore, of paramount importance when operating a laser device. Any person that may possibly be exposed to optic radiation must wear appropriate protective eyewear, including the laser operator, support staff, patients, and visitors. Protective eyewear is chosen based on the wavelengths of light emitted by the laser. Each pair of laser safety eyewear is designated with a central wavelength of rejection or a rejection band of wavelengths, as well as the optical density (OD) afforded by the lens. The OD parameter is defined by the log of the attenuation of light transmitted through the lens. For example, protective laser goggles with an OD of 4 allow $1/10^4$ of the laser energy to penetrate. Thus, the higher the OD value, the greater the protection afforded by the eyewear. On the other hand, higher OD laser eyewear may significantly reduce visible transmittance and impair color discrimination, factors that may challenge a laser surgeon [2, 6].

Two common laser eyewear options include those constructed with coated glass and those made up of polymeric materials. Some of the disadvantages of the glass eyewear include their greater weight and the ability of the reflective coating to become scratched, thus compromising protection. On the other hand, polymeric material-constructed eyewear is lightweight and absorbs, rather than reflects, radiation. A disadvantage of the latter, however, is that it may crack or melt. Protection against indirect exposure should be maintained with wrap-around goggles or side shields. Patient eye protection should be provided with corneal shields or mini-goggles that ensure complete light-proofing. Only heat-proof stainless steel corneal shields should be employed, as they reflect away the laser beam without generating heat, in contrast to plastic eye shields that heat up and may melt, causing ocular injury. Corneal shields must have a smooth polished concave surface and an anodized convex surface [2, 6].

Ocular injury may occur not only from a direct laser beam exposure, but also from exposure to the light reflected or scattered off glass, glossy metal, or plastic surfaces. Thus, all windows and mirrors in the room should be covered with opaque material, all jewelry removed, and all instruments anodized, roughened, and ebonized with black fluoropolymeric coating [5, 7].

A non-beam related ocular hazard associated with laser surgery that cannot be overlooked involves preoperative disinfectant chlorhexidine, which may cause keratitis and corneal opacification. Thus, this agent should not be used to disinfect corneal eyeshields or periorbital areas lest the substance be aerosolized into the laser plume. Instead, corneal shields should be autoclave-sterilized. They should be rinsed in sterile water and lubricated with an ophthalmic ointment prior to insertion [8].

A warning sign should clearly be displayed on the door of the room where laser surgical procedures take place informing potential visitors and staff of the potential ocular hazard inside.

Skin and Teeth Hazards

Cutaneous injury from a laser beam may be significant and its spectrum may range from redness to overt burns and scarring. Lasers may also be a hazard to oral health. Dental enamel, in particular, is vulnerable to ultraviolet and infrared light. Thus, lasers operating in these wavelengths, including the CO_2 and Er:YAG, pose a particular threat. Melting and resolidification of enamel at high fluences and cracking, charring, flaking, and discoloration at lower fluences has been reported. Therefore, oral protection is important in order to avoid injury. This may be achieved by keeping the mouth closed or by covering it with a moistened gauze or a protective mouthpiece [9].

Fire Hazard

Laser fires may result from ignition of combustible material in the vicinity of a laser procedure, resulting in burns. Common sources of fuel include flammable materials used during a laser surgical procedure, including gauze, towels or drapes, and respiratory devices, including face masks and nasal cannulae. Flammable materials, including makeup and hair spray, should be removed prior to a laser procedure. Alcohol should never be utilized to prep a laser surgical field. Caution should be exercised when performing laser procedures in hair-bearing areas, as hair can ignite and cause skin burns. In order to mitigate the risk of ignition of potentially combustible materials, the prepping with saline of the surgical area and supplies – including drapes, gauze, and tubing – is appropriate. The patient's hair should also be kept moist during the procedure to reduce its incendiary potential [2, 10, 11].

A water reservoir, such as a bowl or an irrigation syringe should always be within easy reach during a laser procedure to combat combustion fires on the surgical field. A halon fire extinguisher should also always be readily available in the event that an electrical fire erupts within the device circuitry. Unlike the carbon dioxide fire extinguisher, this fluorohydrocarbon fire extinguisher does not generally damage the laser components [12].

Electrical Hazard

Given the high-voltage high-current electrical nature of lasers, these devices carry a significant hazard of electrocution, and should only be maintained and repaired by specially trained and authorized personnel [2].

Plume Hazards

The interaction between a heat-producing device and the treated tissue has the potential to produce surgical smoke or plume. In fact, laser is the second most common heat-generating device employed by surgeons, second only to electrosurgical units. The evaluation of laser plume as a hazard has centered on its mutagenic and carcinogenic capacity, as well as the possible role it may play in disease transmission. Plappert et al. [13] have found substances released in CO_2 laser pyrolysis of tissue to be cytotoxic, genotoxic, clastogenic, and mutagenic. Numerous chemical substances, some carcinogenic, have been detected in the laser plume, including carbon monoxide, hydrogen cyanide, ammonia, formaldehyde, acrolein, toluene, and benzene [2, 14]. Substances found in the surgical smoke have been showed to be respiratory irritants in laboratory animals, resulting in the development of pulmonary and bronchiolar inflammation [14].

Laser procedures utilizing lower irradiance energies carry the risk of liberating cellular clumps and red blood cells that may be carried in the laser plume. It is believed that laser plume may harbor a greater infectious potential compared to electrosurgical smoke [15]. The presence of viral particles in the laser plume has been documented in the literature. For instance, intact human papillomavirus DNA has been isolated from laser plume generated in the CO_2 laser treatment of verruca plantaris and respiratory tract papillomatosis [16]. Human immunodeficiency virus (HIV) DNA capable of infecting cultured cells has also been recovered from a laser plume [17].

Implications of these findings on viral disease transmission have been a subject of debate, with emerging evidence supporting infectivity potential. Type 6 and 11 human papillomavirus has been isolated in the case of laryngeal papillomatosis developed by a laser surgeon involved in the treatment of anogenital papillomatosis. Furthermore, the development by a laser operator of verrucae in the anterior nares may point to human papillomavirus transmission via smoke plume [14, 18, 19].

In light of these findings and concerns, effective methods should be employed to combat the laser plume hazard. These include a smoke evacuator and laser masks. Most surgical masks only filter particles to approximately 0.5 μm in size. However, about 77% of particles in the laser plume are 1.1 μm or smaller. Therefore, well-fitted high-filtration, or laser, masks that filter particles larger than 0.1 μm by means of electrostatically charged synthetic fibers, should be employed in place of standard surgical masks. These must be frequently changed as condensate moisture eliminates their polarity. The mask should also have a moldable nosepiece to allow it to be worn snuggly [2].

High-filtration masks should not be the only protective method utilized during a laser surgical procedure. A high-efficiency smoke evacuation device with a powerful vacuum should also be used. The device should employ a triple-filter system. This includes a high-efficiency particulate air filter that removes particles larger than 0.3 μm; an ultra-low particular air filter, which is capable of removing particular matter up to 0.1 μm in size; and a charcoal filter to remove toxic chemicals within the smoke. An appropriate smoke evacuation system should have a 30- to 50-ft^3/min capacity [2]. The importance of proper operation of a smoke evacuator cannot be overemphasized. The effectiveness of the device is drastically reduced from 99 to 50% when the distance from the laser-treated site is increased from 1 to 2 cm [15].

Q-switched lasers present a particular challenge. The light energy provided by these systems in the course of treatment generates high-speed tissue fragments and particles that may escape capture by the smoke evacuation device. Safety measures that have been recommended in these situations include splatter shields or collecting cones, as well as appropriate eye protection, gloves, gowns, and laser surgical masks. Treatment through a transparent membrane has also been recommended [5, 20].

Conclusion

As lasers are becoming commonplace in medical practices, is it crucial more than ever to emphasize their safe use to protect all those involved in laser procedures. Potential laser hazards, including ocular and skin hazards, laser plume hazards, and fire hazards, should all be appreciated and appropriately addressed in order to ensure maximal safety for the surgeon, the staff, and the patient.

References

1 Friedman NR, Saleeby ER, Rubin MG, Sandu T, Krull EA: Safety parameters for avoiding acute ocular damage from the reflected CO_2 (10.6 microns) laser beam. J Am Acad Dermatol 1987;17(part 1):815–818.
2 Youker SR, Ammirati CT: Practical aspects of laser safety. Facial Plast Surg 2001;17:155–163.
3 Barkana Y, Belkin M: Laser eye injuries. Surv Ophthalmol 2000;44:459–478.
4 Goldman AI, Ham WT Jr, Mueller HA: Mechanisms of retinal damage resulting from the exposure of rhesus monkeys to ultrashort laser pulses. Exp Eye Res 1975;21:457–469.
5 Cao LY, Taylor JS, Vidimos A: Patient safety in dermatology: a review of the literature. Dermatol Online J 2010;16:3.
6 Teichman JM, Vassar GJ, Yates JT, Angle BN, Johnson AJ, Dirks MS, Thompson IM: Color vision deficits and laser eyewear protection for soft tissue laser applications. J Urol 1999;161:874–880.
7 Wood RL Jr, Sliney DH, Basye RA: Laser reflections from surgical instruments. Lasers Surg Med 1992;12:675–678.
8 Christian MM, Cox DO, Smith CV, Onouye T, Moy RL: Ocular damage due to chlorhexidine versus eyeshield thermal injury. Dermatol Surg 2001;27:153–157.
9 Pogrel MA, Muff DF, Marshall GW: Structural changes in dental enamel induced by high energy continuous wave carbon dioxide laser. Lasers Surg Med 1993;13:89–96.
10 Epstein RH, Brummett RR Jr, Lask GP: Incendiary potential of the flash-lamp pumped 585-nm tunable dye laser. Anesth Analg 1990;71:171–175.
11 Rohrich RJ, Gyimesi IM, Clark P, Burns AJ: CO_2 laser safety considerations in facial skin resurfacing. Plast Reconstr Surg 1997;100:1285–1290.
12 American Society of Anesthesiologists Task Force on Operating Room Fires, Caplan RA, Barker SJ, Connis RT, Cowles C, de Richemond AL, Ehrenwerth J, Nickinovich DG, Pritchard D, Roberson D, Wolf GL: Practice advisory for the prevention and management of operating room fires. Anesthesiology 2008;108:786–801.
13 Plappert UG, Stocker B, Helbig R, Fliedner TM, Seidel HJ: Laser pyrolysis products – genotoxic, clastogenic and mutagenic effects of the particulate aerosol fractions. Mutat Res 1999;441:29–41.
14 Barrett WL, Garber SM: Surgical smoke: a review of the literature. Is this just a lot of hot air? Surg Endosc 2003;17:979–987.
15 Sawchuk WS, Weber PJ, Lowy DR, Dzubow LM: Infectious papillomavirus in the vapor of warts treated with carbon dioxide laser or electrocoagulation: detection and protection. J Am Acad Dermatol 1989;21:41–49.
16 Ferenczy A, Bergeron C, Richart RM: Human papillomavirus DNA in CO_2 laser-generated plume of smoke and its consequences to the surgeon. Obstet Gynecol 1990;75:114–118.
17 Baggish MS, Poiesz BJ, Joret D, Williamson P, Refai A: Presence of human immunodeficiency virus DNA in laser smoke. Lasers Surg Med 1991;11:197–203.
18 Andre P, Orth G, Evenou P, Guillaume JC, Avril MF: Risk of papillomavirus infection in carbon dioxide laser treatment of genital lesions. J Am Acad Dermatol 1990;22:131–132.
19 Hallmo P, Naess O: Laryngeal papillomatosis with human papillomavirus DNA contracted by a laser surgeon. Eur Arch Otorhinolaryngol 1991;248:425–427.
20 Pay AD, Kenealy JM: Laser transmission through membranes using the Q-switched Nd:YAG laser. Lasers Surg Med 1999;24:48–54.

Jacob Dudelzak, MD
Accredited Dermatology, Laser & Cosmetic Surgery
2615 East 16th Street, 2nd Floor
Brooklyn, NY 11235 (USA)
E-Mail ydudelzak@hotmail.com

Bogdan Allemann I, Goldberg DJ (eds): Basics in Dermatological Laser Applications.
Curr Probl Dermatol. Basel, Karger, 2011, vol 42, pp 40–47

Ablative Skin Resurfacing

Jennifer Chwalek[a] · David J. Goldberg[b,c]

Skin Laser & Surgery Specialists of New York & New Jersey, [a]Hackensack, N.J., and [b]New York, N.Y., and [c]Mount Sinai School of Medicine, New York, N.Y., USA

Abstract

Ablative skin resurfacing has remained the gold standard for treating photodamage and acne scars since the development of the first CO_2 lasers. CO_2 and Er:YAG lasers emit infrared light, which targets water resulting in tissue contraction and collagen formation. The first ablative laser systems created significant thermal damage resulting in unacceptably high rates of scarring and prolonged healing. Newer devices, such as high-energy pulsed lasers and fractional ablative lasers, are capable of achieving significant improvements with fewer side effects and shorter recovery times. While ablative resurfacing has become safer, careful patient selection is still important to avoid post-treatment scarring, dyspigmentation, and infections. Clinicians utilizing ablative devices need to be aware of possible side effects in order to maximize results and patient satisfaction. This chapter reviews the background of ablative lasers including the types of ablative lasers, mechanism of action, indications for ablative resurfacing, and possible side effects.

Ablative laser resurfacing has remained the gold standard for treating photodamage since its inception in the 1960s when CO_2 lasers were first introduced in dermatology to treat actinic cheilitis [1, 2]. The early CO_2 lasers operated with a continuous-wave mode producing nonspecific cutaneous injury, which was better suited for tissue cutting [2]. These devices produced significant thermal damage to the surrounding skin, increasing the risk of scarring and limiting their utility. The possible dangers of resurfacing led many practitioners to abandon ablative lasers in favor of non-ablative technology when it became available because it was safer and associated with less downtime. However, non-ablative lasers have been unable to produce the dramatic improvements achievable with ablative lasers. Over the past two decades, the development of high-energy pulsed and scanned CO_2 lasers, Er:YAG lasers, and fractional ablative devices has allowed for a more controlled approach to resurfacing. This newer generation of lasers is associated with fewer side effects and a quicker recovery. As a result, ablative laser resurfacing has once again become the treatment of choice for photodamage and scarring.

In 1983, Parish and Anderson introduced the theory of selective photothermolysis, which revolutionized cutaneous laser surgery [3]. Their premise was that a target molecule or chromophore in the skin could be selectively destroyed when the irradiating wavelength of light was preferentially absorbed by the target tissue rather than the surrounding tissue. In order to avoid damaging surrounding tissue, the pulse duration of the laser has to be less than the thermal relaxation time of skin

(200–600 ms), which is defined as the time it takes for the chromophore to cool to one half of its peak temperature following irradiation. Longer pulse durations increase the risk of thermal damage, which can result in hyperpigmentation and scarring. Additionally, the energy density or fluence delivered must be great enough to effectively ablate the target [2, 3].

The first ablative laser approved by the Food and Drug Administration in 1996 was the CO_2 laser [3]. CO_2 lasers emit light at a wavelength of 10,600 nm in the infrared spectrum targeting water-containing tissue. The CO_2 laser vaporizes the epidermis and part of the dermis leaving a zone of thermal injury responsible for collagen shrinkage and remodeling [2, 4]. Previous studies have shown that a fluence of at least 5 J/cm^2, the vaporization threshold of tissue, is required to ablate tissue and minimize thermal injury. Failure to reach this fluence results in a charred wound with surrounding thermal damage similar to electrocautery [5]. The earliest CO_2 lasers functioned in a continuous-wave mode, capable of ablating 400–500 μm of tissue. These lasers produced significant thermal damage making them more effective for selective lesional destruction rather than resurfacing. In contrast, the newer high-energy short-pulsed CO_2 lasers, such as the UltraPulse (Coherent Laser Corp., Palo Alto, Calif., USA), ablate 20–30 μm after 1 pass (at 4.5–5 J/cm^2, pulse duration <1 ms). Residual thermal damage extends only to 100–150 μm even after successive passes [2, 5–7]. CO_2 laser resurfacing can achieve up to a 50% improvement in rhytides and scars as well as immediate collagen contraction [3].

The Er-YAG, a flash-pumped crystal laser system, was developed in an effort to design an ablative laser that would have less downtime and risks of scarring than the CO_2 [3]. The Er:YAG emits a wavelength of light at 2,940 nm, which more closely approximates the peak absorption of the coefficient of water (3,000 nm). This wavelength is absorbed by tissue 12–18 times more efficiently than the CO_2 laser, confining energy to the epidermis and papillary dermis [4]. The Er:YAG ablates 20–25 μm (at 5 J/cm^2) with thermal damage extending to only 10–40 μm [2, 5]. While this results in few side effects and faster reepithelialization, the minimal thermal injury produces insufficient coagulation and reduced collagen contraction [4]. Long-pulse and variable-pulse Er:YAG systems with pulse durations between 500 and 10 ms have improved hemostasis and induce greater collagen remodeling at the expense of increased thermal injury [2, 3, 7]. Although it is possible to achieve deeper ablation with the Er:YAG with multiple passes, the CO_2 is still felt to produce superior results, especially for deeper rhytides [8].

Patient demand has driven the development of newer fractionated ablative devices that are effective and associated with less downtime. Fractionated technology creates microthermal treatment zones, i.e. columns of thermally denatured skin. Instead of completely ablating the epidermis, fractionated ablative lasers leave intact surrounding islands of skin from which wound healing occurs [9]. These devices can improve rhytides, photodamage, and scars with shorter healing times. Whereas the average time to reepithelialization is 5–7 days for ablative devices, it has been noted within 24 h of fractional resurfacing [10]. The shorter recovery and improved side effect profile is at the expense of clinical efficacy, which does not match a full ablative procedure. To achieve a clinically significant improvement with a fractionated ablative device multiple sessions may be required.

The histological changes following ablative resurfacing have been well studied and compared to other resurfacing modalities such as dermabrasion and chemical peels. Fitzpatrick et al. [11] compared the histological changes of a pulsed CO_2 laser, 35% trichloroacetic acid, Baker-Gordon phenol peel, and dermabrasion using a pig model. The UltraPulse 5000 (Coherent Laser Corp.) laser was used at increasing energy increments (150, 250, 350, 450 mJ per pulse). For each energy level, 1, 2, and 3 passes were performed to remove increasing amounts of epidermis and dermis. Saline-soaked gauze was used to remove any tissue debris between successive passes. Biopsy specimens were obtained immediately after treatment and on days 7, 21, and 42 from the laser-irradiated tissue, and 1 week postoperatively following chemical peels. Reepithelialization occurred by day 7 in all sites except the phenol-treated site, which took approximately 3 weeks to heal. The trichloroacetic acid peel was clinically and histologically comparable to the laser treatment sites that received 3 passes of 150 mJ per pulse or 1–2 passes of 250 mJ per pulse. The dermabrasion site was clinically and histologically similar to the laser treatment sites that received 250–450 mJ per pulse and 2–3 passes. The Baker-Gordon phenol peel produced a deeper wound than any combination of laser energies and passes. Partial or complete epidermal loss with no dermal vaporization was achieved with a single pulse with pulse energies of 150–450 mJ. Higher energies were more effective at producing a more complete epidermal ablation. The degree of dermal vaporization and necrosis depth increased with increasing energies and passes [6, 11]. Other studies have confirmed that short-pulse CO_2 lasers produce histological changes similar to medium-depth chemical peels [12].

The depth of laser ablation varies not only with pulse energies and the number of passes, but also depending on the device used. Kauvar et al. [6] examined three CO_2 systems (UltraPulse, SilkTouch, and SurgiPulse). The SilkTouch (Sharplan Laser Inc., Allendale, N.J., USA) and SurgiPulse (Sharplan Laser Inc.) lasers ablated to a depth of 30–50 μm, through the epidermis with 1 pass, whereas the UltraPulse (Lumenis, Santa Clara, Calif., USA) ablated 20–30 μm with some dermoepidermal junction remaining following 1 pass.

The degree of injury, in particular the amount of thermal damage, as well as the type of tissue injury determines the extent of neocollagenesis. Ross et al. [13] examined the gross and microscopic effects of thermal and mechanical ablation using a pig model. They evaluated the immediate and long-term wound histology of the CO_2 laser (UltraPulse, SilkTouch), Er:YAG laser (Con Bio, Livermore, Calif., USA), dermabrasion, and dermatome. Three histological patterns of injury were noted. Type-one wounds had minimal-to-moderate dermal ablation and minimal thermal damage (1-pass CO_2 or Er:YAG wounds, superficial dermabrasion and dermatome wounds). The second type had minimal dermal ablation and greater thermal damage (3-pass CO_2 wounds); the third wound type had greater dermal ablation with no thermal damage (deep dermabrasion). CO_2-laser-induced wounds showed immediate and final tissue contraction that increased with successive passes and increasing residual thermal damage. While there was slight tissue contraction immediately following Er:YAG treatment, versus none observed with mechanical injuries, in the end the Er:YAG sites healed in a manner similar to mechanically induced wounds. In wounds with more than 70 μm of residual thermal damage (3-pass CO_2 sites), there was greater compaction of collagen [13].

In another trial using the UltraPulse CO_2 laser, skin biopsies from patients 12 weeks after treatment showed a grenz zone of new collagen and elastin in the superficial papillary dermis. Collagen thickness was not associated with the number of laser passes or the amount of preceding photodamage [6].

Indications

Ablative lasers have been successfully utilized to treat a number of cutaneous conditions.

By far the most common indication is photoaging, including dyspigmentation, rhytides, actinic keratoses, and vascular changes. Scarring – either from acne, trauma, or surgery – remains the second most common indication. Acne scars that are soft atrophic scars with sloping edges respond well to laser resurfacing. Ice pick or deep scars with sharp borders that are fixed do not respond as well and often require punch excisions [1]. In addition, ablative lasers have been used to successfully treat verruca, benign skin neoplasms (such as syringoma), and acantholytic diseases [1, 2]. The level of improvement when laser resurfacing is used for photodamage or acne scarring can range from 20 to 90%. Since collagen remodeling occurs over time, patients may continue to notice improvements up to 6 months after treatment [1].

Patient Selection

Careful patient selection is required to avoid complications and optimize outcomes. Patient expectations, skin type, treatment area, and medical history should be carefully reviewed. While ablative resurfacing can be performed on all skin types, darker skinned patients almost universally experience hyperpigmentation following treatment [1]. For this reason, the ideal patient for ablative resurfacing is Fitzpatrick skin types I–III with moderate-severe photodamage, rhytides, or soft/atrophic acne scars. The past medical history should be reviewed for a history of hypertrophic scars, keloids, and dermatologic diseases that can possibly koebnerize (such as vitiligo, lichen planus, and psoriasis). Resurfacing depends on reepithelialization from stem cells found in skin appendages. Patients with conditions with decreased appendages, such as scleroderma, or areas previously treated with radiation therapy may experience delayed wound healing, and are not ideal candidates for ablative resurfacing. Likewise, some anatomical areas are at increased risk of delayed healing or scarring, such as the neck and chest, and should be avoided or treated conservatively. Recent treatment with isotretinoin is also a contraindication to ablative resurfacing. There are reports of patients previously treated with isotretinion developing hypertrophic scarring following dermabrasion and chemical peels [5]. Because intensive wound care is required following resurfacing, the physician must assess whether the patient is capable of tolerating and following treatment guidelines. An inability to change dressings and take medications as directed can result in infections, delayed wound healing, and scarring.

Perioperative Management

Anesthesia

Anesthesia is required for laser resurfacing to minimize discomfort during the procedure. The degree of ablation oftentimes dictates the type of anesthesia required. The thermal damage produced with an Er:YAG for instance, is less than that with a CO_2 resulting in less discomfort with the former [3]. In many cases, topical anesthesia – such as topical lidocaine and prilocaine cream – can be used alone or combined with local nerve blocks. In pain sensitive or anxious patients, IV sedation or general anesthesia may be warranted [1, 5].

Prophylaxis

Post-ablative skin is most at risk of bacterial, viral, and fungal infections during the first week following treatment. Reactivation of herpes simplex virus (HSV) is the most common infection following resurfacing. Given the high incidence of HSV carriers, it is essential to pretreat any patients undergoing full facial or perioral resurfacing with antivirals. Even with antiviral prophylaxis, 2–7% of patients develop HSV reactivation. Oral antivirals, such as acyclovir, famcyclovir, or valacyclovir, should be initiated 1–2 days prior to the resurfacing procedure and continued for 7–10 days [14]. Many clinicians also choose to use prophylaxis for secondary bacterial and antifungal infections, though the necessity of the later is more controversial. In these cases, antibiotics (such as penicillins, macrolides, or quinolones) and an antifungal (such as fluconazole) are commenced before the procedure [1].

Postoperative Wound Care
The wound care regimen following laser resurfacing is important to ensure appropriate healing and minimize side effects. Open and closed wound care techniques have been described. The open technique involves application of emollients such as aquaphor or petrolatum, as well as cool compresses to reduce edema. Closed wound care approaches utilize semi-occlusive biosynthetic dressings (e.g. Biobrane, Flexzan, Second Skin, Silon-TSR, Vigilon) for a few days postoperatively [14]. The open technique is often associated with more discomfort than the closed technique, especially during the first few days following treatment; however, in general it is associated with fewer infections, possibly because it is easier to monitor the healing process. The closed wound technique requires less effort for patients, but can be associated with increased infections when the dressings are left in place for several days. For this reason, some ascribe to a combined closed and open regimen. In these cases, patients are initially treated with semi-occlusive dressing for the first day or two followed by the open technique [1]. Dilute acetic acid soaks in the postoperative period can decrease bacterial counts and prevent infection.

Complications
The possible serious side effects associated with ablative resurfacing – including scarring and delayed-onset hypopigmentation – have led many clinicians to avoid it in favor of safer albeit less effective modalities. The risk of complications is related to the depth of ablation, the skill of the surgeon and his or her technique, as well as the patient's genetic background. Even in the most experienced hands with the ideal candidate, complications can occur. Most untoward effects if caught early can be successfully managed, and for this reason being able to identify potential side effects quickly is necessary. While using higher fluences and increasing treatment passes (pulse stacking) produces more dramatic outcomes, these increase the risk of complications [5].

Erythema
All patients experience erythema following ablative resurfacing. This is directly proportional to the depth of ablation. Erythema is most intense in the first month after resurfacing and is more pronounced with CO_2 resurfacing compared to Er-YAG resurfacing. Erythema can last up to 6 months with CO_2 resurfacing, whereas erythema following Er-YAG treatment usually resolves within weeks. [14]. Alster [15] noted that in patients treated with L-ascorbic acid in an aqueous formulation following reepithelialization, erythema was less intense and resolved quicker. The use of light-emitting diodes following resurfacing may also reduce the severity and duration of erythema [16].

Acne/Milia
Acne/milia formation can be a common side effect in patients following laser resurfacing, occurring by some estimates in as many as 80% of patients. Acne usually presents within the first few weeks and is likely due to the occlusive nature of petrolatum-based emollients and biosynthetic dressings. Aberrant follicular reepithelialization and thermal injury to the sebaceous gland resulting in increased sebum production may contribute to acne formation [1, 14]. Patients with a prior history of acne seem to be at greater risk. For mild flares, treatment is not necessary. Acne usually resolves once the occlusive ointments are discontinued. In more severe cases, courses of oral antibiotics may be necessary and once reepithelialization is complete, topical retinoids, antibiotics, and α-hydroxy acids can be instituted [1, 14].

Contact Dermatitis
Debrided skin is more susceptible to irritation. Worsening erythema or pruritus can signal contact dermatitis, either allergic or irritant, due to

topical preparations used in the perioperative period. When it occurs following laser resurfacing, it is usually irritant in nature. The most common offending agents are topical antibiotics, soaps, emollients and cosmetics with fragrances. These should be avoided while the skin is reepithelializing. Emphasize should be on using bland emollients in the postoperative period and avoiding nonessential products or self-prescribed remedies. If contact dermatitis is suspected, all possible offending agents should be discontinued. The use of topical steroids and oral antihistamines can decrease the inflammatory response and control pruritus [14].

Petechiae/Purpura

Petechiae are an unusual complication that has been reported following ablative resurfacing. These appear during the reepithelialization phase and resolve spontaneously without treatment [1]. Delayed purpura following fractional laser resurfacing can occur within the first week. Avoiding nonsteroidal anti-inflammatory medications, aspirin, and blood thinners in the postoperative period may reduce the risk [9].

Infection

Following ablative resurfacing, the debrided skin is at risk of bacterial, fungal, and viral infections. The risk of infection is greatest the first week following treatment when the skin is reepithelializing. Patients presenting with active infections commonly complain of acute onset of increased pain (noted in up to 50%) following by burning or pruritus. Other signs of infection include increasing erythema, crusting, exudates, pustules, and superficial erosions [1, 14]. The most common bacterial infection is due to *Staphylococcus aureus* followed by *Pseudomonas*. Fungal infections are usually due to *Candida* species and present with beefy erythema and pustule formation. HSV reactivation has been reported to occur despite prophylaxis. Failure to recognize infections can result in prolonged healing and scarring. Patients presenting with the aforementioned symptoms should be cultured and started on broad antimicrobial coverage. In the case of HSV reactivation despite prophylaxis, maximizing treatment doses or changing antiviral agents may be necessary. Rarely in cases of HSV dissemination is intravenous acyclovir required [9].

Post-Inflammatory Hyperpigmentation

Post-inflammatory hyperpigmentation occurs in 20–30% of patients with Fitzpatrick skin type II and 90–100% of patients with Fitzpatrick skin type V–VI [1]. Hyperpigmentation usually appears 1 month after treatment and can take several months to resolve without intervention. Pretreating with topical agents such as retinoic acid or glycolic acid to lessen the incidence of hyperpigmentation has been examined in a few studies. However, no studies have shown a reduction in the rate of post-inflammatory hyperpigmentation in patients pre-treated with topicals. West and Alster [17] examined the effects of pretreating 100 patients with either glycolic acid cream, hydroquinone 4% with tretinoin 0.025%, or no treatment prior to CO_2 resurfacing. No significant differences were noted among the 3 groups. Topical medications target the superficial epithelium rather than the melanocytes in deeper adnexal structures where hyperpigmentation is stimulated. As a result, topical lightening medications are unlikely to protect against hyperpigmentation [14]. It is important to screen patients carefully prior to laser resurfacing in order to determine those at greatest risk. If post-inflammatory hyperpigmentation does occur, it can be treated with a variety of topical bleaching agents (hydroquinone, glycolic acid, retinoic acid, or kojic acid). Hydroquinone 4% cream initiated at approximately 3 weeks may halt the progression [1]. Glycolic acid peels every 2–4 weeks can also expedite the resolution [14]. Patients should be instructed to use sunscreen regularly before and after the procedure. Patients with suntans are at increased risk of post-inflammatory hyperpigmentation [14].

Delayed-Onset Hypopigmentation

Delayed-onset hypopigmentation is a rare phenomenon developing in patients after 6 months or longer once post-resurfacing erythema has resolved. Gradual skin lightening develops in the laser-treated areas. It has been noted more commonly in skin types I–II and in individuals previously treated with dermabrasion and phenol peels. It is believed to be related to the depth of tissue injury and the degree of thermal necrosis since patients treated with the Er:YAG appear to be at less risk for this complication. The excimer laser and topical Oxsoralen with ultraviolet exposure may result in some improvement [1, 14].

It is important to differentiate delayed-onset hypopigmentation from relative or pseudo-hypopigmentation. Pseudo-hypopigmentation is skin lightening due to rejuvenation of photodamaged skin. Treated skin appears lighter than the surrounding non-treated skin. Treating the entire face or complete cosmetic units and feathering into the surrounding areas can minimize the contrast of treated versus non-treated skin [1].

Hypertrophic Scarring

One of the more severe and dreaded complications of ablative resurfacing is hypertrophic scarring. The use of excessively high fluences, pulse stacking, and failure to remove desiccated tissue between laser passes increases the risk of scarring. Other factors that can increase the risk of scarring are a recent history of isotretinoin use, a history of keloids, or postoperative complications such as infection and contact dermatitis [5].

Certain anatomic locations are more prone to scarring, such as the periorbital area, mandible, and neck. Focal areas of erythema or induration can signal scar formation and should be treated aggressively with strong topical steroids and intralesional corticosteroid and/or 5-fluorouracil. The pulsed dye laser can be used to improve erythema and induration once scar tissue has formed [1, 7, 14].

Ectropion Formation

Ectropion formation of the eyelids is a rare but serious complication that can occur with ablative resurfacing. Infraorbital scars contract leading to exposure of conjunctiva and possibly excessive tearing, dryness, and photophobia [1]. Patients over the age of 50 years and those with a history of prior eyelid surgery such as lower blepharoplasty are most at risk. Performing a snap lid test to assess for lid laxity can aid in identifying those patients most at risk. Using lower energy densities and few passes in the infraorbital region can reduce the risk of scar formation [14]. Mild ectropion of the lower lids is usually reversible, whereas severe cases require surgical intervention [1].

Conclusions

Ablative laser resurfacing provides the best treatment for rhytides, photodamaged skin, and scarring. Ablative resurfacing continues to evolve providing the laser surgeon with devices that can more precisely and safely ablate to the necessary depth based on clinical indication. Although newer ablative lasers are safer than the early CO_2 lasers, even these can be associated with risks of infection, scarring, and pigmentary changes. Increasing the energy and number of passes can result in more dramatic results, but also increases the risk of complications. With careful patient selection, good technique, and postoperative monitoring for complications, the laser surgeon can maximize results and patient satisfaction using ablative technology.

References

1 Zachary CB, Rokhsar CK, Fitzpatrick RE: Laser skin resurfacing; in David Goldberg (ed): Procedures in Cosmetic Dermatology: Lasers and Lights. Amsterdam, Elsevier, 2007, vol 2.

2 Riggs K, Keller M, Humphreys T: Ablative laser resurfacing: high-energy pulsed carbon dioxide and erbium:yttrium-aluminum-garnet. Clin Dermatol 2007;25:462–473.

3 Alster TS, Lupton JR: Erbium: YAG cutaneous laser resurfacing. Dermatol Clin 2001;19:453–466.

4 Alster TS, Tanzi EL: Single-pass carbon dioxide versus multiple-pass Er:YAG laser resurfacing: a comparison of postoperative wound healing and side effect rates. Dermatol Surg 2003;29:80–84.

5 Brightman LA, Brauer JA, Anolik R, Weiss E, Karen J, Chapas A, Hale E, Bernstein L, Geronemus RG: Ablative and fractional ablative lasers. Dermatol Clin 2009;27:479–489.

6 Kauvar ANB, Geronemus RG: Histology of laser resurfacing. Dermatol Clin 1997;15:459–467.

7 Alexiades-Armenakas MR, Dover JS, Arndt KA: The spectrum of laser skin resurfacing: nonablative, fractional, and ablative laser resurfacing. J Am Acad Dermatol 2008;58:719–737.

8 Pozner JM, Goldberg DJ: Histologic effect of a variable pulsed Er:YAG laser. Dermatol Surg 2000;26:733–736.

9 Metelitsa AI, Alster TS: Fractionated laser skin resurfacing treatment complications: a review. Dermatol Surg 2010;36:299–306.

10 Dierickx CC, Khatri KA, Tannous ZS, Childs JJ, Cohen RH, Erofeev A, Tabatadze D, Yaroslavsky IV, Altshuler GB: Micro-fractional ablative skin resurfacing with two novel erbium laser systems. Lasers Surg Med 2008;40:113–123.

11 Fitzpatrick RE, Tope WE, Goldman MP, Satur NM: Pulsed carbon dioxide laser, trichloroacetic acid, Baker-Gordon phenol, and dermabrasion: a comparative clinical and histologic study of cutaneous resurfacing in a porcine model. Arch Dermatol 1996;132:469–471.

12 Cotton J, Hood AF, Gonin R, Beeson WH, Hanke WC: Histologic evaluation of preauricular and postauricular human skin after high-energy, short-pulse carbon dioxide laser. Arch Dermatol 1996;132:425–428.

13 Ross VE, Naseef GS, McKinlay JR, Barnette DJ, Skrobal M, Grevelink J, Anderson RR: Comparison of carbon dioxide laser, erbium: YAG laser, dermabrasion, and dermatome. J Am Acad Dermatol 2000;42:92–105.

14 Alster TS, Lupton JR: Prevention and treatment of side effects and complications of cutaneous laser resufacing. Plast Reconstr Surg 2002;109:308–316.

15 Alster TS, West TB: Effect of topical vitamin C on postoperative carbon dioxide laser resurfacing erythema. Dermatol Surg 1998;24:331–334.

16 Alster TS, Wanitphakdeedecha R: Improvement of postfractional laser erythema with light-emitting diode photomodulation. Dermatol Surg 2009;35:813–815.

17 West TB, Alster TS: Effect of pretreatment on the incidence of hyperpigmentation following cutaneous CO_2 laser resurfacing. Dermatol Surg 1999;25:15–17.

Jennifer Chwalek, MD
Skin Laser & Surgery Specialists of New York & New Jersey
20 Prospect Street, Suite 702
Hackensack, NJ 07601 (USA)

Bogdan Allemann I, Goldberg DJ (eds): Basics in Dermatological Laser Applications.
Curr Probl Dermatol. Basel, Karger, 2011, vol 42, pp 48–55

Non-Ablative Lasers

David H. Ciocon[a] · Daven Doshi[b] · David J. Goldberg[a,c]

[a]Skin Laser & Surgery Specialists of New York & New Jersey, [b]Division of Dermatology, Albert Einstein College of Medicine, and [c]Department of Dermatology, Mount Sinai School of Medicine, New York, N.Y., USA

Abstract

Non-ablative laser devices exert their effects by inducing dermal collagen remodeling while sparing the epidermis. The remodeling process is achieved by producing thermal energy in a controlled and precise fashion, which recruits fibroblasts and stimulates collagen production. Although these devices do not produce the same degree of improvement as traditional CO_2 or er:YAG resurfacing techniques, they are an excellent alternative for people seeking gradual aesthetic improvement with minimal downtime. They are also extremely well suited to patients with darker skin types. In this chapter, we review the various non-ablative devices currently available on the market, with an emphasis on safety profiles and efficacy.

Non-ablative laser rejuvenation refers to the use of laser or light-based technology to affect improvements in cutaneous signs of photoaging, including dyspigmentation, wrinkles, and skin laxity. In contrast to traditional ablative rejuvenation procedures, such as carbon dioxide or erbium:YAG resurfacing, non-ablative rejuvenation technologies induce dermal heating and collagen remodeling without significant injury to the epidermis, particularly its upper layers such as the stratum corneum. While the advantage of such procedures is the shorter recovery times, the results tend to be subtle, and multiple sessions rather than single treatments are often required to achieve expected goals [1].

The exact mechanisms of non-ablative dermal remodeling are still under investigation; however, a subthreshold laser-induced injury to the dermis and/or the dermal vasculature theoretically results in a wound repair response, fibroblast recruitment and stimulation, and collagen reformation [2]. Because thermal energy is the source of this injury, most non-ablative devices target water, the major component of the dermis, which is heated through the process of selective photothermolysis. To reach these dermal targets, non-ablative devices typically rely on wavelengths in the mid-infrared regions of the electromagnetic spectrum. However, visible light devices, such as intense pulsed light (IPL) and light-emitting diodes (LED), have also been shown to induce non-ablative dermal remodeling.

This chapter focuses on the mechanisms, indications, and efficacy of a variety of visible light and mid-infrared lasers or light-based technologies that affect improvements in wrinkles, scars, and skin laxity through dermal remodeling, while sparing the epidermis.

Non-Fractionated Mid-Infrared Laser Devices

Chief among the family of non-ablative lasers are devices with wavelengths in the mid-infrared range that induce collagen remodeling but spare the epidermis [1, 2]. Most of these devices – which include the 1,064 nm Nd:YAG, the 1,320 nm Nd:YAG, and the 1,450 nm diode – have handpieces coupled to a thermal sensor and/or a cooling cryogen blast that optimize control of the surface temperature of the skin. Consequently, thermal stimulation of dermal fibroblasts is achieved while the epidermis is protected from direct thermal injury. These devices use non-fractionated flat beams with large spot sizes (3–10 mm) that cause bulk dermal heating without inducing tissue coagulation or vaporization. In this manner, the temperature of the dermis can increase to 60–70°C, while the temperature of the entire epidermal surface can be safely maintained at 40–48°C. As observed clinically, effective cooling of the skin minimizes the risks of scarring and hyperpigmentation. With strict control of the surface temperature, safety profiles increase, particularly in individuals with darker skin types (Fitzpatrick skin types IV and V).

Q-Switched 1064 nm Nd:YAG

The Q-switched Nd:YAG laser emits a 1,064 nm wavelength that targets water within the papillary and reticular dermis, and is believed to promote collagen remodeling through selective photothermolysis [1]. The short pulse duration in the nanosecond range is thought to confine injury to the intended target and minimize heat diffusion to other structures. To our knowledge, this was the first laser system examined for non-ablative rejuvenation. We conducted an early study examining its benefits as a non-ablative modality for facial rhytides in a comparison study of 11 patients between the scanned and pulsed CO_2 laser and the 1064 nm Q-switched Nd:YAG laser [3]. By using a 3-mm spot size and a fluence of 5.5 J/cm^2, perioral and periorbital rhytides were treated. Three patients had results comparable to those of the CO_2 laser, 6 had some improvement, and 2 had no improvement. Erythema lasted up to 3 months in patients. A disadvantage of this modality in this early study was the resultant pinpoint bleeding.

A subsequent study investigated the use of a Q-switched Nd:YAG laser with a topical carbon suspension that was believed to potentiate its efficacy [4]. The suspension provided a chromophore target to enable the use of a lower fluence and to reduce epidermal damage. Four treatments at 7- to 10-day intervals at 2.5 J/cm^2, using a 6- or 7-mm spot size, were administered to 12 patients. An average improvement of 25% was seen. Two patients with type VI skin, however, had focal areas of temporary hypopigmentation.

More recently, low fluence treatments with the Q-switched 1,064 nm Nd:YAG laser have been shown to be effective for the treatment of melasma, particularly in Asian patients, who historically have been poor candidates for ablative laser treatments because of an increased risk of post-inflammatory pigmentation [5]. In a study by Suh et al. [5], 23 Korean patients (skin types III–V) with melasma were treated with the 1,064 nm Q-switched Nd:YAG laser at 1-week intervals for 10 weeks. The melasma area and severity index (MASI) score, lightness of melasma, patient satisfaction score and side effects were assessed at baseline, weeks 4, 7, and 10, as well as at 1, 2, and 3 months after the last treatment. A decreased MASI score and increased lightness of melasma were statistically significant at visits on weeks 7 and 10. Furthermore, clinical improvement was maintained at 1, 2, and 3 months after the last treatment.

Example photographs taken before and after laser treatment are presented in figures 1 and 2.

Long-Pulse 1,064 nm Nd:YAG

The long-pulse Nd:YAG laser emits energy in the infrared spectrum at a wavelength of 1,064 nm with extended pulse durations in the millisecond domain [1]. Diffuse heating of dermal tissue at the

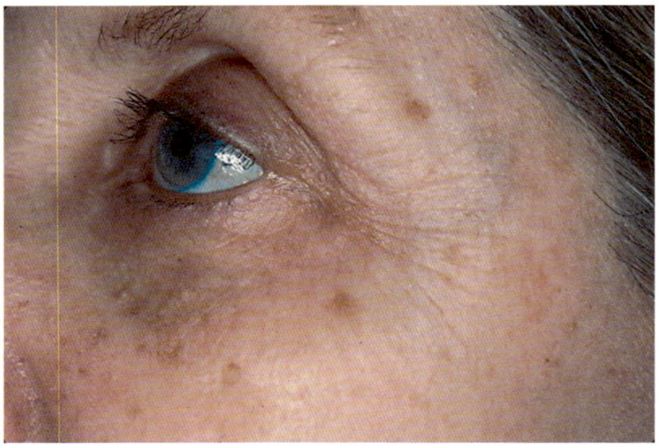

Fig. 1. Before laser treatment.

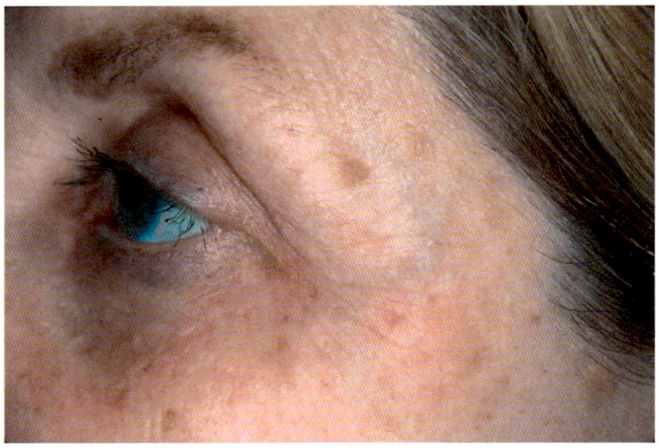

Fig. 2. After laser treatment.

1,064 nm wavelength penetrates to depths of 5–10 mm and permits slow heat diffusion with low energy absorption by melanin. In one study, 34 patients completed at least 7 treatments using the long-pulse Nd:YAG laser (Lyra; Laserscope, San Jose, Calif., USA) [6]. Skin types I–V were treated at 22 J/cm^2, with a 50-ms pulse width, using a 10-mm spot size. A contact cooling device provided epidermal protection. Three passes were administered over the full face. Treatments were repeated weekly. By using photographic documentation, statistically significant improvements in coarse and fine wrinkles and skin laxity were found in all patients based on masked objective analysis of pretreatment and post-treatment photographs by 3 independent evaluators. Similarly, subjects also reported improvements in facial rhytides. No significant adverse events were noted.

We subsequently conducted a study examining the safety and efficacy of a sequential dual-wavelength 595 nm and 1,064 nm Nd:YAG laser device for treatment of photoaged facial skin [7]. We hypothesized that the 595 and 1,064 nm wavelengths would offer improvement in superficial and deep telangiectasias from photoaging, while the 1,064 nm wavelength would stimulate collagen remodeling. Fifteen individuals, between the ages of 38 and 66 years, with various stigmata of facial photoaging, were entered into the study. Five sequential combined PDL/Nd:YAG (Cynergy, Cynosure Inc., Westford, Mass., USA) treatments were performed using a 10-mm handpiece at monthly intervals. PDL parameters included a 10-ms pulse duration and a fluence set at 1 J/cm^2 below the purpura threshold. Nd:YAG parameters were set at a 50-ms pulse duration with fluences varying between 35 and 50 J/cm^2, depending on patient comfort. Improvement was determined by evaluation of photographs taken before the first treatment and at 1 and 3 months following the last treatment, as well as the individual's self-assessment.

All individuals tolerated treatments well with no serious, long-term, or permanent adverse effects. Improvement was most pronounced in telangiectasias and diffuse erythema, followed by epidermal dyspigmentation and lentigines. Some individuals were also noted to have improved smoothness, radiance, or pore size. The improvement was generally maintained at the 3-month follow-up. In addition, the average self-reported improvement was 1.7 (on a 0–4 scale) at 1 month and 2.4 at 3 months following the last treatment.

1,320 nm Nd:YAG Laser

The 1,320 nm Nd:YAG laser system (CoolTouch, Roseville, Calif., USA) has been specifically designed for non-ablative resurfacing. Epidermal protection is provided by cryogen cooling. The initial study by Kelly et al. [8] reported its use with fluences ranging from 28 to 36 J/cm^2 and a 5-mm spot size. A cryogen spurt of 20–40 ms was delivered followed by a 10-ms delay prior to each laser pulse. Among 35 subjects treated, a statistically significant improvement was seen in mild, moderate, and severe facial rhytides at 3-month follow-up; however, at 6-month follow-up, only patients with severe rhytides retained their previously observed improvements. We conducted a subsequent study with the 1,320 nm Nd:YAG laser for facial rhytides consisting of 10 subjects with class I–II rhytides who were treated in 5 sessions and assessed clinically 6 months after their final treatment. All subjects demonstrated new collagen formation in the papillary dermis, and all 10 showed subjective improvement in their skin [9].

1,450 nm Diode Laser

The 1,450 nm diode laser is another mid-infrared wavelength laser that penetrates to a depth of about 500 μm [1, 2]. Peak powers are in the range of 10–15 W, and pulse durations range from 150 to 250 ms. Epidermal cooling is also provided by a cryogen spray. To determine whether the improvements seen in non-ablative therapy using this mid-infrared device are attributable to the laser itself or to nonspecific physical trauma from cryogen, we conducted a split-face study examining the effect of the 1,450 nm diode laser (Candela, Wayland, Mass., USA) compared with cryogen alone. Twenty patients with class I–II rhytides were treated in 2–4 treatment sessions, with a 6-month follow-up. Patients were treated with the laser and cryogen spray on one side of the face [10]. The contralateral side was treated with cryogen spray only. Seven subjects showed no improvement, 10 had mild improvement, and 3 had moderate improvement on the laser-treated side. The cryogen-treated side reflected no improvement.

In a separate study, Tanzi et al. [11] evaluated the efficacy of the 1,450 nm diode laser for the treatment of facial rhytides in a split-face controlled trial. By using a fluence range of 15–20 J/cm^2 (average, 17.5 J/cm^2), a 4-mm spot size, a pulse duration of 210 ms, and a dynamic cooling device set at 10 ms for pre-cooling, 30 ms for intra-cooling, and 20 ms for post-treatment cooling, a single pass of non-overlapping pulses was delivered. Four consecutive treatments were administered at 3- to 4-week intervals. A clinical evaluation demonstrated modest improvements in the treated areas. Improvements peaked at the 6-month follow-up visit. The 12-month visit revealed no additional benefit. A histological examination noted new collagen formation in the treated areas. In a subsequent meta-analysis by Doshi and Alster [12] of 5 studies in which this device was used for photoaging or acne scars, the authors concluded that: 'This laser is a good choice for patients with minimal epidermal damage who desire improvement of mild to moderate rhytides without the postoperative morbidity common to the ablative laser systems. Additionally, the results in the treatment of active acne and atrophic scars are promising.'

Intense Pulsed Light

Intense pulsed light (IPL) sources are widely utilized in medicine for various cosmetic problems, including irregular pigmentation, vascular lesions, telangiectasias, hypertrichosis, and rhytides [13, 14]. The core technology involves the use of polychromatic broadband flash lamps equipped with optical filters that allow non-coherent visible to infrared wavelengths (500–1,200 nm) to reach the skin. Because multiple wavelengths are delivered, multiple chromophores including hemoglobin, melanin, water, and collagen, can be targeted with the same exposure. From a treatment standpoint, the multi-target capacity makes for great therapeutic versatility. In a single session of IPL, a wide range of problems including telangiectasias, pigmentation, and dermal atrophy can be addressed.

However, due to the complexity of selecting the appropriate cutoff filters and wavelength range, there is also a greater risk of developing side effects from non-specific thermal damage.

IPL has been reported to be one of the most effective light-based methods for the treatment of melanocytic hyperpigmentation associated with aging and long-term sun exposure. Solar lentigines are the most common example of this [15]. The mechanism of action of IPL for treating pigmentary lesions is thought to be the result of rapid differentiation of pigmented keratinocytes induced by thermal heating. This process induces an upward transfer of melanosomes along with necrotic keratinocytes, resulting in their elimination as the microcrusts are removed from the skin surface. These effects have been demonstrated in vivo using both reflectance-mode confocal microscopy and optical coherence tomography, which allow precise visualization of melanosomes in the skin in horizontal and vertical dimensions, respectively [16]. The surrounding non-pigmented keratinocytes are spared by keeping the pulse duration of these devices above the thermal relaxation time of the epidermis and by using contact cooling with chilled handpiece tips.

In recent years, multiple groups including our own have also demonstrated the efficacy of IPL technology for the treatment of textural changes associated with photoaging including wrinkles and lax skin. A series of 4–6 IPL treatments termed 'photorejuvenation' is available in many dermatology practices nationwide [17, 18]. The principle behind tightening of the skin using IPL rests on the theory that heating collagen fibers with high intensity light energy leads to their contracture, while sparing the epidermis. This may account for the textural change described in skin treated with IPL that has been reported as a secondary observation in several studies [19]. Furthermore, the thermal stimulation of dermal fibroblasts by the higher wavelengths within the IPL spectrum has been shown to result in increased synthesis of collagen and extracellular matrix proteins, leading to at least partial replacement of the lost dermal volume [20, 21]. Specifically, wavelengths in the 1,200 nm spectrum are absorbed by water in the dermis, triggering a cytokine reaction which in turn stimulates the formation of new collagen I, III, and elastin. Histological evaluation of the effects of 5 subsequent IPL treatments with 570–645 nm wavelengths showed epidermal thickening of 100–300 μm, a decrease in horny plugs, new rete ridge formation, a decrease in proportion of degenerated elastic fibers, and new dermal collagen formation [22, 23].

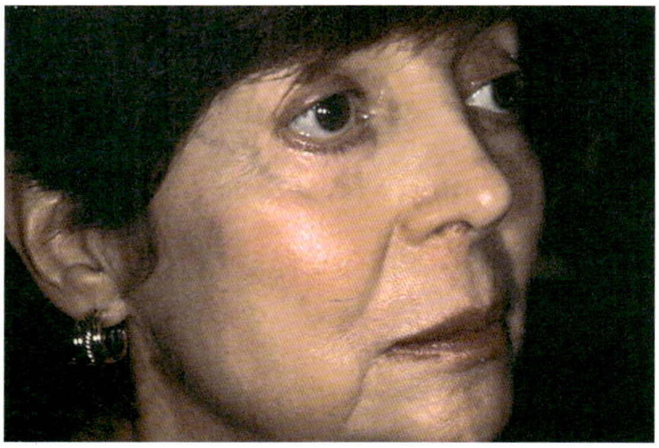

Fig. 3. Before IPL treatment.

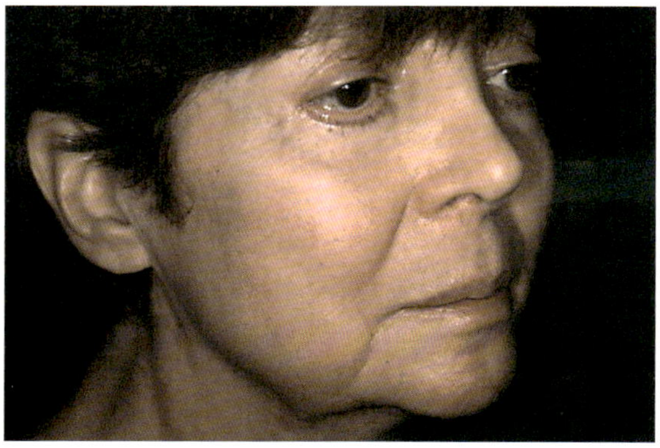

Fig. 4. Decreased erythema after IPL treatment.

Example photographs taken before and after IPL treatment are presented in figures 3 and 4.

Light-Emitting Diodes

LED are devices that emit narrow bands of low-intensity visible light without inducing dermal or epidermal damage. It is theorized that cellular receptors are activated by LED irradiation and enable the regulation of cellular and subcellular function based on the settings of specific LED modules [24]. This notion of modifying cellular activity by narrow light bands is termed photomodulation. Available LED devices are arranged as panels, so that large surface areas can be treated simultaneously. LED light pulsation has been shown to induce fibroblast stimulation in cell culture [25]. The mechanism of action is believed to be the stimulation of the mitochondria, which activates the electron transport chain and thereby upregulates cellular production. New adenosine triphosphate formation immediately after LED exposure has also been demonstrated to increase fibroblast activity [26]. Fibroblast stimulation in this manner has been hypothesized to upregulate the production of collagen and elastin and downregulate the production of metalloproteinase, an enzyme responsible for the degradation of collagen. This is corroborated by immunohistochemical staining of pretreatment and post-treatment biopsy specimens.

We previously published our results of the treatment of 36 subjects with a combined 633- and 830 nm LED device [27]. Subjects were treated for 20 min using each wavelength (total treatment time 40 min) twice weekly for 5 weeks and evaluated 12 weeks after treatment. Five of the subjects were biopsied to assess the ultrastructural impact of the treatments. The majority of subjects showed improvements in smoothness, firmness, and softness at the final visit, and ultrastructural analysis showed thickening of collagen fibers in all 5 biopsy specimens, consistent with clinical evidence of rejuvenation.

Example photographs taken before and after IPL treatment are presented in figures 5 and 6.

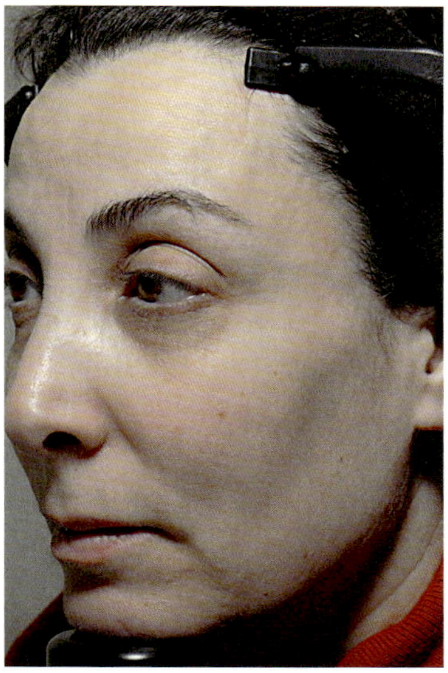

Fig. 5. Before LED treatment.

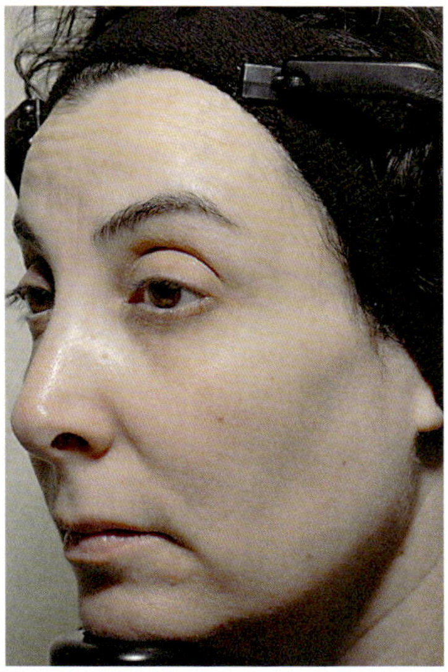

Fig. 6. After LED treatment.

Conclusion

Non-ablative laser devices exert their effects by inducing dermal collagen remodeling while sparing the epidermis. The remodeling process is achieved by producing thermal energy in a controlled and precise fashion, using mechanisms that control epidermal temperature and wavelengths that penetrate deeply enough to reach their appropriate dermal targets. Although these devices do not produce the same degree of improvement as traditional CO_2 or Er:YAG resurfacing techniques, they are an excellent alternative for people seeking gradual aesthetic improvement with minimal downtime. There is abundant evidence demonstrating the clinical efficacy of these devices, but there are those who remain skeptical because the macroscopic results tend to be subtle compared to microscopic improvements. At present multiple non-ablative devices are available, but their efficacies relative to one another have yet to be tested. Finally, non-ablative rejuvenation shows great potential in the treatment of darker skin types, who are prone to pigmentary alterations from classic resurfacing modalities, because of its emphasis on epidermal sparing.

References

1 Kim KH, Geronemus RG: Nonablative laser and light therapies for skin rejuvenation. Arch Facial Plast Surg 2004; 6:398–409.
2 Vartanian J, Dayan SH: Nonablative facial skin tightening (eMedicine website). http://emedicine.medscape.com/article/844068-overview (updated March 18, 2009; accessed July 4, 2010).
3 Goldberg DJ, Whitworth J: Laser skin resurfacing with the Q-switched Nd:YAG laser. Dermatol Surg 1997;23:903.
4 Newman J, Lord J, McDaniel DH: Nonablative laser therapy in skin types I–VI: clinical evaluation of facial treatment using QS 1064 nm Nd:YAG laser combined with carbon suspension lotion (abstract). Lasers Surg Med Suppl 2000;12:17
5 Suh KS, Sung JY, Roh HJ, Jeon YS, Kim YC, Kim ST: Efficacy of the 1064-nm Q-switched Nd:YAG laser in melasma. J Dermatolog Treat 2010, E-pub ahead of print.
6 Dayan SH, Vartanian J, Menaker G, Mobley SR, Dayan AN: Nonablative laser resurfacing using the long-pulse (1064 nm) Nd:YAG laser. Arch Facial Plast Surg 2003;5:310.
7 Berlin AL, Hussain M, Goldberg DJ: Cutaneous photoaging treated with a combined 595/1064 nm laser. J Cosmet Laser Ther 2007;9:214–217.
8 Kelly KM, Nelson JS, Lask GP, Geronemus RG, Bernstein LJ: Cryogen spray cooling in combination with nonablative laser treatment of facial rhytides. Arch Dermatol 1999;135:691–694.
9 Goldberg DJ: Full face nonablative dermal remodeling with a 1320 nm Nd:YAG laser. Dermatol Surg 2000;26:915–918.
10 Goldberg DJ, Rogachefsky AS, Silapunt S: Non-ablative laser treatment of facial rhytides: a comparison of 1450 nm diode laser treatment with dynamic cooling as opposed to treatment with dynamic cooling alone. Lasers Surg Med 2002;30:79–81.
11 Tanzi EL, Williams CM, Alster TS: Treatment of facial rhytides with nonablative 1450 nm diode laser: a controlled clinical and histologic study. Dermatol Surg 2003;29:124–128.
12 Doshi SN, Alster TS: 1,450 nm long-pulsed diode laser for nonablative skin rejuvenation. Dermatol Surg 2005; 31(part 2):1223–1226.
13 Goldberg DJ: Laser- and light-based hair removal: an update. Expert Rev Med Devices 2007;4:253–260.
14 Goldberg DJ: Laser removal of pigmented and vascular lesions. J Cosmet Dermatol 2006;5:204–209.
15 Ortonne JP, Pandya AG, Lui H, Hexsel D: Treatment of solar lentigines. J Am Acad Dermatol 2006;54(suppl 2):S262–S271.
16 Yamashita T, Negishi K, Hariya T, Kunizawa N, Ikuta K, Yanai M, Wakamatsu S: Intense pulsed light therapy for superficial pigmented lesions evaluated by reflectance-mode confocal microscopy and optical coherence tomography. J Invest Dermatol 2006;126:2281–2286.
17 Goldberg DJ, Cutler KB: Nonablative treatment of rhytids with intense pulsed light. Lasers Surg Med 2000;26:196–200.
18 Bitter PH: Noninvasive rejuvenation of photodamaged skin using serial, full-face intense pulsed light treatments. Dermatol Surg 2000;26:835–842.
19 Weiss RA, Weiss MA, Beasley KL: Rejuvenation of photoaged skin: 5 years results with intense pulsed light of the face, neck, and chest. Dermatol Surg 2002;28:1115–1119.
20 Goldberg DJ: New collagen formation after dermal remodeling with an intense pulsed light source. J Cutan Laser Ther 2000;2:59–61.
21 Wong WR, Shyu WL, Tsai JW, Hsu KH, Pang JH: Intense pulsed light effects on the expression of extracellular matrix proteins and transforming growth factor beta-1 in skin dermal fibroblasts cultured within contracted collagen lattices. Dermatol Surg 2009;35:816–825.
22 Hernández-Pérez E, Ibiett EV: Gross and microscopic findings in patients submitted to nonablative full-face resurfacing using intense pulsed light: a preliminary study. Dermatol Surg 2002;28:651–655.
23 Zelickson B, Kist D: Pulsed dye laser and photoderm treatment stimulates production of type-1 collagen and collagenase transcripts in papillary dermis fibroblasts. Lasers Surg Med Suppl 2001;13:132.
24 Barolet D: Light-emitting diodes in dermatology. Semin Cutan Med Surg 2008;27:227–238.

25 Weiss RA, McDaniel DH, Geronemus RG, Weiss MA, Beasley KL, Munavalli GM, Bellew SG: Clinical experience with light-emitting diode (LED) photomodulation. Dermatol Surg 2005;31(part 2):1199–1205.

26 Weiss RA, Weiss MA, Geronemus RG, McDaniel DH: A novel non-thermal non-ablative full panel LED photomodulation device for reversal of photoaging: digital microscopic and clinical results in various skin types. J Drugs Dermatol 2004;3:605–610.

27 Goldberg DJ, Amin S, Russell BA, Phelps R, Kellett N, Reilly LA: Combined 633-nm and 830-nm led treatment of photoaging skin. J Drugs Dermatol 2006;5:748–753.

David J. Goldberg, MD
Skin Laser & Surgery Specialists of New York & New Jersey
115 E. 57th Street, Suite 710
New York, NY 10022 (USA)
Tel. +1 212 750 8090, E-Mail drdavidgoldberg@skinandlasers.com

Bogdan Allemann I, Goldberg DJ (eds): Basics in Dermatological Laser Applications.
Curr Probl Dermatol. Basel, Karger, 2011, vol 42, pp 56–66

Fractional Photothermolysis

Inja Bogdan Allemann[a] · Joely Kaufman[b]

[a]Klinik für Plastische Chirurgie und Handchirurgie, Universitätsspital Zürich, Zürich, Switzerland;
[b]University of Miami Miller School of Medicine and Private Practice Coral Gables, Miami, Fla., USA

Abstract

The novel concept of 'fractional photothermolysis' was introduced to the market in 2003, and has since gained tremendous popularity and importance for numerous clinical indications. Its development and introduction were driven by the need for an effective yet safe and low-risk resurfacing technique, in contrast to the ablative and non-ablative lasers available at the time. The first device that was introduced to the market used non-ablative wavelengths. In 2007, the technology was further developed with ablative wavelengths leading to vaporization of tissue. This new technology has revolutionized the field of laser skin resurfacing and resulted in the development of numerous non-ablative and ablative fractional devices. There is little doubt that fractional technology is here to stay, as (when used properly) it allows the operator to obtain significant clinical results with minimal post-treatment recovery and hence fewer complications.

To understand the development of fractional technology, it is best to start with the clinical indication of photoaged skin and the treatment of skin resurfacing. The first lasers used for this indication were ablative lasers. The ablative lasers, such as the Er:YAG and CO_2, are highly effective lasers that ablate the entire epidermis and parts of the papillary dermis, leading to a considerable wound area with slow reepithelialization from the base of the wound over 7–10 days. Although, in skilled hands, this procedure is considered to be the most clinically efficacious, downtimes are long and risks of serious side effects – such as infection, persistent erythema, dyspigmentation, and scarring – are rather high [1, 2]. These facts eventually made these original ablative lasers less attractive treatment alternatives.

The subsequent development of non-ablative lasers and intense pulsed light devices – such as the long-pulse 1,320 nm or 1,064 nm Nd:YAG, or the long-pulse 1,450 nm diode laser – led to a controlled thermal hyperthermia deep in the skin, without destruction of the epidermis, which was protected by cooling. The dermal thermal wound led to neocollagenesis. With this approach, recovery, and tolerability were greatly improved, and side effects were reduced, but at the same time clinical efficacy was markedly limited [3, 4]. This was the main factor that finally led to the development of fractional photothermolysis (FP).

Fractional Photothermolysis

As the name states, the concept of FP is based on treating only a fraction of the skin [5]. Unlike the conventional ablative and non-ablative lasers,

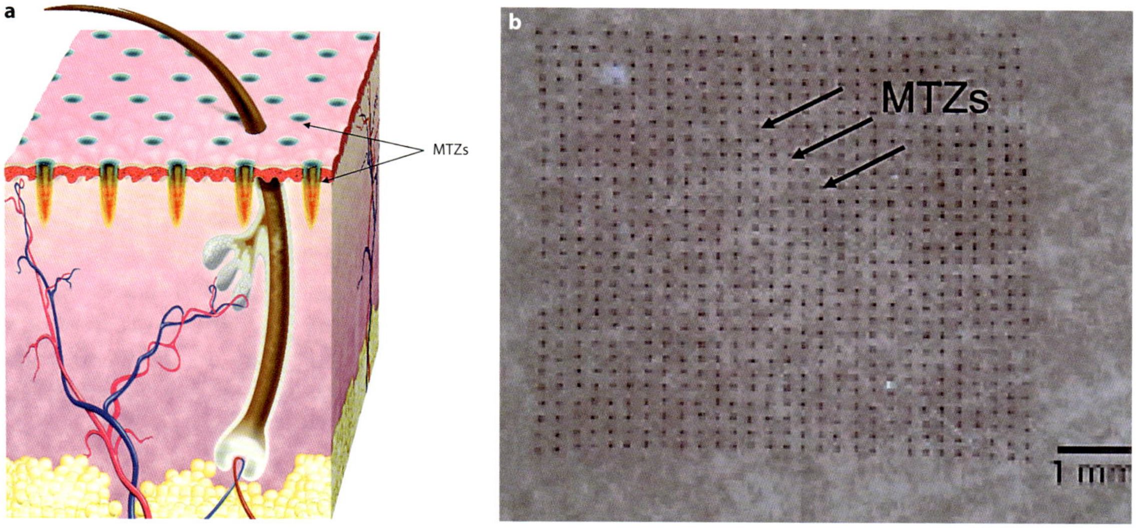

Fig. 1. a Schematic picture of FP depicting MTZs surrounded by unharmed tissue. Courtesy of Palomar Medical. **b** Clinical picture depicting MTZs surrounded by unharmed tissue.

which treat large uninterrupted areas of the skin, leading to correspondingly large wounds and bulk heating, FP only treats microscopic areas of the skin. This is achieved by separating a laser into an array of smaller beams creating multiple small 3D zones of thermal damage in the treated skin, referred to as microscopic treatment zones (MTZs). Those column-like MTZs consist of sharply defined coagulated tissue, each of which is surrounded by unharmed tissue (fig. 1). The tissue adjacent to the MTZs remains intact with viable skin cells. This vital tissue represents the reservoir of cells surrounding the damaged skin, allowing for rapid epidermal repair and hence decreased complication rates. As such, the thermal tissue damage per MTZ is much greater than would be possible in non-fractional procedures.

The size of individual MTZs depends on the energy per MTZ used; the higher the energy, the wider the MTZ and the deeper the thermal damage. The MTZs per cm^2 define the density of the treatment, also referred to as the percentage of coverage (extent of damaged dermis within a defined treatment area). Hence, by increasing the energy, the density is also increased. Typically a density of 1,000–2,000 MTZs/cm^2 is achieved. Usually 10–30% of the skin can be treated, i.e. thermally damaged with 1 treatment pass. With repetitive passes, the density can subsequently be increased.

Thus, the two most important input parameters in the treatment with FP are the energy and the density (table 1).

The target chromophore for all FP devices is water. Hence, a multitude of water-containing structures – such as epidermal keratinocytes and blood vessels located at differing depths throughout the skin – can be targeted, resulting in a wide array of clinical indications [6].

Based on the wavelength's affinity for water, fractional technologies can be divided into 2 main categories. Those with wavelengths that are highly absorbed by water are termed 'ablative', while those wavelengths that are less strongly absorbed by water are termed non-ablative. Examples of non-ablative wavelengths on the market are 1,410, 1,440, 1,540, 1,550, and 1,927 nm. Ablative wavelengths available

Table 1. Important terms used in FP treatment

MTZ	column-like zone of thermal injury; size of individual MTZs depends on the energy per MTZ; the higher the energy, the wider the MTZ and the deeper the thermal damage
Treatment density	MTZs per cm^2
Micro-epidermal necrotic debris	button-shaped accumulation of necrotic debris below the intact stratum corneum over each MTZ which forms after treatment

are the 2,790 nm YSGG, 2,940 nm Er:YAG, and CO_2 (10,600 nm) lasers. The various devices differ in wavelength, pulse energy, and damage pattern, and hence provide numerous combinations of possible treatments. Using various laser parameters allows the customization of the 3D tissue damage pattern, its shape, density, and depth according to the clinical indication.

Non-Ablative Fractional Photothermolysis

In 2003, Manstein et al. [5] introduced the novel concept of FP in its basic application. In the prototype device, they used a non-ablative wavelength of 1,550 nm with an Er-doped fiber laser and named it the Fraxel™ laser [7]. In the beginning, Fraxel laser treatment was synonymous with non-ablative fractional photothermolysis (NAFP). However, the true novelty lay in the treatment of only fractions of the skin, which were surrounded by unharmed tissue, which led to rapid and controlled epidermal and dermal regeneration and repair via the migration of the surrounding viable cells. This allowed for deep thermal damage without compromising safety. The tissue injury created stimulates collagen remodeling and promotes elastic tissue formation, which is postulated to be responsible for the clinical improvements observed with FP.

Technical Data

The small three-dimensional zones of non-ablative thermal damage, referred to as MTZs, are in general smaller than 400 μm in diameter and can penetrate the skin at varying depths of up to 1,400 μm, depending on the wavelength, pulse energy, and device chosen. The distance between the MTZs, also referred to as the pitch, depends on the density and can be anywhere from 200 to 300 μm and larger. As mentioned above, the energy chosen directly influences not only the depth of the MTZ, i.e., of the thermally damaged tissue, but also its diameter and hence the density or percentage of the skin covered. Some devices adjust the density appropriately after the operator chooses the energy to account for the increase in MTZ diameter.

The terminologies may be congruent at low energies, but, as the energy increases and the MTZs increase, these parameters may differ significantly. In order to translate one device's density settings into another, a consistent density parameter should be chosen. Percent coverage is more translatable, as all MTZ sizes and individual spots differ not only between devices but also within a single device.

Depending on the device, energies of up to 70 mJ are available.

There are two different modes of how the handpieces are handled and the MTZs delivered to the skin. The treatment can be delivered either in a scanning mode, where the thermal damage is induced in the skin while simply moving the handpiece over the treatment area with constant contact. These pulses are delivered randomly and individually, theoretically decreasing the possibility of bulk heating. In stamping mode, all microbeams within the handpiece reach the skin at the same time. Cooling with fractional

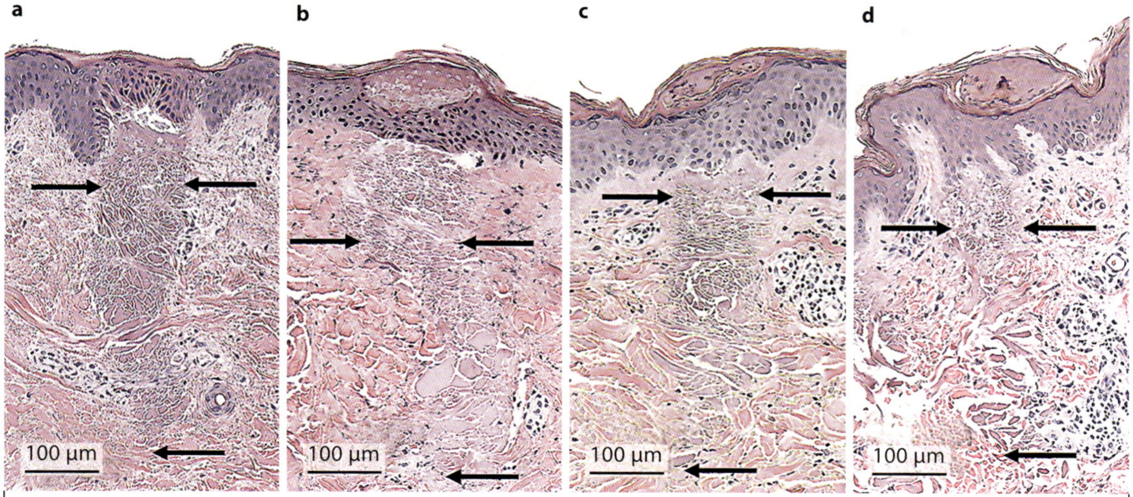

Day 0
Column-like denaturation of epidermis and dermis
Subepidermal clefting
Stratum corneum intact

Day 1
Reepithelialization of epidermis
Formation of MENDs with thermally damaged epidermal and dermal cells

Day 3
Compaction of MENDs
Virtually normalized epidermis

Day 7
Intracorneal MENDs
Beginning exfoliation
No more subepidermal clefting

Fig. 2. Histological damage pattern and healing course after NAFP (Fraxel SR1500). Courtesy of Solta Medical.

devices is only used as a comfort measure to make the treatment more tolerable to the patient. Microscopic epidermal injury leading to epidermal turnover is a desirable effect with all resurfacing devices.

Non-ablative lasers currently on the market have wavelengths of 825–1,350, 1,410, 1,440, 1,540, 1,550, and 1,927 nm. As these wavelengths are not strongly absorbed by water, the wounds induced into the tissue are coagulative in nature.

Histological Damage Pattern and Healing Course

Directly after treatment with NAFP, there is a characteristic histology with 4 main features: a column-like denaturation of the epidermis and dermis which represents the MTZ, and an intact stratum corneum with subepidermal clefting and, surrounding this, vital unharmed tissue (fig. 2a).

The intact stratum corneum maintains and allows the barrier function of the skin, decreasing the risk of post-procedural infections to virtually zero. Also, there is no oozing or bleeding due to the coagulative nature of the non-ablative process.

Within the first 24 h, the thermally destroyed tissue becomes replaced by keratinocytes that migrate from the healthy surrounding tissue. At day 1 after treatment, the epidermis is already fully re-epithelialized and the continuity of the epidermal basal cell layer is restored [8]. A button-shaped structure becomes evident as early as 1 day after FP, which is located below the intact stratum corneum over each MTZ and has been shown to contain necrotic tissue. It therefore has been named microscopic epidermal necrotic debris (MEND; fig. 2b). Melanin, elastic tissue, and other dermal contents can be found within this structure. Within the first week, the MEND becomes more and more compacted, until it finally gets eliminated transepidermally through the stratum corneum, which is facilitated by the subepidermal clefting. This process has been called the MEND shuttle, as it literally 'shuttles' dermal content out of the skin.

Around day 3, the epidermis is virtually normalized; at day 7, the subepidermal clefting is not visible anymore and complete epidermal regeneration is obtained [9, 10]. MEND is shed from the epidermis within 7 days (fig. 2c, d). This thermal damage induced in the epidermis and dermis ultimately leads to collagen induction and epidermal turnover. MTZs become replaced with new collagen within 3–6 months, which means that the clinical result continues to improve over 6 months.

In one study, tissue reactions concordant to a wound healing response by NAFP could be shown by immunohistochemical examination. At day 1, expression of heat shock protein 70 was shown; at day 7, increased collagen III production as well as the presence of myofibroblasts could be demonstrated by immunohistochemistry [10].

Clinical Indications

Currently, NAFP is FDA approved for the treatment of pigmented lesions, periorbital rhytides, skin resurfacing, melasma and soft tissue coagulation, acne and surgical scars, striae, and actinic keratoses. However, its off-label use is clearly more extended with case reports on the successful treatment of striae distensae, Becker nevus, nevus of Ota, poikiloderma of Civatte, residual hemangioma, burn scars, hyper- and hypopigmented scars, telangiectatic matting, enlarged pores, and even persistent pigment and hypopigmentation after QS tattoo removal.

Current experience shows that with 5–7 sessions, safe and effective treatments can be achieved for facial acne scars, mild-to-moderate rhytides and dyschromia in photoaged skin. Deep rhytides, especially around the mouth, are difficult to treat with a non-ablative approach, whereas fine lines and wrinkles around the eyes respond more consistently.

Energy settings of each treatment have to be chosen according to clinical indications and, more precisely, according to the desired depth. Conditions extending deep into the dermis, such as scars and wrinkles for example, have to be treated with high energies for thermal damage to penetrate deeply into the dermis and induce collagen remodeling. Superficial conditions on the other hand, such as epidermal dyschromia, should be treated with low-energy settings. Melasma is a difficult-to-treat condition, and it should be treated very conservatively, if at all, with any fractional device. Results from fractional resurfacing for treatment of melasma are unpredictable at best and should not be used as a first-line therapy.

Depending on the indication, multiple treatments are usually necessary, which can be performed in intervals of at least 2 weeks. In patients who are at higher risk of post-inflammatory hyperpigmentation (PIH), these intervals should be increased to at least 4–6 weeks.

An important finding in NAFP is that there have also been good results in off-face locations and locations with fragile skin (such as the eyelids, neck, chest, and extremities), which are known to usually be more prone to side effects when treated with lasers. Examples for off-face skin conditions that can be treated with FP are dyschromia on the hands or chest, poikiloderma of Civatte, and surgical scars. When performing NAFP off the face, post-treatment erythema may last slightly longer, roughly up to 1 week.

As melanin does not directly absorb the various non-ablative wavelengths, the risk of pigmentary side effects is generally low. Post-inflammatory pigment alteration is a risk in any laser procedure where an inflammatory response is created. Despite

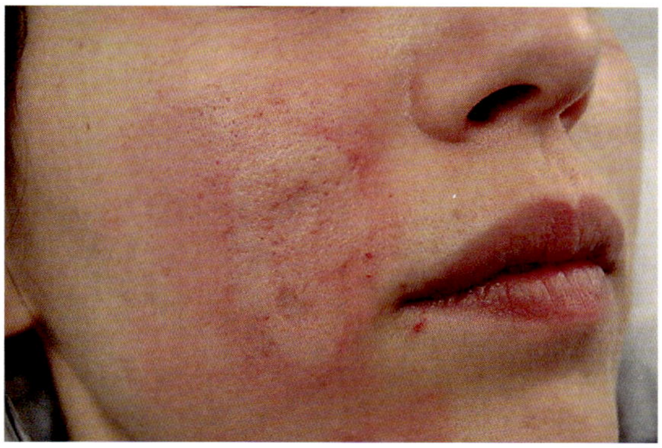

Fig. 3. Erythema and edema immediately after treatment with NAFP (Palomar LUX 1540 fractional laser).

this, NAFP is safe for Fitzpatrick skin types I–VI, when used appropriately. However, pigmentary alterations can still occur, and clearly patients with a Fitzpatrick skin type >III are at higher risk. Thus, it is advisable to treat such patients with conservative settings, i.e., lower density settings. As the rate of PIH mainly depends on the density, a decreased percent coverage should be targeted.

The treatment itself is only slightly painful, and to ease the pain topical anesthesia can be used.

Clinical Healing Course

Directly after treatment, the treated area shows erythema and edema, which usually lasts up to 3 days depending on the energy and density used, as well as the location treated (fig. 3). Periorbital swelling is usually more pronounced than in other areas. However, to speak of downtime with NAFP seems almost exaggerated, as the symptoms are mild and redness can be covered with makeup after 24 h. Thereafter, the patient may experience some xerosis or bronzing and peeling of the skin, which represents the MEND being exfoliated. Usually after 7 days, the skin has virtually normalized.

Side Effects of NAFP

Non-ablative fractional resurfacing is a very safe procedure with an almost complete absence of prolonged side effects. The most common side effect is post-inflammatory pigment alteration. These pigmentary changes are more prominent in darker skin types. In patients with melasma, hyperpigmentation can occur in more than 30% of cases [11]. If side effects do occur, they are typically mild and transient [12].

Mild Side Effects

Immediate post-treatment erythema and swelling as well as slight darkening of the skin (bronzing) as the MEND desquamates are expected consequences of the treatment and usually last 1–3 days. If the erythema persists longer than 4 days, it is defined as prolonged erythema, which is a rare mild complication occurring in less than 1% of cases [13]. Acneiform eruptions are the most frequently observed mild side effect. They are presumed to occur due to disruption of the follicular units during treatment and reepithelialization.

Moderate Side Effects

The most common type of infections that occur after NAFP are herpes simplex virus (HSV) infections. They have been reported to occur in 0.3–2% of cases. Bacterial infections are very rarely observed and usually present 1–3 days after treatment.

PIH can also be observed. It depends on the energy and density used, and patients with darker skin phototypes (Fitzpatrick IV–VI) are at higher risk. It seems that high densities are more likely to produce PIH. Thus, it is recommended to use lower densities and longer treatment intervals in those at risk of hyperpigmentation [14]. Whilst keeping this in mind, NAFP can also be performed on dark skin types [15]. Hypopigmentation has not been reported.

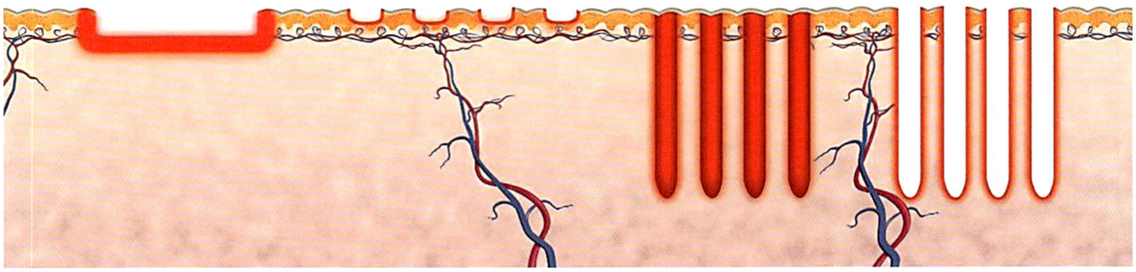

Fig. 4. Patterns of tissue damage. Courtesy of Palomar Medical. NAFR = Non-ablative fractional resurfacing; AFR = ablative fractional resurfacing.

Severe Side Effects
To date, severe side effects such as scarring have not been reported for NAFP.

Ablative Fractional Photothermolysis

The next logical step in the evolution of the fractional technology was to expand it into ablative wavelengths. Hence, ablative FP (AFP) was introduced into the market in 2007. The basic principle is the same as for non-ablative wavelengths, only the water absorption curve is not a linear one but an exponential one; hence, the ablative wavelengths have a thousand times stronger absorption in water than the non-ablative ones. This fact ultimately changes the pattern of tissue damage (fig. 4).

Technical Data

With the ablative approach, the resulting histological damage pattern in the skin shows an ablation zone, i.e. micro-columns of missing tissue. Again, depending on the energy and wavelength used, those ablated micro-columns vary in width and depth and can penetrate up to 1.5 mm deep. Due to the ablative character of the procedure, the stratum corneum is mostly disrupted, contrary to the damage pattern in non-ablative fractionated technology (fig. 5a) [16]. A thin layer of eschar or necrotic tissue lines the ablated cavity. The width of this coagulated tissue lining the ablative column will vary depending on the laser wavelength used. Around these cavities, annular coagulation zones of 10–150 μm, representing denatured collagen, can be observed. Finally, there is unharmed tissue surrounding each of these ablation-necrosis-coagulation damage zones.

The fact that there is actual tissue ablation results in observable tissue tightening. The layer of coagulation promotes hemostasis and tissue shrinkage. The ablation/coagulation ratio is different for each of these devices and may also change within the same device based on the pulse width chosen.

With AFP, we can also choose between scanning and stamping devices. The ablative fractional devices on the market at time of writing are the 2,940 nm Er:YAG, 2,790 nm YSGG, and 10,600 nm CO_2 lasers.

With regard to how deep a penetration devices actually obtain, it has been suggested to divide the ablative fractional devices into two categories, namely the ablative ones and the micro-ablative ones (superficial ablative fractional resurfacing). The ablative ones are those devices that have enough power to truly penetrate and ablate the

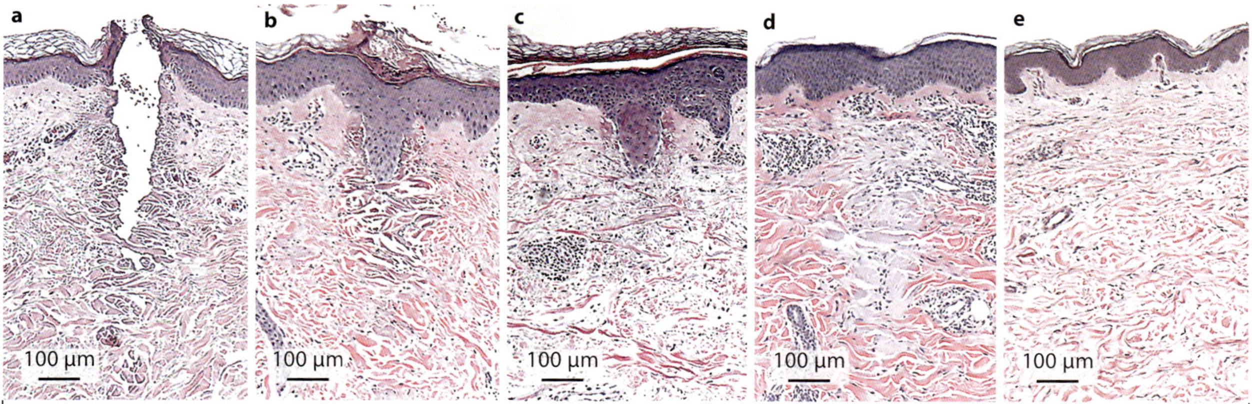

Day 0	Day 2	Week 1	Month 1	Month 3
Epidermal and dermal ablation	Invagination of the epidermis into ablated dermis Formation of MENDs with thermally damaged epidermal and dermal cells	Complete reepithelialization of epidermis MEND exfoliated	Invagination of epidermis decreasing Coagulation zone in the dermis still visible	New compact collagen

Fig. 5. Histological damage pattern and healing course after treatment with 20 mJ CO_2 AFP (Fraxel re:pair). Courtesy of Solta Medical.

tissue down to 1.5 mm. The micro-ablative ones are devices with limited power and hence limited penetration depth. Not all ablative fractional devices are the same. Pulse width also has a role in the penetration of ablative wavelengths, with the energy being constant, longer pulse widths typically result in more superficial heating with more coagulative than ablative damage. In some instances this may be desirable and in some cases detrimental to a good clinical outcome.

Histological Damage Pattern and Healing Course

Immediately after treatment, one can see the typical pattern of fractional ablative tissue damage with a disrupted stratum corneum, a central ablation zone (which is lined with a thin eschar layer followed by a coagulation zone), and then unharmed tissue (fig. 5a). Thanks to the vital tissue surrounding each damaged zone, reepithelialization of the coagulation zones occurs rapidly, within 48 h. Around day 2, one can observe an invagination of the epidermis into the ablated dermis and again – the formation of MEND (fig. 5b). After 1 week, there is complete reepithelialization of epidermis and the MEND is exfoliated (fig. 5c). The invagination of the epidermis into the dermis starts to decrease around 1 month after treatment; after 3 months new collagen formation can be observed [17] (fig. 5d, e). Immunohistochemical studies have indicated a prolonged stimulation of the wound repair induced by AFP, which is believed to lead to prolonged clinical improvements for at least 3 months.

Clinical Indications

Currently, ablative fractional devices are FDA cleared to treat wrinkles, rhytides, furrows, fine lines, textural irregularities, pigmented lesions, and dyschromia.

Photodamaged skin, skin laxity, facial rhytides, acne scarring can all be clinically improved after a single treatment. However, with severe rhytides or acne scarring, up to three treatments can sometimes be necessary. Intervals between treatments can be anywhere from 6 to 12 months.

Split-face studies have confirmed that with ablative fractionated a higher degree of clinical improvement per session can be achieved than with non-ablative fractionated treatments [18]. However, downtime is longer and the rate of side effects is slightly higher.

Unlike conventional ablative lasers, treatments are tolerable with only topical anesthesia with or without the addition of air-cooling.

AFP is a safe treatment for lighter skin types, such as Fitzpatrick I–III. Regarding safety for Fitzpatrick skin types IV–VI, we still need more experience and data. However, for darker skin types (Fitzpatrick IV–VI) there is clearly a higher risk of PIH; thus, indications have to be evaluated carefully and settings should be chosen more conservatively.

True care has to be taken when performing AFP in areas where the skin is thin or relatively devoid of pilosebaceous units, such the neck and chest, as well as around the eyes. There are reports of scarring after CO_2 AFP on the neck, chest and face. These are possible complications in these areas, even when performed by experienced operators [19, 20]. A very recent interesting and promising new indication is drug delivery by AFP [21].

Clinical Healing Course

When using AFP, a wound is created. Hence, there is downtime associated with swelling, oozing, and crusting. The duration of the downtime strongly varies according to wavelength, energy, and density chosen. Usually, downtime can be anywhere from 3 to 7 days, with high energies and high densities being associated with longer downtimes. Generally AFP with Er:YAG has somewhat shorter downtimes (up to 3 days), whereas those of AFP with CO_2 devices are longer (3–7 days). Theoretically though, by increasing the density to a maximum, one can virtually ablate such high percentages of the skin that the downtime would approximate the one of traditional ablative resurfacing.

Downtime is defined by oozing and serosanguineous discharge and some edema, during which proper wound care is essential. Once the wound is healed, one can observe erythema anywhere from 2–4 weeks, but with aggressive settings even up to 6–8 weeks. Off-face treatment areas typically result in longer periods of erythema. During this time and thereafter, sun avoidance is crucial.

Side Effects of AFP

In contrary to non-ablative fractional resurfacing, the more invasive procedure of ablative fractional resurfacing is associated with the possibility of more severe and frequent side effects and complications [22]. We have to bear in mind though that this procedure has been on the market for a shorter period of time than NAFP, and hence there are fewer publications on side effects and complications with smaller patient numbers than have been published for NAFP.

Mild Side Effects

In ablative FP, prolonged erythema is defined as post-treatment erythema persisting longer than 1 month and occurs quite frequently in more than 12.5% of treated patients. In most cases, it resolves within 3 months [23].

Acneiform eruptions and milia formation are further mild side effects that can be observed more frequently. Again, occlusive wound care will increase the risk and severity of such eruptions; thus, non-comedogenic wound care is recommended.

Moderate Side Effects
HSV infections are the most common infections and due to the ablative, hence more invasive, character of the procedure, seem to occur more frequently than with NAFP. Thus, with a prior history of facial HSV, it is recommended to administer an oral antiviral prophylaxis, initiated 1 day prior to the treatment and continued 5–7 days thereafter.

Excessive wound care with too occlusive agents after AFP can lead to wound occlusion, which will result in prolonged healing times and can increase the risk of bacterial overgrowth. Nonetheless, it is still controversial whether prophylactic systemic antibiotics should be used. It is recommended for high-risk patients, such as immunosuppressed patients.

As in NAFP, PIH can be observed and it occurs more frequently in patients with Fitzpatrick III–VI skin phototypes.

Severe Side Effects
AFP can induce scarring, especially on locations off the face such as the neck and chest, as well as on locations of fragile skin, such as around the eyes. It seems to be more frequent with CO_2 AFP [24–26]. Such scar-prone anatomic locations should – if treated at all – be treated only by experienced physicians with conservative treatment settings and strict postoperative wound care and follow-up.

Cicatricial ectropion formation of the lower eyelid is another rare severe side effect observed with CO_2 AFP as a result of treatment of the lower eyelids [20].

When to Use Which Device

Thanks to the numerous devices on the markets – using different wavelengths, energies, and damage patterns – there is a large choice of which device to use. Not all offices have multiple devices, so in many instances the choice becomes more of a proper patient selection as opposed to a device selection. A more risk-adverse patient or a patient that cannot accept any downtime will be treated with repetitive non-ablative treatments, taking into account the less significant improvement in one treatment. Such patients will accept the decreased success rate and increased treatment sessions in exchange for the shorter downtime and lower risk of side effects.

A patient that wants results rather fast and can accept several days of downtime will be treated with an ablative approach, with dramatic improvements in 1 or 2 treatments, at the same time accepting a higher risk profile. Patient skin type and clinical indication will also steer this decision process.

From a clinical indication point of view, penetration depth (energy) should be matched to the depth of the targeted structure. Pigmented lesions, actinic damage, and actinic keratoses should be treated with superficial-to-medium penetration depths (100–200 μm), textural irregularities and wrinkles with medium-to-deep penetration depths (400–700 μm), and scars and deep furrows with deep penetration depths (1,000–1,200 μm). However, the importance of depth in achieving excellent clinical results is still controversial and being discussed.

Preliminary data suggest that the combination of AFP and NAFP might deliver promising results. FP has changed the field of laser resurfacing. With more sophisticated systems in development and greater understanding of the skin's reactions to these treatments, we are expanding our ability to safely resurface the skin.

References

1 Manuskiatti W, Fitzpatrick RE, Goldman MP: Long-term effectiveness and side effects of carbon dioxide laser resurfacing for photoaged facial skin. J Am Acad Dermatol 1999;40:401–411.
2 Schwartz RJ, Burns AJ, Rohrich RJ, Barton FE, Byrd HS: Long-term assessment of CO_2 facial laser resurfacing: aesthetic results and complications. Plast Reconstr Surg 1999;103:592–601.
3 Hohenleutner S, Koellner K, Lorenz S, Landthaler M, Hohenleutner U: Results of nonablative wrinkle reduction with a 1,450-nm diode laser: difficulties in the assessment of "subtle changes". Lasers Surg Med 2005;37:14–18.
4 Tan MH, Dover JS, Hsu TS, Arndt KA, Stewart B: Clinical evaluation of enhanced nonablative skin rejuvenation using a combination of a 532 and a 1,064 nm laser. Lasers Surg Med 2004;34:439–445.
5 Manstein D, Herron GS, Sink RK, Tanner H, Anderson RR: Fractional photothermolysis: a new concept for cutaneous remodeling using microscopic patterns of thermal injury. Lasers Surg Med 2004;34:426–438.
6 Khan MH, Sink RK, Manstein D, Eimerl D, Anderson RR: Intradermally focused infrared laser pulses: thermal effects at defined tissue depths. Lasers Surg Med 2005;36:270–280.
7 Huzaira M, Anderson RR, Sink K, Manstein D: Intradermal focusing of near-infrared optical pulses: a new approach for non-ablative laser therapy. Lasers Surg Med 2003;32(suppl 15):17–38.
8 Stumpp OF, Bedi VP, Wyatt D, Lac D, Rahman Z, Chan KF: In vivo confocal imaging of epidermal cell migration and dermal changes post nonablative fractional resurfacing: study of the wound healing process with corroborated histopathologic evidence. J Biomed Opt 2009;14:024018.
9 Jih MH, Kimyai-Asadi A: Fractional photothermolysis: a review and update. Semin Cutan Med Surg 2008;27:63–71.
10 Laubach HJ, Tannous Z, Anderson RR, Manstein D: Skin responses to fractional photothermolysis. Lasers Surg Med 2006;38:142–149.
11 Wind BS, Kroon MW, Meesters AA, Beek JF, van der Veen JP, Nieuweboer-Krobotová L, Bos JD, Wolkerstorfer A: Non-ablative 1,550 nm fractional laser therapy versus triple topical therapy for the treatment of melasma: a randomized controlled split-face study. Lasers Surg Med 2010;42:607–612.
12 Graber EM, Tanzi EL, Alster TS: Side effects and complications of fractional laser photothermolysis: experience with 961 treatments. Dermatol Surg 2008;34:301–305.
13 Fisher GH, Geronemus RG: Short-term side effects of fractional photothermolysis. Dermatol Surg 2005;31:1245–1249.
14 Kono T, Chan HH, Groff WF, Manstein D, Sakurai H, Takeuchi M, Yamaki T, Soejima K, Nozaki M: Prospective direct comparison study of fractional resurfacing using different fluences and densities for skin rejuvenation in Asians. Lasers Surg Med 2007;39:311–314.
15 Mahmoud BH, Srivastava D, Janiga JJ, Yang JJ, Lim HW, Ozog DM: Safety and efficacy of erbium-doped yttrium aluminum garnet fractionated laser for treatment of acne scars in type IV to VI skin. Dermatol Surg 2010;36:602–609.
16 Hantash BM, Bedi VP, Chan KF, Zachary CB: Ex vivo histological characterization of a novel ablative fractional resurfacing device. Lasers Surg Med 2007;39:87–95.
17 Jih MH, Kimyai-Asadi A: Fractional photothermolysis: a review and update. Semin Cutan Med Surg 2008;27:63–71.
18 Cho SB, Lee SJ, Cho S, Oh SH, Chung WS, Kang JM, Kim YK, Kim DH: Non-ablative 1550-nm erbium-glass and ablative 10 600-nm carbon dioxide fractional lasers for acne scars: a randomized split-face study with blinded response evaluation. J Eur Acad Dermatol Venereol 2010;24:921–925.
19 Avram MM, Tope WD, Yu T, Szachowicz E, Nelson JS: Hypertrophic scarring of the neck following ablative fractional carbon dioxide laser resurfacing. Lasers Surg Med 2009;41:185–188, erratum in Lasers Surg Med 2009;41:398.
20 Fife DJ, Fitzpatrick RE, Zachary CB: Complications of fractional CO_2 laser resurfacing: four cases. Lasers Surg Med 2009;41:179–184.
21 Haedersdal M, Sakamoto FH, Farinelli WA, Doukas AG, Tam J, Anderson RR: Fractional CO(2) laser-assisted drug delivery. Lasers Surg Med 2010;42:113–122.
22 Metelitsa AI, Alster TS: Fractionated laser skin resurfacing treatment complications: a review. Dermatol Surg 2010;36:299–306.
23 Rahman Z, MacFalls H, Jiang K, Chan KF, Kelly K, Tournas J, Stumpp OF, Bedi V, Zachary C: Fractional deep dermal ablation induces tissue tightening. Lasers Surg Med 2009;41:78–86.
24 Ross RB, Spencer J: Scarring and persistent erythema after fractionated ablative CO_2 laser resurfacing. J Drugs Dermatol 2008;7:1072–1073.
25 Fife DJ, Fitzpatrick RE, Zachary CB: Complications of fractional CO_2 laser resurfacing: four cases. Lasers Surg Med 2009;41:179–184.
26 Avram MM, Tope WD, Yu T, Szachowicz E, Nelson JS: Hypertrophic scarring of the neck following ablative fractional carbon dioxide laser resurfacing. Lasers Surg Med 2009;41:185–188.

I. Bogdan Allemann, MD
Klinik für Plastische Chirurgie und Handchirurgie, Universitätsspital Zürich
Rämistrasse 100
CH–8091 Zürich (Switzerland)
Tel. +41 44 255 1111, E-Mail Inja.bogdan@usz.ch

Bogdan Allemann I, Goldberg DJ (eds): Basics in Dermatological Laser Applications.
Curr Probl Dermatol. Basel, Karger, 2011, vol 42, pp 67–80

Vascular Lesions

Patricia T. Ting · Jaggi Rao

Division of Dermatology and Cutaneous Sciences, University of Alberta, Edmonton, Alta., Canada

Abstract

Advances in laser and light-based technology have increased their potential applications, safety and efficacy for the management of vascular lesions in dermatology. Light devices for vascular lesions include the variable-pulse 532 nm potassium titanyl phosphate laser, 577 to 595 nm pulsed dye laser, intense pulsed light devices, and 800 to 940 nm diode, long-pulse 755 nm alexandrite and 1,064 nm Nd:YAG lasers. This review will discuss the various different laser and light-based devices, and provide a focused treatment approach for the management of common congenital and acquired vascular lesions.

Significant advances in the technology and availability of pulsed light devices and lasers for the treatment of vascular lesions have occurred over the past 25–30 years.

In 1983, Anderson and Parish [1] described the principles of selective photothermolysis outlining the use of specific pulsed wavelengths of light to target cells, organelles and pigmented structures in vivo. The same basic principle of lasers and noncoherent light sources applies in that both emit electromagnetic radiation containing photons that target a specific chromophore and the subsequent absorption of photons induces thermal injury to the target tissue [1].

The three main chromophores in human skin are hemoglobin, melanin and water. Intravascular oxyhemoglobin (red lesions), deoxygenated hemoglobin (blue lesions) and methemoglobin are the intended chromophores for vascular lesions. Absorption peaks for oxyhemoglobin are 418, 542 and 577 nm, while deoxyhemoglobin ranges from 800 to 1,200 nm [1–3]. Variations in wavelength (nanometers, nm), fluence (energy density, J/cm^2) and pulse duration (milliseconds) result in controlled thermal injury to target tissue and alter the penetration depth [3]. It is also important that the pulse duration is less than or equal to the target tissue thermal relaxation time (i.e. time for half of the thermal energy to dissipate from a structure into surrounding tissue) to minimize damage to adjacent tissue [1]. The optimal pulse duration is dependent upon the vessel diameter; the estimated pulse duration for 30- to 100-μm diameter vessels is between 1 and 10 ms (i.e. thermal relaxation time for cutaneous vasculature) [4].

Table 1 summarizes the numerous vascular lesions that can be treated with pulsed light devices and lasers, including hemangiomas and port wine

Table 1. Vascular lesions treated by laser and light devices [3, 5, 7, 8, 17, 38, 73]

Congenital	Acquired	Vascular malformations	Venous	Other dermatological conditions with vascular component
Hemangioma Port wine stain	Angiomas (Campbell de Morgan, spider, senile) Angiofibroma Angiokeratoma Cutaneous Kaposi sarcoma Poikiloderma of Civatte Pyogenic granuloma Rosacea – erythematotelangiectatic Telangiectasia – facial, leg	Arteriovenous fistula Arteriovenous malformation Capillary malformation Venous malformation Lymphatic malformation	Venulectasia Venous angioma Venous lake Blue reticular veins	Acne Hypertrophic scars Psoriasis Striae atrophicae distensae (early) Verruca vulgaris

stains (PWS), facial telangiectasia, rosacea, angiomas, angiokeratomas, poikiloderma of Civatte, venous lakes, venulectasias and blue reticular veins [2, 3, 5–9]. Safe and effective light devices for vascular lesions are the variable-pulse 532 nm potassium titanyl phosphate (KTP) laser, 577 to 595 nm pulsed dye laser (PDL), intense pulsed light (IPL) devices, 800 to 940 nm diode, as well as the long-pulse 755 nm alexandrite and 1,064 nm Nd:YAG lasers (table 2). Endovascular lasers that can be used for varicose veins include the 810, 940 and 980 nm diodes as well as the 1,043 and 1,320 nm Nd:YAG [10–14].

Continuous- and Quasi-Continuous-Wave Lasers

Some of the first lasers used in the treatment of vascular lesions were continuous (CW) and quasi-continuous wave (quasi-CW). CW lasers include the 488 and 514 nm argon, 488 to 638 nm argon-pumped tunable dye, 511 and 578 nm copper vapor/bromide, 532 nm KTP and 568 nm krypton lasers. Between the 1970s and 1980s, the argon laser (418–514 nm, 0.1- to 1.0-mm spot size, 0.5-mm penetration depth) with an absorption spectrum that matched oxyhemoglobin was preferentially used for face and neck telangiectasias [15], capillary hemangiomas, angiokeratomas, congenital vascular anomalies (such as hemangiomas and PWS) and other smaller cutaneous vascular lesions such as venous lakes and spider angiomas [16–19]. The continuous mode of CW lasers (e.g. argon laser) exceeds the vascular thermal relaxation time and frequently results in nonspecific thermal damage and post-procedural pigmentary changes (i.e. hypopigmentation or hyperpigmentation), thereby limiting their use to Fitzpatrick I–III skin types; other adverse effects include atrophic or hypertrophic scarring [20–22].

Quasi-CW lasers such as copper bromide, krypton and KTP have also been used to treat many types of vascular lesions similar to those of CW lasers [21, 23, 24]. The variable-pulse KTP laser utilizes a 1,064 nm Nd:YAG energy source via a KTP crystal to emit a 532 nm green light with variable pulse duration (1–100 ms) and spot size (1–10 mm). The KTP laser is effective for superficial vascular lesions, including facial telangiectasia, erythematotelangiectatic rosacea, cherry and spider angiomas, venous malformations and leg telangiectasia [25–28]. It has also

Table 2. Lasers and light-based treatments used for the management of vascular lesions

Light device/ laser	Wave-length nm	Fluence J/cm²	Pulse duration ms	Applications	Advantages/ disadvantages
Argon [18, 20] Argon pumped tunable [19]	488, 514 488–638	–	CW quasi-CW	Face and neck telangiectasia, capillary hemangiomas, PWS, other small cutaneous vascular lesions (e.g. angiokeratomas, venous lakes, spider angiomas), leg telangiectasia	Disadvantages: – Superficial depth approximately 0.5 mm – CW exceeds thermal relaxation time – Non-specific thermal damage to non-target tissue – Post-treatment pigmentary changes – Atrophic or hypertrophic scarring – Limited use in Fitzpatrick I–III skin types
KTP laser [25–28]	532	up to 240	1–100	Facial telangiectasia, erythematotelangiectatic rosacea, cherry and spider angiomas, venous malformations, leg telangiectasia	Advantages: – Longer pulse durations – Heating rather than blood vessel rupture – Less post-treatment purpura Disadvantages: – Superficial penetration (0.5 mm), small spot size (1–10 mm) – Significant procedural pain – Edema, crusting, atrophic scars (nasal region) – Post-treatment pigmentary changes in ethnic or tanned skin
PDL [2, 9, 32, 34, 35, 37, 38, 44, 64]	577*, 585, 595, 600	up to 40	0.45-40	Congenital vascular anomalies (hemangioma, PWS), facial telangiectasia, erythematotelangiectatic rosacea, leg telangiectasia and veins (<0.5 mm), poikiloderma of Civatte, cutaneous lesions with vascular component (e.g. acne, hypertrophic scars, psoriasis, early striae distensae), pyogenic granuloma	Advantages: – Treatment of choice for PWS – Safe and effective for use in infants and children – Reduced thermal epidermal injury with dynamic cooling – Deeper penetration to dermal vessels with longer wavelength Disadvantages: – Post-treatment purpura and erythema (with 577 and 585 nm) – Post-treatment pigmentary changes (usually hyperpigmentation)
IPL [35, 45– 47]	400–1,200	10–80	2– 200	Facial telangiectasia, erythematotelangiectatic rosacea, hemangioma, PWS, poikiloderma of Civatte	Advantages: – Versatility: range of wavelength, fluence and pulse duration – Large spot size, high skin coverage rate – Reduced thermal epidermal injury with dynamic cooling – Minimal erythema, edema, post-treatment pigmentary changes – Rapid recovery (usually < 48 hours) Disadvantages: – Inconsistent wavelength and fluence from noncoherent light – Optical coupling gel application required – Handpiece: heavy, requires direct skin contact

Table 2. Continued

Light device/ laser	Wave-length nm	Fluence J/cm²	Pulse duration ms	Applications	Advantages/ disadvantages
Alexandrite (long-pulsed) laser [3, 66]	755	up to 100	3–100	Face and leg telangiectasia, blue reticular veins, PWS (various stages), vascular malformations	Advantages: – Minimal pain and purpura – Reduced thermal epidermal injury with dynamic cooling – Deeper penetration of dermal vessels Disadvantages: – Transient hyperpigmentation – Post-treatment pigmentary changes
Diode [63, 74]	800	10–100	5–400	Facial telangiectasia, erythematotelangiectatic rosacea, leg telangiectasia, blue reticular veins, venous lakes, varicose veins	Advantages: – Reduced thermal epidermal injury with dynamic cooling – Deeper penetration of dermal vessels – May be used endovascularly Disadvantages: – Post-treatment pigmentary changes
Nd:YAG laser [6, 52–55, 72, 73]	1,064	5–900	0.25–500	Leg telangiectasia, blue reticular veins up to 4 mm, venous lakes, PWS, hemangioma, venous malformation, angioma, facial telangiectasia	Advantages: – Deeper penetration to dermal vessels – Treats small and medium leg veins up to 4 mm – Less likely to cause post-treatment pigmentary changes – Safer for ethnic or darkly pigmented skin Disadvantages: – Requires dynamic cooling to reduce thermal epidermal injury – Less absorption by hemoglobin – Crusting up to 2–3 weeks
Endovascular lasers [10–14] Diode Nd:YAG	 810–940 1,320	Variable	Variable	Varicose veins, chronic venous insufficiency	Advantages: – High success rates (>90%) and patient satisfaction Disadvantages: – Requires intravascular fiberoptic catheter – Higher risk of complications – Risk of DVT, pulmonary embolus and nerve injury (<1%)

CW = Continuous wave; quasi-CW = quasi-continuous wave; IPL = intense pulsed light; KTP = potassium titanyl phosphate; PDL = pulsed dye laser; Nd:YAG = neodymium:yttrium aluminium garnet; DVT = deep venous thrombosis; PWS = port wine stain; *orginal PDL 577 nm.

been successfully used to treat microvascular lesions on the vocal cords [29, 30]. Advantages of KTP laser are shorter procedural time, longer pulse durations with resultant heating rather than the rupture of vessels and therefore less post-procedural purpura [26]. Disadvantages of KTP are superficial penetration (approximately 0.5 mm), small spot size, significant procedural pain, edema and crusting, thermal injury and post-treatment pigmentary changes particularly in ethnic or tanned skin, multiple treatment sessions required for larger vessels, and possible development of atrophic scars in the nasal region [26, 28]. All quasi-CW lasers, which still have relatively long pulse durations, are associated with the same adverse effects as CW lasers, but to a lesser extent [22].

Pulsed Dye Laser

Introduced in 1989 for the treatment of PWS in children, the pulsed dye laser (PDL) generates a noncoherent pulsed yellow light using a high-power flashlamp pump to activate an organic rhodamine dye [9]. The original PDL emitted 577 nm (pulse duration 0.45 ms), corresponding to the third absorption peak of oxyhemoglobin (approximately 577 nm) [2]. Newer PDL systems emit longer wavelengths of 585, 590, 595 and 600 nm, which allows for greater tissue penetration depth and more variable pulse duration (0.45–40 ms), fluence (5–10 J/cm^2) and spot size (2–12 mm) [31, 32]. Treatment endpoints indicating the optimal photophysical parameters that have been used for PDL are mild purpura for 585–600 nm PDL or vessel clearance for 595 nm PDL [33].

New-generation PDL are considered the standard of care for congenital vascular anomalies, such as hemangioma and PWS in infants and children [9, 34, 35].

PDL is also effective in acquired vascular lesions, including facial telangiectasia, erythematotelangiectatic rosacea, leg telangiectasia and veins (<0.5 mm), poikiloderma of Civatte, cutaneous lesions with a vascular component (e.g. acne, hypertrophic scars, psoriasis and early striae distensae) and pyogenic granuloma [36–38]. Newer PDL devices equipped with dynamic cooling systems reduce the risk of thermal epidermal injury (even with higher fluence settings), reduce procedural pain and minimize post-procedural pigmentation (even in ethnic skin) [39–41]. Furthermore, long-pulse 595 nm PDL with pulse durations up to 40 ms can reduce or ameliorate post-procedural purpura [42, 43]. Disadvantages of PDL are immediate post-treatment purpura (more prominent with shorter wavelengths and high fluence) lasting for up to 3 weeks, edema and serous crusting (up to 7 days), hyperpigmentation (in ethnic skin), and rarely atrophic scarring and formation of pyogenic granuloma [44].

Intense Pulsed Light

Intense pulsed light (IPL) devices use flashlamps (usually xenon) and bandpass filters to generate polychromatic noncoherent pulsed light with a spectral distribution ranging from 500 to 1,200 nm in the visible to near-infrared range with variable fluence and pulse durations [45, 46]. Cutoff filters of 515, 550, 560, 570, 590, 615, 645, 690 and 755 nm shift the short end of the spectrum to target specific chromophores such as oxyhemoglobin, melanin or water; shorter wavelengths (i.e. 515–550 nm) and longer wavelengths (i.e. >570 nm) result in more superficial and deeper penetration, respectively [45, 46].

IPL has a vast number of applications in dermatology for vascular lesions, pigmented lesions, hair removal and photorejuvenation [46]. Numerous vascular lesions such as facial telangiectasia, erythematotelangiectatic rosacea, hemangioma, PWS as well as poikiloderma of Civatte can be treated effectively with IPL [35, 45–47]. The treatment endpoint for IPL is an immediate dark blue to light gray discoloration or vessel clearance [33].

Advantages of IPL include: (1) the versatility of a range of wavelengths (500–1,200 nm), fluence (10–18 J/cm^2) and pulse duration (2–200 ms), which target vessels at different depths; (2) large spot size coverage; (3) rapid recovery (usually within 48 h). Disadvantages of IPL are: inconsistent spectral shift distribution due to variable current to the flashlamp [48], a heavy hand piece requiring direct skin contact, and the requirement of an optical coupling gel to the skin. There is also a risk of erythema, but minimal risk of edema and post-treatment pigmentary changes, and a virtual absence of purpura (except with high fluence) [35]. Newer IPL devices have integrated cooling mechanisms with chilled sapphire or quartz crystal windows to reduce epidermal thermal injury; alternatively exogenous cooling may be used (e.g. contact cooling with ice or ice gels, refrigerated air, cryogen spray, etc.) for the same reasons as with PDL [46, 49].

Infrared Lasers

Lasers such as the 755 nm alexandrite, 800 to 940 nm diode and 1,064 nm Nd:YAG operate in the red or near-infrared range of 700–1,200 nm. These devices emit radiation relatively close to the third absorption peak of oxyhemoglobin, but their longer wavelengths allow for deeper penetration into the dermis. Longer wavelengths are also less absorbed by melanin (i.e. 351 nm) [1], and therefore these devices are considered safer for use in ethnic skin. The primary uses for these lasers are telangiectasia, venulectasias such as spider veins and blue reticular veins (small to medium leg veins up to 4 mm), PWS (various stages) and vascular malformations [3].

755 nm Alexandrite Laser
The longer wavelength of the 755 nm alexandrite has been reported to improve leg telangiectasia and blue reticular veins among its more popular uses in hair and pigment removal. Optimal parameters for Fitzpatrick I–III with leg telangiectasia (0.2–1.3 mm) are as follows: fluence 89–90 J/cm^2 and pulse duration 60–80 ms [50, 51]. Treatment endpoints are mild purpura (i.e. persistent bluing) and/or vessel clearance [33, 51]. Reported clearance of leg telangiectasia following a single treatment varies between 26 and 75% at 12-week follow-up [51]. Minimal pain and purpura as well as transient hyperpigmentation are noted with 755 nm alexandrite laser treatment of leg telangiectasia [50, 51].

1,064 nm Nd:YAG Laser
The 1,064 nm Nd:YAG laser has a high absorption coefficient of deoxyhemoglobin and methemoglobin (i.e. the main components of veins), and can penetrate vessels 4–6 mm in depth [6]. The primary uses of this device include leg telangiectasia, reticular veins and (less often) hair removal. The 1,064 nm Nd:YAG can also be used to treat vascular lesions such as hemangioma, PWS, vascular malformations, telangiectasia and other small vascular lesions with good results [52–55]. Moreover, the longer wavelength of Nd:YAG is less likely to affect epidermal melanin, and therefore may potentially minimize the risk of post-inflammatory hyperpigmentation in ethnic or tanned skin. With longer pulse durations, the Nd:YAG will slowly coagulate the vessel rather than cause vessel rupture, which results in purpura and post-procedural pigmentary alterations [6].

Optimal parameters for the management of leg veins (0.2–2 mm) in the mid-dermis (1.5 mm deep) are fluence 100–200 J/cm^2 and pulse duration 10–100 ms [6]. Prevention of inadvertent epidermal thermal damage and subsequent post-inflammatory hyperpigmentation requires the use of cooling systems such as refrigerated air (–32°C), cryogen spray and cooled sapphire windows. Adverse effects of 1,064 nm ND:YAG include crusting for up to 2 weeks and post-treatment pigmentary changes. The treatment endpoint for Nd:YAG is vessel clearance [33].

Photodynamic Therapy

Photodynamic therapy (PDT) involves the application of a photosensitizer (e.g. 5-aminolevulinic acid or methyl aminolevulinate) which is subsequently converted to the endogenous photosensitizer protoporphyrin IX upon exposure to the appropriate absorption spectrum (i.e. 410, 504, 538, 576 and 630 nm) [46]. Activation of protoporphyrin IX generates reactive oxygen species and free radicals leading to subsequent cell death. Traditionally, a coherent light source (e.g. 410 nm blue light) is used, but monochromatic PDL [56] and broad spectrum polychromatic IPL encompassing the absorption spectra of protoporphyrin IX may be even more effective [46]. Primary uses of PDT include actinic keratoses followed by superficial basal cell and squamous cell in situ, inflammatory acne and photorejuvenation [46].

In vivo studies in chick chorioallantoic membrane models using a benzoporphyrin derivative monoacid ring A (BPD) photosensitizer with 576 nm CW laser and 585 nm PDL reported that PDT causes coagulation, vasoconstriction, angiostasis and hemorrhage [57]. A mouse angiosarcoma model created by intradermal injection of immortalized murine endothelial cells (i.e. intended to mimic the proliferative phase of hemangiomas) reported intravenous BPD combined with 690 nm red light (50–150 J/cm^2) resulted in eradication of the tumor with higher fluences [58]. The use of PDT for photorejuvenation may also yield some benefit for improvement of facial telangiectasia. Based on in vivo animal studies, PDT certainly has potential applications for vascular lesions in dermatology, but further studies are needed.

Common Congenital Vascular Lesions

Infantile Hemangioma

Infantile hemangiomas are common vascular tumors presenting in the neonatal period. While some forms of infantile hemangiomas are formed at birth, most lesions appear as a small macule or patch of erythema, telangiectasia or hypopigmentation, and subsequently undergo a rapid proliferative growth phase. These lesions may cause functional impairments, bleed and ulcerate with growth or minor trauma, develop secondary infection and result in cosmetic disfigurement [3, 8, 9, 17, 59].

Treatment should begin in the early macule patch stage. PDL, particularly long-pulse PDL, is the treatment of choice [60]. Other treatment options include IPL, 1,064 nm Nd:YAG and 532 nm KTP. Discomfort is minimal with these devices, but a local anesthetic either in the form of a topical application or nerve block may be used. Caution is necessary with the use of topical anesthetics in infants due to the risk of methemoglobinemia [3]. It is important to note that all treatment modalities suppress superficial component growth, but the effects on the deeper component are controversial [8]. Systemic therapy, rather than light or laser therapy, is recommended for life-threatening infantile hemangioma [3].

A retrospective chart analysis of 90 children (median age 3 months) with 105 hemangiomas reported near-complete to complete clearance of color in 81% (n = 85 lesions) and thickness in 64% (n = 67 lesions) using long-pulse 595 nm PDL with dynamic epidermal cooling (11.5 J/cm^2, range 7.5–14.0 J/cm^2, for 7-mm spot size versus 8.6 J/cm^2, range 6.2–11.5 J/cm^2, for 10-mm spot size) [61]. Furthermore, the use of the same 595 nm PDL has been reported to be safe and effective in 239 Korean patients with Fitzpatrick III–V with hemangioma (n = 37), telangiectasia (n = 100) and nevus flammeus (n = 102) [62]. The hemangioma group showed good and excellent responses in 21.5% (8 patients) and 32.4% (12 patients), respectively, particularly for superficial lesions and younger patients following a mean of 5.95 treatments with fluence at 11.67 J/cm^2 (range 5–15 J/cm^2) and pulse duration of 1.6–6 ms [60].

Diode lasers may also be adapted for endolesional use. An interdisciplinary group from surgery, radiology, dermatology and pediatrics used an endolesional 980 nm diode laser to treat 250 pediatric patients with hemangiomas (n = 160), vascular malformations (n = 50) and lymphatic malformations (n = 40) of the head and neck [63]. All lesions were successfully treated with the exception of 34 patients who required a second session [63].

Port Wine Stains

Port wine stains (PWS) are capillary malformations often found on the face and upper limbs that arise from small vessels in the papillary and upper reticular dermis. Generally, treatment of PWS should commence at age <1 year with multiple treatments for better chance of clearance [3]. Mature (hypertrophic and nodular) PWS tend to respond poorly to treatment [3, 8]. Analgesia is required as the treatment area is usually quite extensive [3].

PDL was originally developed for PWS and remains the treatment of choice [59]. A randomized side-by-side trial of 20 patients with PWS treated with 595 nm PDL, 500 to 670 nm and 870 to 1,400 nm IPL determined that PDL demonstrated better clinical improvement in 65% of those treated (vs. 30% with IPL), higher clearance rates of 75% (vs. 30% with IPL) and improved skin reflectance (i.e. lightening) in 33% (vs. 12% with IPL) [64].

A significant proportion, up to 20%, of PWS fail initial PDL treatment, and therefore other laser and light-based treatment options include IPL, 755 nm alexandrite, 810 nm diode and 1,064 nm Nd:YAG [65]. PDL may also be used in combination with other lasers. A retrospective review of 20 patients with either hypertrophic or PDL-resistant PWS treated with 755 nm alexandrite laser in combination with PDL found significant lightening of hypertrophic PWS and only moderate improvement in PDL-resistant PWS [66]. In the future, the use of reflectance spectrophotometry may offer a more tailored treatment approach for PWS [67].

Common Acquired Vascular Lesions

Facial Telangiectasia and Rosacea

Facial telangiectasia are dilated arterioles, capillaries and venules that present as erythematous small-diameter (0.1- to 1-mm) vessels in a linear (simple), arborizing (spider), punctiform or papular configuration, and are usually distributed on the central and lateral areas of the face [3]. The cause of facial telangiectasia is likely multifactorial – with genetics, hormonal influences, photoaging, medications, alcohol and trauma all potentially responsible [3]. Any device that emits a wavelength coinciding with the absorption spectrum of oxyhemoglobin can be use to treat facial telangiectasia. Overall, PDL, IPL and 532 nm KTP are considered safe and efficacious treatment options [37, 68].

Long-pulse PDL is considered the treatment of choice for diffuse telangiectasia, erythema and rosacea. The use of longer 6- to 10-ms pulse durations and relatively high fluence (7–9 J/cm^2) minimize post-treatment purpura [69]. However, 532 nm KTP is preferential for more discrete telangiectasia. Using 532 nm KTP in 204 patients (Fitzpatrick I–III) affected by facial telangiectasia (n = 102) or spider angioma (n = 102), 90% (44/49) of facial telangiectasia and 98% (57/58) of spider angioma cases markedly improved or cleared in those patients who completed follow-up [28]. A split-face comparison study of 532 nm KTP to IPL in 17 patients (Fitzpatrick I–IV) with facial telangiectasia and/or dyschromia (i.e. vascular and pigmented) found KTP and IPL resulted in a 41.99 versus 38.16% improvement respectively for vascular lesions at 1-month follow-up [26]. Patient global self-evaluations of vascular and pigmented lesion improvements were rated as 60.88 and 65.59% with 532 nm KTP and IPL, respectively, thereby suggesting that these devices may have equivocal outcomes [26]. Among many other uses, IPL is also efficacious for facial telangiectasia and rosacea. Sixty patients (Fitzpatrick I–IV) with rosacea (n = 580 sites) treated with 515 to 1,200 nm IPL (4.3 and 6.5 ms, 25–35 J/cm^2) achieved a 77.8% clearance rate with mean of 4.1 treatments per site [47].

Small Cutaneous Vascular Lesions

Spider angiomas consist of arterioles that flow into a more superficial capillary bed; these occur most commonly in children, but can also occur with pregnancy and liver disease. In contrast, cherry angiomas are dilated capillary loops in the superficial dermis; these occur most often in the middle-aged to elderly [3]. In the context of vascular lasers and light-based devices, many treatment options are available, including 532 nm KTP, PDL, IPL and 1,064 nm Nd:YAG lasers [3]. The use of 532 nm KTP for angiomas has been reported to be efficacious with 98% clearance of spider angiomas [28], usually only requiring a single treatment for clearance compared to an average of 2 treatments with hyfrecator electrosurgery for spider and cherry angiomas [25]. Spider angiomas appear to be more predictable for successful treatment with laser and light-based devices [27].

Pyogenic granulomas (also known as lobular capillary hemangiomas) are benign vascular tumors that tend to bleed profusely with minor trauma. These lesions can develop spontaneously, particularly in pregnancy, or at the site of a recent trauma. Options for treatment with vascular devices include 1,064 nm Nd:YAG (1–14 pulses, 100–130 J/cm^2) [36], 585 nm PDL [70] and IPL. In approximately 1% of patients, treatment with PDL can induce the formation of a pyogenic granuloma [44].

Poikiloderma of Civatte

Characterized by atrophy, reticulate pigmentary irregularity and telangiectasia, poikiloderma of Civatte occurs in sun-exposed areas of the upper mid-chest, lateral neck and cheeks. Treatment options include IPL and PDL. Rusciani et al. [71] conducted the largest single-centre study on 175 Italian patients (Fitzpatrick I–III) with this condition treated by 3 sessions of IPL separated by 3 weeks intervals using variable pulse durations (proportional to vessel size) and fluence (according to skin type). These authors reported only transient adverse effects in 5% of patients and 80% clearance of the vascular and pigmentation component at 3-month follow-up. While multiple (usually 3–5) treatments are required, IPL appears to be more effective, associated with less adverse effects and allows for larger spot size treatment compared to PDL [35, 46]. Caution should be used in areas of hyperpigmentation and scar-prone areas of the neck and chest [3].

Venous Lakes

Venous lakes are composed of dilated venules likely from degeneration of connective tissue, which appear on the lip, ear and face of elderly individuals. Minor trauma can precede the development of these lesions. Treatment options include 1,064 nm Nd:YAG, PDL, IPL, 532 nm KTP and 808 nm diode lasers.

Bekhor [72] described 35 adults with venous lakes on the lips and cheeks treated with Nd:YAG (250 J/cm^2 for the 3-mm spot size versus 140–180 J/cm^2 for the 5-mm spot size) and demonstrated a 94% clearance rate following a single treatment. A combination of laser modalities is also possible. A case series of 30 patients (with 39 venous lakes) treated with 595 nm PDL (20 ms, 10 J/cm^2) followed by 1,064 nm Nd:YAG (20 ms, 70 J/cm^2) reported a 95% clearance rate [73].

Diode lasers have also been used. Complete resolution of venous lakes on the lip was demonstrated in 17 patients treated with a single session of 808 nm diode laser treatment (2–3 W power output, CW) [74]. Additionally, 1 treatment session with the 980 nm diode laser (150 ms, 300–500 J/cm^2) yielded at least a 50% improvement in 60% of patients with venous lakes [75].

Venulectasia and Varicose Veins

Leg telangiectasia are superficial small 0.2- to 1-mm diameter vessels in the superficial dermis. Spider veins are slightly larger 0.2- to 2-mm diameter vascular ectasias. Reticular veins are subcutaneous veins up to 4 mm in diameter [3]. There are numerous treatment modalities available for venulectasia and varicose veins ranging from sclerotherapy and phlebectomy to lasers and light-based devices such as PDL (long pulsed), 532 nm KTP (pulsed), IPL, 755 nm alexandrite as well as endovascular diode and Nd:YAG lasers [76]. Nevertheless, sclerotherapy for superficial telangiectasia and spider veins, sclerotherapy and phlebectomy for reticular veins, and surgical ligation with stripping and phlebectomy for long saphenous veins are still considered the standards of treatment [76]. Advances in endovascular lasers have led to high success rates with minimal adverse effects.

Diode lasers such as the 810 , 940 and 980 nm are used in endovenous laser ablation (EVLA). Mackenzie et al. [13] treated 631 patients (704 incompetent truncal veins) with a 810 nm diode laser (12–14 W, 70–85 J/cm^2) and adjuvant foam sclerotherapy (129/704, 18%), phlebectomies (53/704, 8%) or both (3 limbs), and reported complete occlusion in 610 veins (96%) at 3 month follow-up. Desmyttere et al. [14] conducted EVLA of incompetent small saphenous veins measuring 2–10 mm in 128 patients (147 limbs) with a 980 nm diode laser (10W, 50–90 J/cm cm^2) and demonstrated 100% initial technical success followed by 97% occlusion at 1-year follow-up and 100% occlusion at 2- and 3-year follow-ups. Overall success rates for EVLA are greater than 90%. Complications of EVLA include ecchymoses and postoperative pain (100%), temporary paresthesia (0–22%), superficial thrombophlebitis (0–25%), skin burns (<1%), deep venous thrombosis and/or pulmonary embolus (0–6%), and permanent nerve damage (<1%) [10, 13, 14].

Practical Considerations and Conclusion

Laser and light-based devices can be very effective in the treatment of vascular lesions. A few important principles to remember are: (1) procedural pain is expected, but should be tolerable; (2) pre- or parallel cooling can prevent thermal injury to the epidermis, aid with analgesia and allow for higher energy levels to be used without increasing the risk of post-treatment pigmentary changes; (3) pulse duration should be directly proportional to vessel size; (4) deeper vessels require longer wavelengths, pulse durations and larger spot sizes; (5) ethnic or darker skin often requires longer pulse durations and pulse intervals; (6) a 10–20% fluence reduction should be made for areas with a thin dermis (eyelids) and scar-prone areas of the chest; (7) multiple treatment sessions are often required [3, 49].

The initial consultation should include options for treatment and an open discussion about patient expectations and adverse effects of treatment. If the patient decides on treatment, then a detailed treatment plan and informed consent should be obtained [3]. For the procedure, adequate analgesia and pre- and postoperative photographs are required. The appropriate settings for the device – including wavelength, fluence, pulse duration and spot size – must be chosen. The patient should be aware of common complications, such as pain, purpura and ecchymoses, edema, bleeding, blistering, pigmentary alterations, scarring and the potential for non-response. Post-treatment care should outline conservative comfort measures (i.e. ice), gentle skin care, application of ointment and dressings (if required), infection prevention and avoidance of sun exposure [3]. A thorough overview of the management options and treatment considerations of common vascular lesions with pulsed light and laser is presented in table 3.

Table 3. Management options and treatment considerations of common vascular lesions with pulsed light and laser (Adapted from Adamic et al. [3])

Type of lesion	Description	Management options	Notes
Congenital			
Infantile hemangioma [8, 60–63]	Common vascular tumors presenting in the neonatal period. These lesions may cause functional impairments, bleed and ulcerate with growth or minor trauma, develop secondary infection and result in cosmetic disfigurement.	PDL (long-pulsed), IPL, 1,064 nm Nd:YAG, 532 nm KTP, endolesional 980 nm diode	Treatment should start early in macular patch stage. All treatment modalities suppress superficial component growth, but are unlikely to affect the deeper component. Caution with topical anesthesia in infants <6 months due to risks with methemoglobinemia (particularly with prilocaine). General anesthesia is required with Nd:YAG in children. Systemic therapy (not PDL or laser) is required for life-threatening hemangioma.
PWS [8, 64, 66]	Capillary malformation of small vessels in the papillary and reticular dermis. PWS become more hypertrophic and nodular with time, as deeper dermal vessels are more involved.	Early PWS: PDL (long-pulsed), IPL, 1,064 nm Nd:YAG Mature PWS: PDL (long-pulsed), IPL, 755 nm alexandrite, 1,064 nm Nd:YAG, diode	PDL (long-pulsed) remains the treatment of choice. Treatment should begin at age <1 year for better chance of clearance. Multiple treatments are required. Mature (nodular) PWS respond poorly to treatment. Topical or intralesional anesthesia is required.
Acquired			
Facial telangiectasia [2, 28, 35, 37, 47]	Dilated arterioles, capillaries and venules present as small diameter 0.1- to 1-mm vessels on the central and lateral face.	PDL, IPL, 532 nm KTP, diode	Topical anesthetics can cause vasoconstriction and suboptimal outcome. Pre- or parallel cooling can aid in analgesia and inadvertent epidermal thermal injury.
Rosacea [2, 35, 47]	Proliferation of telangiectasia with or without papules and pustules with chronic facial erythema.	PDL, 532 nm KTP (pulsed), IPL	Multiple treatments often required.
Angioma [7, 25, 27, 28]	Spider angiomas consist of arterioles that flow into a more superficial capillary bed. Cherry angiomas are dilated capillary loops in the superficial dermis.	532 nm KTP, PDL, IPL, 1,064 nm Nd:YAG	Most angiomas tend to respond well to a single treatment.

Table 3. Continued

Type of lesion	Description	Management options	Notes
Pyogenic granuloma [36, 44, 70]	Benign vascular tumors that tend to bleed profusely. They can occur following minor cutaneous trauma.	1,064 nm Nd:YAG, PDL, IPL	Multiple treatments are often required (with the exception of Nd:YAG). The use of PDL has been reported to cause pyogenic granulomas.
Poikiloderma of Civatte [35, 46, 71]	Atrophy, reticulate pigmentary irregularity and telangiectasia that occurs in sun-exposed areas of the upper mid-chest, lateral neck and cheeks.	IPL, PDL	Several treatments are often required. Neck and upper chest may require lower fluence settings. Use caution in hyperpigmented regions.
Venous lakes [72–75]	Dilated venules likely from degeneration of connective tissue on the lip, ear and face of elderly individuals.	1,064 nm Nd:YAG, PDL, IPL, 532 nm KTP, 808 nm diode	High clearance rate >94% reported with single treatment by Nd:YAG or combination 595 nm PDL + Nd:YAG [73].
Leg telangiectasia and veins [6, 7, 42, 51, 75]	Telangiectasia are superficial, small 0.2- to 1-mm diameter vessels in the superficial dermis. Spider veins are slightly larger 0.2- to 2-mm diameter vascular ectasias. Reticular veins are subcutaneous veins up to 4 mm in diameter.	Small (<0.5-mm) superficial veins: PDL (long-pulsed), 532 nm KTP (pulsed), IPL; Reticular veins: 1,064 nm Nd:YAG, 755 nm alexandrite, diode Endovascular lasers: 810, 940 and 980 nm diodes, 1,043 and 1,320 nm Nd:YAG	Sclerotherapy still remains the gold standard. Longer wavelengths and pulse durations are required for larger vessels. Endovascular laser ablation with adjuvant phlebectomy have >90% success rates.

References

1 Anderson RR, Parrish JA: Selective photothermolysis: precise microsurgery by selective absorption of pulsed radiation. Science 1983;220:524–527.

2 Railan D, Parlette EC, Uebelhoer NS, Rohrer TE: Laser treatment of vascular lesions. Clin Dermatol 2006;24:8–15.

3 Adamic M, Troilius A, Adatto M, Drosner M, Dahmane R: Vascular lasers and IPLS: guidelines for care from the European Society for Laser Dermatology (ESLD). J Cosmet Laser Ther 2007;9:113–124.

4 Dierickx CC, Casparian JM, Venugopalan V, Farinelli WA, Anderson RR: Thermal relaxation of port-wine stain vessels probed in vivo: the need for 1–10-millisecond laser pulse treatment. J Invest Dermatol 1995;105:709–714.

5 Bucci J, Goldberg D: Past, present, and future: vascular lasers/light devices. J Cosmet Laser Ther 2006;8:149–153.

6 Baumler W, Ulrich H, Hartl A, Landthaler M, Shafirstein G: Optimal parameters for the treatment of leg veins using Nd:YAG lasers at 1064 nm. Br J Dermatol 2006;155:364–371.

7 Galeckas KJ: Update on lasers and light devices for the treatment of vascular lesions. Semin Cutan Med Surg 2008;27:276–284.

8 Stier MF, Glick SA, Hirsch RJ: Laser treatment of pediatric vascular lesions: port wine stains and hemangiomas. J Am Acad Dermatol 2008;58:261–285.

9 Alster TS, Railan D: Laser treatment of vascular birthmarks. J Craniofac Surg 2006;17:720–723.

10 Van Den Bos RR, Neumann M, De Roos KP, Nijsten T: Endovenous laser ablation-induced complications: review of the literature and new cases. Dermatol Surg 2009;35:1206–1214.
11 Sadick NS, Wasser S: Combined endovascular laser with ambulatory phlebectomy for the treatment of superficial venous incompetence: a 2-year perspective. J Cosmet Laser Ther 2004;6:44–49.
12 Sharif MA, Lau LL, Lee B, Hannon RJ, Soong CV: Role of endovenous laser treatment in the management of chronic venous insufficiency. Ann Vasc Surg 2007;21:551–555.
13 Mackenzie RK, Cassar K, Brittenden J, Bachoo P: Introducing endovenous laser therapy ablation to a national health service vascular surgical unit – the Aberdeen experience. Eur J Vasc Endovasc Surg 2009;38:208–212.
14 Desmyttere J, Grard C, Stalnikiewicz G, Wassmer B, Mordon S: Endovenous laser ablation (980 nm) of the small saphenous vein in a series of 147 limbs with a 3-year follow-up. Eur J Vasc Endovasc Surg 2010;39:99–103.
15 Achauer BM, Vander Kam VM: Argon laser treatment of telangiectasia of the face and neck: 5 years' experience. Lasers Surg Med 1987;7:495–498.
16 Landthaler M, Haina D, Waidelich W, Braun-Falco O. A three-year experience with the argon laser in dermatotherapy. J Dermatol Surg Oncol 1984;10:456–461.
17 Diwan R: Laser therapy in the treatment of congenital vascular abnormalities. Md Med J 1990;39:343–346.
18 Arndt KA: Argon laser therapy of small cutaneous vascular lesions. Arch Dermatol 1982;118:220–224.
19 Scheibner A, Wheeland RG: Use of the argon-pumped tunable dye laser for port-wine stains in children. J Dermatol Surg Oncol 1991;17:735–739.
20 Parkin JL, Dixon JA: Argon laser treatment of head and neck vascular lesions. Otolaryngol Head Neck Surg 1985;93:211–216.
21 McCoy S, Hanna M, Anderson P, McLennan G, Repacholi M: An evaluation of the copper-bromide laser for treating telangiectasia. Dermatol Surg 1996;22:551–557.
22 Nanni CA, Alster TS: Complications of cutaneous laser surgery: a review. Dermatol Surg 1998;24:209–219.
23 Dinehart SM, Waner M, Flock S: The copper vapor laser for treatment of cutaneous vascular and pigmented lesions. J Dermatol Surg Oncol 1993;19:370–375.
24 Pickering JW, Walker EP, Butler PH, van Halewyn CN: Copper vapour laser treatment of port-wine stains and other vascular malformations. Br J Plast Surg 1990;43:273–282.
25 Dawn G, Gupta G: Comparison of potassium titanyl phosphate vascular laser and hyfrecator in the treatment of vascular spiders and cherry angiomas. Clin Exp Dermatol 2003;28:581–583.
26 Butler EG 2nd, McClellan SD, Ross EV: Split treatment of photodamaged skin with KTP 532 nm laser with 10 mm handpiece versus IPL: a cheek-to-cheek comparison. Lasers Surg Med 2006;38:124–128.
27 Goodman GJ, Roberts S, Bezborodoff A: Studies in long-pulsed potassium titanyl phosphate laser for the treatment of spider naevi and perialar telangiectasia. Australas J Dermatol 2002;43:9–14.
28 Clark C, Cameron H, Moseley H, Ferguson J, Ibbotson SH: Treatment of superficial cutaneous vascular lesions: experience with the KTP 532 nm laser. Lasers Med Sci 2004;19:1–5.
29 Hsiung MW, Kang BH, Su WF, Pai L, Wang HW: Clearing microvascular lesions of the true vocal fold with the KTP/532 laser. Ann Otol Rhinol Laryngol 2003;112:534–539.
30 Zeitels SM, Akst LM, Bums JA, Hillman RE, Broadhurst MS, Anderson RR: Pulsed angiolytic laser treatment of ectasias and varices in singers. Ann Otol Rhinol Laryngol 2006;115:571–580.
31 Tan OT, Murray S, Kurban AK: Action spectrum of vascular specific injury using pulsed irradiation. J Invest Dermatol 1989;92:868–871.
32 Saafan AM, Salah MM: Using pulsed dual-wavelength 595 and 1064 nm is more effective in the management of hemangiomas. J Drugs Dermatol 2010;9:310–314.
33 Astner S, Anderson RR: Treating vascular lesions. Dermatol Ther 2005;18:267–281.
34 Geronemus RG. Pulsed dye laser treatment of vascular lesions in children. J Dermatol Surg Oncol 1993;19:303–310.
35 Campolmi P, Bonan P, Cannarozzo G, Bruscino N, Troiano M, Prignano F, Lotti T: Intense pulsed light in the treatment of non-aesthetic facial and neck vascular lesions: report of 85 cases. J Eur Acad Dermatol Venereol 2011;25:68–73.
36 Bedard MS, Boulanger J: Treatment of lobular capillary hemangioma with the Nd:YAG laser: retrospective case series of 25 patients. J Cutan Med Surg 2009;13:181–182.
37 Smit JM, Bauland CG, Wijnberg DS, Spauwen PH: Pulsed dye laser treatment, a review of indications and outcome based on published trials. Br J Plast Surg 2005;58:981–987.
38 Bouzari N, Davis SC, Nouri K: Laser treatment of keloids and hypertrophic scars. Int J Dermatol 2007;46:80–88.
39 Tunnell JW, Nelson JS, Torres JH, Anvari B: Epidermal protection with cryogen spray cooling during high fluence pulsed dye laser irradiation: an ex vivo study. Lasers Surg Med 2000;27:373–383.
40 Chang CJ, Nelson JS: Cryogen spray cooling and higher fluence pulsed dye laser treatment improve port-wine stain clearance while minimizing epidermal damage. Dermatol Surg 1999;25:767–772.
41 Waldorf HA, Alster TS, McMillan K, Kauvar AN, Geronemus RG, Nelson JS: Effect of dynamic cooling on 585-nm pulsed dye laser treatment of port-wine stain birthmarks. Dermatol Surg 1997;23:657–662.
42 Tanghetti E, Sherr E: Treatment of telangiectasia using the multi-pass technique with the extended pulse width, pulsed dye laser (Cynosure V-Star). J Cosmet Laser Ther 2003;5:71–75.
43 Galeckas KJ, Ross EV, Uebelhoer NS: A pulsed dye laser with a 10-mm beam diameter and a pigmented lesion window for purpura-free photorejuvenation. Dermatol Surg 2008;34:308–313.
44 Wlotzke U, Hohenleutner U, Abd-El-Raheem TA, Baumler W, Landthaler M: Side-effects and complications of flashlamp-pumped pulsed dye laser therapy of port-wine stains: a prospective study. Br J Dermatol 1996;134:475–480.
45 Babilas P, Schreml S, Szeimies RM, Landthaler M: Intense pulsed light (IPL): a review. Lasers Surg Med 2010;42:93–104.

46 Ciocon DH, Boker A, Goldberg DJ: Intense pulsed light: what works, what's new, what's next. Facial Plast Surg 2009;25:290–300.
47 Schroeter CA, Haaf-von Below S, Neumann HA: Effective treatment of rosacea using intense pulsed light systems. Dermatol Surg 2005;31:1285–1289.
48 Eadie E, Miller P, Goodman T, Moseley H: Time-resolved measurement shows a spectral distribution shift in an intense pulsed light system. Lasers Med Sci 2009;24:35–43.
49 Raulin C, Greve B, Hammes S: Cold air in laser therapy: first experiences with a new cooling system. Lasers Surg Med 2000;27:404–410.
50 Brunnberg S, Lorenz S, Landthaler M, Hohenleutner U: Evaluation of the long pulsed high fluence alexandrite laser therapy of leg telangiectasia. Lasers Surg Med 2002;31:359–362.
51 Ross EV, Meehan KJ, Gilbert S, Domankevitz Y: Optimal pulse durations for the treatment of leg telangiectasias with an alexandrite laser. Lasers Surg Med 2009;41:104–109.
52 Ulrich H, Baumler W, Hohenleutner U, Landthaler M: Neodymium-YAG Laser for hemangiomas and vascular malformations – long term results. J Dtsch Dermatol Ges 2005;3:436–440.
53 Goldman A, Rossato F, Prati C: Stretch marks: treatment using the 1,064-nm Nd:YAG laser. Dermatol Surg 2008;34:686–691, discussion 91–92.
54 Dias Coelho J, Serrao V: Treatment of vascular lesions of the tongue with Nd:YAG laser. Case Report Med 2009;2009:795363.
55 Civas E, Koc E, Aksoy B, Aksoy HM: Clinical experience in the treatment of different vascular lesions using a neodymium-doped yttrium aluminum garnet laser. Dermatol Surg 2009;35:1933–1941.
56 Channual J, Choi B, Osann K, Pattanachinda D, Lotfi J, Kelly KM: Vascular effects of photodynamic and pulsed dye laser therapy protocols. Lasers Surg Med 2008;40:644–650.
57 Kelly KM, Kimel S, Smith T, Stacy A, Hammer-Wilson MJ, Svaasand LO, Nelson JS: Combined photodynamic and photothermal induced injury enhances damage to in vivo model blood vessels. Lasers Surg Med 2004;34:407–413.
58 Middelkamp-Hup MA, Sanchez-Carpintero I, Kossodo S, Waterman P, González S, Mihm MC Jr, Anderson RR: Photodynamic therapy for cutaneous proliferative vascular tumors in a mouse model. J Invest Dermatol 2003;121:634–639.
59 Cordisco MR: An update on lasers in children. Curr Opin Pediatr 2009;21:499–504.
60 Kono T, Sakurai H, Groff WF, et al: Comparison study of a traditional pulsed dye laser versus a long-pulsed dye laser in the treatment of early childhood hemangiomas. Lasers Surg Med 2006;38:112–115.
61 Rizzo C, Brightman L, Chapas AM, et al: Outcomes of childhood hemangiomas treated with the pulsed-dye laser with dynamic cooling: a retrospective chart analysis. Dermatol Surg 2009;35:1947–1954.
62 Woo SH, Ahn HH, Kim SN, Kye YC: Treatment of vascular skin lesions with the variable-pulse 595 nm pulsed dye laser. Dermatol Surg 2006;32:41–48.
63 Angiero F, Benedicenti S, Benedicenti A, Arcieri K, Berne E: Head and neck hemangiomas in pediatric patients treated with endolesional 980-nm diode laser. Photomed Laser Surg 2009;27:553–559.
64 Faurschou A, Togsverd-Bo K, Zachariae C, Haedersdal M: Pulsed dye laser vs. intense pulsed light for port-wine stains: a randomized side-by-side trial with blinded response evaluation. Br J Dermatol 2009;160:359–364.
65 Jasim ZF, Handley JM: Treatment of pulsed dye laser-resistant port wine stain birthmarks. J Am Acad Dermatol 2007;57:677–682.
66 Izikson L, Nelson JS, Anderson RR: Treatment of hypertrophic and resistant port wine stains with a 755 nm laser: a case series of 20 patients. Lasers Surg Med 2009;41:427–432.
67 Lister T, Wright P, Chappell P: Spectrophotometers for the clinical assessment of port-wine stain skin lesions: a review. Lasers Med Sci 2010;25:449–457.
68 Clementoni MT, Gilardino P, Muti GF, et al: Facial teleangectasias: our experience in treatment with IPL. Lasers Surg Med 2005;37:9–13.
69 Jasim ZF, Woo WK, Handley JM: Long-pulsed (6-ms) pulsed dye laser treatment of rosacea-associated telangiectasia using subpurpuric clinical threshold. Dermatol Surg 2004;30:37–40.
70 Gonzalez S, Vibhagool C, Falo LD Jr, Momtaz KT, Grevelink J, Gonzalez E: Treatment of pyogenic granulomas with the 585 nm pulsed dye laser. J Am Acad Dermatol 1996;35:428–431.
71 Rusciani A, Motta A, Fino P, Menichini G: Treatment of poikiloderma of Civatte using intense pulsed light source: 7 years of experience. Dermatol Surg 2008;34:314–319; discussion 319.
72 Bekhor PS: Long-pulsed Nd:YAG laser treatment of venous lakes: report of a series of 34 cases. Dermatol Surg 2006;32:1151–1154.
73 Roncero M, Canueto J, Blanco S, Unamuno P, Boixeda P: Multiwavelength laser treatment of venous lakes. Dermatol Surg 2009;35:1942–1946.
74 Azevedo LH, Galletta VC, Eduardo Cde P, Migliari DA: Venous lake of the lips treated using photocoagulation with high-intensity diode laser. Photomed Laser Surg 2010;28:263–265.
75 Levy JL, Berwald C: Treatment of vascular abnormalities with a long-pulse diode at 980 nm. J Cosmet Laser Ther 2004;6:217–221.
76 Leopardi D, Hoggan BL, Fitridge RA, Woodruff PW, Maddern GJ: Systematic review of treatments for varicose veins. Ann Vasc Surg 2009;23:264–276.

Patricia Ting, MD
2–104 Clinical Sciences Building
University of Alberta
Edmonton, AB, T6G 2B3 (Canada)
Tel. +1 780 708 3886, E-Mail patriciating@gmail.com

Bogdan Allemann I, Goldberg DJ (eds): Basics in Dermatological Laser Applications.
Curr Probl Dermatol. Basel, Karger, 2011, vol 42, pp 81–96

Benign Pigmented Lesions

Inja Bogdan Allemann[a] · David J. Goldberg[b,c]

[a]Klinik für Plastische Chirurgie und Handchirurgie, Universitätsspital Zürich, Zürich, Switzerland;
[b]Skin Laser & Surgery Specialists of New York & New Jersey, and [c]Mount Sinai School of Medicine, New York, N.Y., USA

Abstract

Benign pigmented lesions are a frequent complaint in dermatological patients, especially those seeking advice and therapy in a laser or cosmetic practice. Significant advances in laser technology over the last decades now allow us to effectively and safely treat various benign pigmented lesions. However, a thorough understanding of the biology of the lesion to be treated, the physical properties of the lasers to be used, and laser-tissue interactions is crucial for a successful and safe treatment. This chapter will give an overview of the types of benign pigmented lesions that can be treated with lasers and the specific lasers used to treat them.

Our skin color is determined by the activity of our pigment-producing melanocytes. The distribution of melanin varies from skin phototype to skin phenotype. Individuals with darker skin types have the same amount of melanocytes as lighter-skinned people; however, they have larger melanosomes that contain more melanin pigment. These melanosomes are also more evenly dispersed throughout the cytoplasm.

Melanin production takes place within the melanosome, which is an intracytoplasmic organelle within the melanocyte. The enzyme tyrosinase is the key enzyme in the biosynthesis of melanin. Melanocytes are found in the basal layer of the epidermis. In normal skin, roughly every tenth cell in the basal layer is a melanocyte. With its dendrites, one melanocyte can reach as far as the stratum spinosum, contacting roughly 30–40 keratinocytes, which is called an epidermal melanin unit. The melanocyte interacts with the keratinocytes by transferring melanosomes from the dendrites of the melanocyte into neighboring keratinocytes.

Pigmented lesions show either an increase in melanin production with excessive melanin deposition in melanosomes or an increase in the density of active melanocytes.

Selective Photothermolysis

The principle of selective photothermolysis first described by Anderson and Parrish [1] more than 2 decades ago led to a revolution in the treatment of pigmented lesions. As described in chapter explaining the physics of lasers in more detail [Bogdan Allemann and Kaufman, pp. 7–23], the principle states firstly that a target chromophore can selectively be destroyed, sparing the surrounding tissue from damage, by choosing the wavelength that corresponds to the

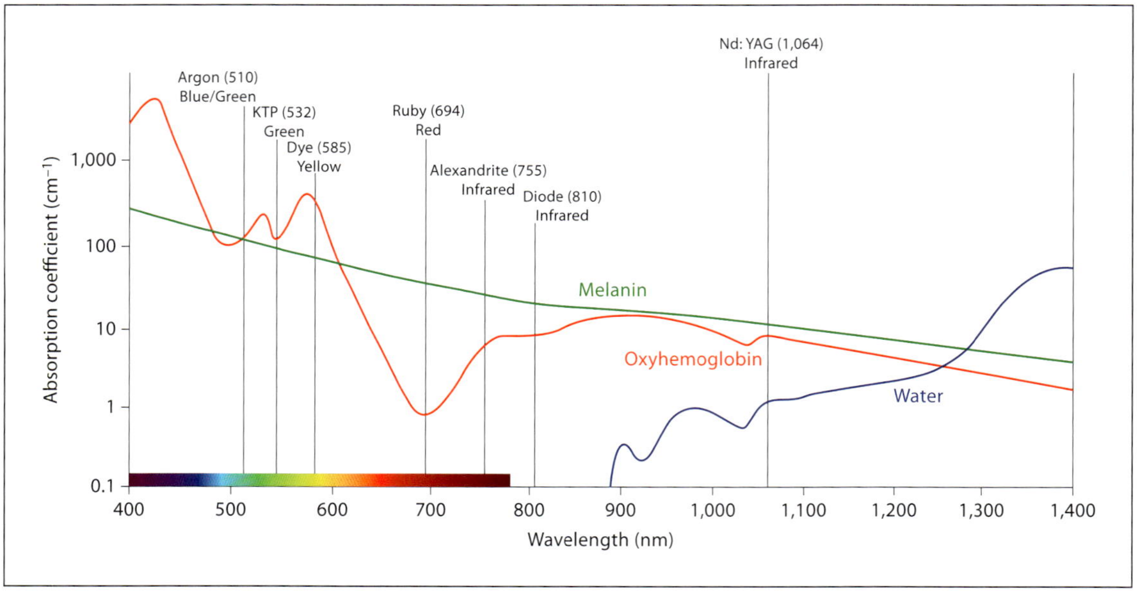

Fig. 1. Absorption spectrum of melanin.

absorption maximum/spectrum of that target chromophore. Secondly, by adapting the pulse width to be equal to or shorter than the thermal relaxation time (TRT, the time needed by a structure to cool down to half the temperature to which it was heated) of the target chromophore, heat is selectively generated in the target and does not harm the surrounding tissue. Finally, the energy of the laser pulse must be high enough to actually destroy the target chromophores. These three parameters, wavelength, pulse duration and energy have to be chosen with respect to the target in order to selectively destroy it and leave surrounding tissue unharmed [2–4].

Physical Principles in the Treatment of Pigmented Lesions

The target chromophore in the treatment of pigmented lesions is the melanin-containing melanosome. Experimental studies have demonstrated that Q-switched (QS) lasers can target and selectively destroy melanosomes [5, 6].

Hence, the three parameters defining selective destruction by the principle of selective photothermolysis – namely wavelength, pulse duration and fluence – have to be chosen according to the characteristics of the melanosome.

Wavelength

The absorption spectrum of melanin is broad and ranges from 351 to 1,064 nm, with its maximum in the ultraviolet region, a gradual decrease throughout the visible region, and an extension into the near-infrared region [7] (fig. 1). Therefore, the wavelength of the chosen laser should correspond to the absorption spectrum of melanin, with an optimum between 600 and 1,100 nm, as light within these wavelengths is able to penetrate into the dermis without significant absorption by competing chromophores, such as hemoglobin and water [37].

The exact choice of the laser wavelength used depends upon the depth of the pigmented lesion as well as on the patient's skin type. Within the visible range, shorter wavelengths penetrate less deeply but show stronger melanin absorption, while long wavelengths penetrate deep into the tissue due to decreased scatter but have poorer melanin absorption; hence, treatment efficacy decreases with increasing wavelength [8–10]. Consequently, epidermal pigment can be targeted with wavelengths that penetrate more superficially, while dermal pigment needs to be targeted with wavelengths that can successfully penetrate deeply into the tissue.

However, as the laser light does not differentiate between pathological (i.e., excess pigment) and normal melanin, defining the patients' skin type, special care should be taken when treating darker skin types. Lasers with high melanin absorption clearly carry an increased high risk of side effects, such as pigment alterations, and should thus be avoided in darker-skinned patients. Patients with Fitzpatrick skin phototype >IV are safest treated with longer wavelengths, such as the 1,064 nm Nd:YAG laser, in order to prevent too much absorption of epidermal melanin and consecutive pigmentary alterations.

Pulse Duration
Melanin is found within the melanosomes. Melanosomes are very small organelles, measuring roughly 0.5–1 μm, and consequently have a short TRT. The TRT is directly proportional to the square of the size of the target; thus, for melanosomes it lies between 0.25 and 1 μs. Consequently, lasers with pulses in the nanosecond range, as is the case with QS lasers, should be used in order to confine the heat within the target and not damage surrounding tissue.

Fluence
Wavelength and pulse duration are both important factors in treating pigmented lesions. Fluence (energy delivered over treated area) is also a highly important treatment parameter. A minimal fluence is required for treatment success in epidermal pigmented lesions. Although at the requisite fluence, epidermal pigmented lesions can be treated with a small treatment spot size, dermal pigmented lesions are different. With dermal pigmented lesions, the larger the delivered spot size, the greater the chances of success at every chosen fluence.

Spot Size
When treating pigmented lesions, the selection of the spot size, as described above, is also crucial. Although a larger spot size decreases peripheral scatter and, subsequently, results in a higher percentage of energy being delivered to the target, the spot size should not be larger than the lesion treated in order to prevent pigmentary changes to the surrounding skin. This is especially crucial when treating darker-skinned patients. The spot size should thus be chosen according to the size of the targeted structure. Ideally, the largest spot size that accommodates the treated lesion should be chosen.

Lasers Used in the treatment of Pigmented Lesions

When treating pigmented lesions, 4 categories of lasers or light devices can be used: (1) the pigment-selective QS lasers, (2) the less-selective long-pulsed (lp) lasers and intense pulsed light (IPL), (3) the non-pigment-specific ablative lasers, and (4) the recently described non-pigment-specific fractionated lasers [11].

Based on the concept of selective photothermolysis, the treatment of pigmented lesions should ideally be performed with lasers that are able to emit high-energy pulses within the nanosecond range. These lasers are referred to as QS lasers. The term 'Q-switching' refers to a switch resulting in the release of all the energy in one short powerful pulse, allowing for selective photothermolysis [12]. The method of Q-switching allows one to produce very short laser pulses in the nanosecond domain, in order to treat small micrometer-sized targets, such as melanin. QS lasers currently on the market are the QS ruby at 694 nm, the QS alexandrite at 755 nm, the QS Nd:YAG at 1,064 nm and the frequency-doubled QS Nd:YAG at 532 nm.

The less-pigment-selective lp lasers used to treat pigmented lesions are the lp ruby at 694 nm, the lp alexandrite at 755 nm, the lp Nd:YAG at 1,064 nm and the diode laser at 800 nm. Due to their pulse durations within the millisecond range, which exceed the TRT of the target chromophore melanin, lp lasers bear the risk of causing damage to the surrounding tissue, which can result in scarring. Therefore, the energy chosen must be low. Lp lasers can better treat epidermal pigmented lesions than dermal ones, due to the thermal damage they cause to melanosome-containing keratinocytes. Dermal lesions should not be treated with lp lasers because of the associated higher risk of scarring [13]. It should be noted, however, that lp lasers are clearly less effective than pigment-specific QS lasers, are associated with a higher risk of complications, and require more treatments to achieve optimal clearing. With pulse durations in the millisecond range, the lp lasers are however ideal to treat hair, where the target chromophore still is melanin, but longer pulse durations are necessary to achieve destruction of the hair follicle. This will be explained in greater detail in the chapter on hair removal [Haedersdal and Haak, pp. 111–121].

Polychromatic light generated by IPL sources (515–1,200 nm) can also be used to treat epidermal pigmented lesions. Due to the polychromatic nature of the IPL devices, competing chromophores, such as hemoglobin, need to be taken into account. In order to avoid absorption by hemoglobin, compression may be used when delivering laser pulses. The pulse duration of the light emitted by IPL devices is within the millisecond range; hence, selective photothermolysis does not apply at all. However, for various indications, such as lentigines and ephelides, IPL devices have been shown to be effective and safe. After treatment of epidermal pigmented lesions with IPL, first a darkening of the respective lesion may occur, followed by sloughing and then clearing.

The non-pigment-specific ablative lasers, such as the CO_2 and Er:YAG lasers, may also be used to treat epidermal pigmented lesions by simply ablating the epidermis and hence the unwanted pigmentation therein. Potential risk of scarring and pigment alterations clearly make these lasers less favorable.

Finally, recently described non-ablative fractionated laser devices at 1,320–1,550 nm have been used to treat pigmented lesions, such as lentigines, melasma, and Becker's nevus. The rational for the successful treatment of pigmented lesions with these non-pigment-specific lasers is the histological observation that melanin found within the microscopic-epidermal necrotic debris becomes extruded from the skin after non-ablative fractional laser treatment [14].

Since QS lasers are the primary laser of choice in the treatment of pigmented lesions, they will be discussed in more detail.

Q-Switched Lasers

Four types of QS lasers, with 4 different wavelengths, are currently available: the ruby at 694 nm, the alexandrite at 755 nm, the Nd:YAG at 1,064 nm, and the double-frequency Nd:YAG at 532 nm. Pulse durations vary, and range between 5 and 100 ns.

As discussed before, to selectively destroy the target (in this case the melanosomes) laser wavelength and pulse duration must be appropriately chosen in order to achieve selective absorption, while minimizing the nonspecific thermal effects induced in the non-targeted chromophores such as hemoglobin or water. By

Table 1. Characteristics of QS lasers used for pigmented lesions

	Double-frequency QS Nd:YAG	QS ruby	QS alexandrite	QS Nd:YAG
Wavelength, nm	532	694	755	1,064
Pulse width, ns	5–10	25–40	50–100	5–10
Fluence, J/cm^2	5	8–10	8–10	10–12
Melanin absorption	good	good	moderate	poor
Depth of tissue penetration	superficial	medium	deep	deep
Type of pigmented lesion	epidermal	epidermal/dermal	epidermal/dermal	dermal
Fitzpatrick skin type	I–III	I–III dermal	I–IV	III–VI

choosing appropriate parameters, the risk of hypo- and hyperpigmentation, or even scarring, is markedly reduced.

Characteristics of QS lasers used for pigmented lesions are summarized in table 1.

Q-Switched Ruby Laser

The ruby laser was the first QS laser to be developed. The wavelength emitted by the QS ruby laser is 694 nm, pulse durations are 25–40 ns, and generally used fluences used are up to 8–10 J/cm^2.

The QS ruby laser is an effective therapeutic device for the treatment of pigmented epidermal and dermal lesions, including lentigines, ephelides, café-au-lait spots, flat seborrheic keratoses, and Becker's nevus [15, 16]. Further, nevi of Ota and Ito with melanin located dermally can be treated [17, 18]. The QS ruby laser shows high melanin absorption; however, its wavelength tissue penetration is limited to roughly 2 mm. Due to the strong melanin absorption, hypopigmentation may occur, which fortunately is mostly transient (2–6 months). However, in darker-skinned patients, there is an increased risk of side effects such as hypo- and depigmentation. In these darker-skinned patients, special caution is required and lower fluences should be used [19]. To select the fluence, the clinical endpoint of immediate tissue whitening with no or minimal bleeding should be aimed for. The optimal energy will vary depending on the pigmented lesion itself, the skin type of the patient, as well as the spot size used. To start treatment, an initial fluence of 3.0–5.0 J/cm^2 is recommended.

Q-Switched Alexandrite Laser

The QS alexandrite laser emits light at a wavelength of 755 nm with a pulse width of 50–100 ns, and an energy density of up to 8–10 J/cm^2. The slightly longer pulse duration as compared to the QS ruby and QS Nd:YAG lasers results in less tissue splatter during treatment. The clinical endpoint is immediate epidermal whitening of the treated area. Since the wavelength, and emitted pulse duration of the alexandrite laser is similar to the ruby laser, its indications and results in treating pigmented lesions are similar. Scarring and hypopigmentation rarely occur; transient hyperpigmentation can be observed a little more frequently [20–22]. When treating patients with darker skin, lower fluences than those used for lighter-skinned individuals should be utilized.

Q-Switched Nd:YAG Laser

The Nd:YAG laser emits light in the infrared range at a wavelength of 1,064 nm and pulse duration of 5–10 ns. Spot sizes can be up to 8 mm and emitted fluences often up to 12 J/cm^2. Repetition rates can be fast (1–10 Hz); beam size is generally 4–8 mm. At the same beam diameter, the penetration at 1,064 nm is deeper than that of the ruby or alexandrite lasers and reaches deep into the papillary dermis. Consequently it can, with everything else being equal, target dermal lesions more effectively [21, 22].

Pigmented lesions with deep dermal pigment, such as nevi of Ota and Ito, can be effectively treated [23]. The 1,064 nm wavelength of the Nd:YAG laser is not well absorbed by melanin, and is thus a suboptimal laser to be used when treating epidermal pigmented lesions. The clinical endpoint, as with all QS lasers, is tissue whitening. When smaller spot sizes and/or higher treatment fluences are used, some tissue splatter and pinpoint bleeding can be observed. Side effects such as scarring or hypopigmentation are rare [24].

Because of its poor melanin absorption, the 1,064 nm Nd:YAG laser is the safest laser to be used in patients with darker skin types (Fitzpatrick IV–VI), as the risk of epidermal injury and pigmentary alteration is greatly reduced [25, 26].

Frequency-Doubled Nd:YAG Laser

By doubling the frequency of the 1,064 nm Nd:YAG laser, the wavelength can be halved to 532 nm. This frequency doubling, accomplished with a KTP potassium-titanyl-phosphate crystal, leads to emission of green laser light. With the shortest wavelengths of all QS lasers, the frequency-doubled Nd:YAG laser possesses strong melanin absorption and little penetration depth. Hence, it shows good results and is ideal for epidermal pigmented lesions, such as lentigines and ephelides [27]. However, when treating epidermal pigmented lesions with this laser, purpura can be observed because the wavelength is also absorbed by hemoglobin.

Tissue Reactions

The tissue reactions induced in the targeted structure by QS lasers can be both photothermal and photomechanical. The photothermal reaction occurs because of a rapid rise in temperature (1,000°C) within the target, resulting in evaporation and vacuolization of targeted pigments within the skin. When the temperature gradient between the target chromophore and the surrounding tissue collapses, the target becomes fragmented; this represents the photomechanical reaction [28–31]. This thermomechanical action on melanosomes results in pigment-dispersion to the periphery of the melanin-containing melanocytes and keratinocytes. This reaction creates the histological appearance of a 'ring-cell' formation [32]. Most probably, the clinical correlate of the above-mentioned photothermal effects is the immediate whitening phenomenon that can be observed when treating pigmented lesions with QS lasers (fig. 2). It is a good clinical marker of melanosomal destruction. Moreover, it is a reversible phenomenon, with resolution seen around 20 min after QS laser treatment, and seems to represent rapid localized heating with steam or gas vacuole formation within the laser-treated target structure. If this whitening effect cannot be observed during treatment, most probably too low an exposure energy has been chosen or no pigmented lesion has been targeted [33].

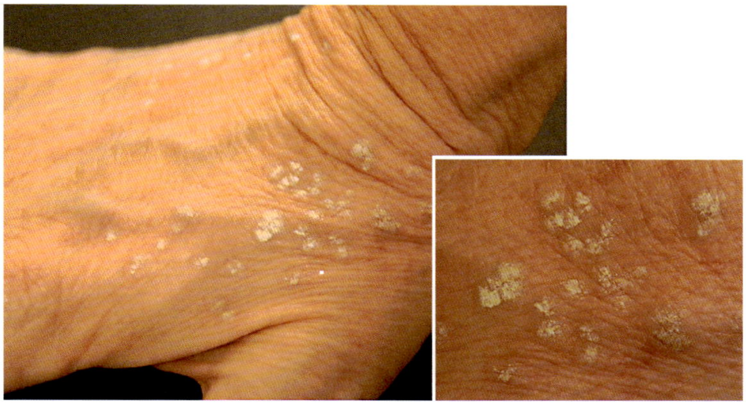

Fig. 2. Immediate whitening after treatment with a QS ruby laser.

Side Effects of Q-Switched Lasers

Although QS lasers can effectively and safely remove melanin whilst sparing the surrounding tissue, complications may still occur. Acute events, which mostly depend on the energy used, are purpura, blistering, and crusting [34]. These are transient and patients need to be informed accordingly, as well as being instructed on post-treatment care.

The most common chronic adverse event observed after treatment with QS lasers is hypopigmentation, which may be transient or permanent, especially when treating darker-skinned patients and using devices with a high rate of melanin absorption, such as the QS ruby or QS alexandrite lasers [35]. The frequency of hypopigmentation reported in various studies ranges anywhere from 0 to 50%, and is dependent on the skin type, with darker-skinned individuals being at higher risk. Post-treatment hyperpigmentation can also occur, generally in darker-skinned patients. This is usually transient. Scarring and textural changes are rare events, reported in most studies to occur in less than 2–5% of cases.

Some professionals have suggested performing an initial test spot prior to treatment to lessen the incidence of complications. However, most clinicians do not use this approach because a normal test spot does not always preclude a rare laser-induced complication. Obtaining patient consent and providing information about typical side effects is prudent.

Clinical Considerations

Two main clinical factors should be considered in choosing the most effective and safe laser for the treatment of a pigmented lesion: the depth of the undesired pigmented lesion and the patient's skin phototype.

Knowing the depth of the excess melanin to be removed will directly influence the choice of the laser, as different laser wavelengths penetrate to various depths within the tissue. Therefore, with regard to laser therapy, pigmented lesions are best classified according to the depth of melanin accumulation in the skin into epidermal or dermal pigmented lesions. Some authors add a third classification of mixed epidermal/dermal pigmented lesions (table 2). Examples of epidermal pigmented lesions that can be treated with lasers and light devices include lentigines (simplex and senilis), ephelides, flat pigmented seborrheic keratoses, café-au-lait macules (CALM), and nevus

Table 2. Types of lesions

Epidermal pigmented lesions	lentigo (simplex and senilis) ephelides pigmented seborrheic keratoses café-au-lait macules
Mixed (epidermal/dermal) pigmented lesions	Becker's nevus nevus spilus melasma post-inflammatory hyperpigmentation
Dermal pigmented lesions	nevi of Ota/Ito acquired bilateral nevus of Ota-like macules congenital nevi

spilus. Mixed epidermal/dermal pigmented lesions include melasma, Becker's nevus, nevocellular nevi, and post-inflammatory hyperpigmentation. Dermal melanocytic lesions that can be treated with lasers are nevus of Ota, nevus of Ito, acquired bilateral nevus of Ota-like macules, and some congenital nevi [36].

As laser light cannot differentiate between pathological pigment and the physiological pigment which is responsible for our skin phototype, it will also be absorbed by the normal physiological pigment. Hence, darker-skinned patients have to be treated with care, ideally using the QS Nd:YAG laser which has the poorest absorption in melanin. Using the QS ruby or QS alexandrite laser, with their strong absorptions in melanin, in a patient with Fitzpatrick skin type >IV bears a higher risk of pigment alterations, especially hypopigmentation.

It is crucial that the diagnosis of the nature of the pigmented lesion be determined by the dermatologist. When there is any suspicion of a malignant lesion, the diagnosis must be confirmed by biopsy. The laser treatment of melanocytic nevi is still a controversial topic and will not be discussed in this chapter as we generally do not advise such treatment. Finally, lasers should not be used to eradicate malignant melanoma or lentigo maligna.

Epidermal Pigmented Lesions

Epidermal lesions with the excess pigment located superficially include lentigines, ephelides (freckles), CALM, and flat pigmented seborrheic keratoses. All can be treated with most QS lasers. Since the pigment is located superficially, the laser light does not have to penetrate deeply into the tissue to destroy the target structure; hence, the more superficially penetrating lasers (double-frequency QS Nd:YAG, QS ruby, QS alexandrite) are preferable to the QS Nd:YAG in lighter-skinned patients. The QS Nd:YAG (532 nm) laser may be associated with purpura when used to treat epidermal lesions. Other than purpura, the risk of side effects and the healing time are similar for all the above-mentioned lasers. It has been shown that QS laser treatment of epidermal pigmented lesions is superior to other superficially destructive methods, such as liquid nitrogen, glycolic acid peels, and trichloroacetic acid [37–39].

As mentioned above, lp lasers, IPL, ablative lasers, and non-ablative fractional lasers can also be used to treat epidermal pigmented lesions.

Lentigines

Lentigines are sharply demarcated irregular- or round-shaped macules with light to dark-brown pigmentation. Solar lentigines, also called senile lentigines, senile freckles, and age or liver spots are acquired due to UV exposure. Hence, they can usually be found on sun-exposed areas, such as the face, neck, and back of the hands. Their number increases with age, and there are usually multiple lesions. Lentigines can also appear as a dermatological finding in various syndromes, such as Peutz-Jeghers syndrome, or LEOPARD

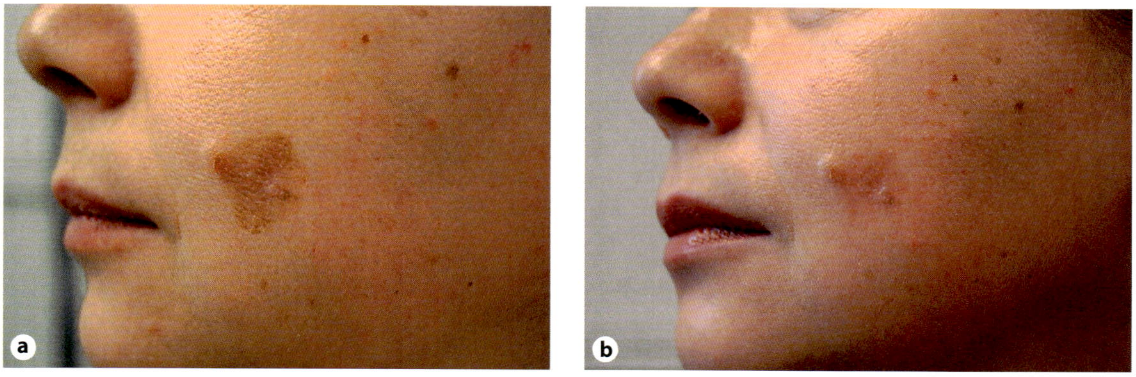

Fig. 3. **a** Histologically verified lentigo before laser treatment. **b** Lentigo after 1 treatment with the 694 nm QS ruby laser (spot size 4 mm, fluence 4.5 J/cm^2).

syndrome. Lentigines have no pre-malignant potential. Histologically, there is an increased pigmentation in the basal layer and often the number of basal melanocytes is increased.

All available QS lasers have been demonstrated to be safe and efficacious in the treatment of lentigines and achieve excellent results – hence any of them can be used [40, 41]. Usually 1 or 2 treatments are enough to achieve satisfactory clearing of most lentigines. As most lentigines are small lesions, spots sizes such as 2–4 mm can be used. The energy should be chosen such that treatment should lead to a brisk whitening of the treated lesion, without much tissue splatter. Crusts, which usually arise after treatment with QS lasers, heal in 7–10 days. Side effects observed are textural changes and hypo- or hyperpigmentation, all of which are usually mild and transient [42, 43]. Due to its stronger melanin absorption, the risk of hypopigmentation is higher for the ruby than for the alexandrite laser. Post-inflammatory hyperpigmentation can be observed more frequently in darker-skinned patients, but also generally improves with time [44].

IPL has also been shown to successfully treat lentigines, although it seems that more than just 1 or 2 treatment sessions are required to achieve optimal clearing [45].

Correctly diagnosing a lentigo senilis is crucial, in order not to miss a lentigo maligna (fig. 3, 4).

Ephelides

Ephelides, commonly called freckles, are light-brown macules of 1–3 mm in diameter. The can be found on sun-exposed areas of the body and histologically show an increased number melanized melanosomes transported from the melanocytes to the keratinocytes in the basal cell layer. Treatment is mostly requested by young adults, who should first be informed that ephelides generally decrease with age. Furthermore, it is important that patients know that ephelides will almost always recur with further UV exposure. If laser intervention is requested, the three pigment-specific QS lasers as well as IPL can be successfully used [46, 47].

Pigmented Seborrheic Keratoses

Flat pigmented seborrheic keratoses can be treated with the QS lasers in the same way as ephelides, and also show an excellent response. The majority of seborrheic keratoses however are of the acantholytic thickened type, and may need additional treatments. The thicker the lesions, the more treatments will be required.

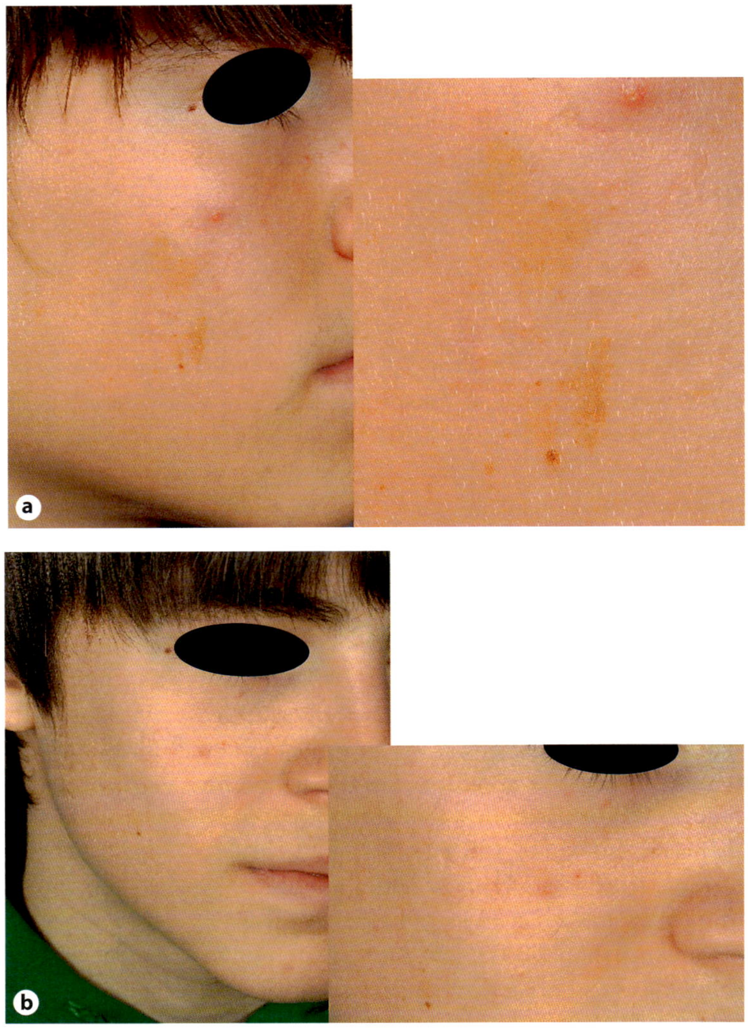

Fig. 4. **a** Segmental lentigo before laser treatment. **b** Segmental lentigo after 1 treatment with the 694 nm QS ruby laser (spot size 4 mm, fluence 4.5 J/cm^2).

Café au Lait Macules

CALM are well-circumscribed macules, uniformly light to dark brown, pigmented, and 2–5 cm. They can be found in the normal population and are usually singular in number. However, they can also be associated with various syndromes, in which case they are often seen in greater numbers. One such syndrome is neurofibromatosis. CALM demonstrate no tendency toward malignancy. Histologically, there is a slightly increased melanin content in the basilar keratinocytes and melanocytes.

Although it seems intuitive that CALM should be an ideal candidate for laser therapy, the treatment of CALM with lasers is not simple and straightforward. Responses to treatment are variable and difficult to predict. Usually, a multitude of treatments are required to achieve clearing, and, once clearing is achieved, recurrence rates are high with numbers of up to 50% at 1 year after clearance [48]. Retreatment can be performed, as CALM are responsive to retreatment. Incomplete clearance or a speckled pattern can also occur. Other side effects observed are the same as with any laser therapy, namely transient or permanent hypo- or hyperpigmentation and scarring. Before treatment, patients have to be informed about the above-mentioned complications. Because of the variable response, it is not unreasonable to perform a test spot prior to treatment. However, as noted above, a test spot may not predict the final response or the possibility of complications from treatment.

Any of the pigment-selective QS lasers can be used to treat CALM; the best response would be expected from the double-frequency 532 nm QS Nd:YAG, the 694 nm ruby and the 755 nm alexandrite lasers [49, 50]. An older, no longer available, 510 nm pulsed dye laser has also successfully been used [51].

As with other pigmented lesions, patients with darker skin types are at higher risk of post-treatment depigmentation, and should thus be evaluated and treated accordingly.

Mixed Epidermal and Dermal Pigmented Lesions

Becker's Nevus

Becker's nevus – also known as Becker's melanosis or pigmented hairy epidermal nevus – usually occurs in adolescence and presents as variably sized (2 to >16 cm) light to dark-brown patches with dark coarse hairs. It is believed to be an organoid nevus with abnormalities in the epidermis, hair follicles, and smooth muscle. Histologically, there is an increased melanin content in the keratinocytes and a normal or slightly increased number of melanocytes, not found in nests.

Due to its clinical presentation, there are 2 targets for laser therapy: the pigmented patch and the excess hairs. Hair removal can be performed by lp pigment-selective lasers as described in the chapter on hair removal [Haedersdal and Haak, pp. 111–121], with the only difference being that the background discoloration of the nevus has to be taken into account as a competing chromophore when removing hairs.

Laser treatment of Becker's nevus is difficult, probably due to its hamartomatous pathophysiology. It is not unusual to see incomplete removal, high recurrence rates, and hyperpigmentation often complicating the treatment. The various QS lasers have been tried with variable therapeutic success [52, 53]. Simultaneous treatment of the pigmented lesion as well as the excess hairs has been reported with the lp alexandrite lasers. Good clearing rates and low recurrence rates have also been reported.

Non-ablative fractional photothermolysis may also represent a valuable and especially safe treatment option; however, more data in regard to this treatment are required [54].

Nevus Spilus

Nevus spilus – also called speckled lentiginous nevus – presents as a sharply demarcated pigmented brown patch with darker 'speckles' that develop over time within it. The larger and lighter brown background, which clinically corresponds to a CALM, histologically corresponds to a lentigo simplex, while the darker macules and papules correspond to junctional or compound melanocytic nevi. The entire lesion can be quite big, up to several centimeters, while the interspersed pigmented lesions are usually only 1–6 mm in diameter.

There are reports of melanoma arising within the nevi of the nevus spilus; thus, special care must be exercised when treating this lesion with lasers. Any atypical areas or clinical changes should be examined by biopsy [55].

The same pigment-selective QS lasers described for CALM have also been reported to successfully clear nevus spilus [56, 57]. In addition, there are reports of successful treatment of this lesion with IPL [58].

The 2 different portions of the nevus – i.e., the CALM-like portion and the nevus portion – react differently to laser treatment. The nevus portion seems to clear much more easily than the CALM part of the nevus [59].

Melasma

Melasma – also referred to as chloasma or 'mask of pregnancy' – is predominantly seen in women and shows a higher prevalence in darker-skinned patients, such as Asians, Hispanics, or Africans. It presents as irregularly shaped and distributed, brown to dark brown macules on sun-exposed areas, such as the face and chest. Based on the location of the excess melanin, melasma can be subdivided into epidermal, dermal, and mixed types [60]. The distinction can be made clinically by using the Wood's light which accentuates epidermal melanin. Corresponding to the clinical presentation, a histological increase in melanin can be found within the epidermis and/or dermis. The number of enlarged melanocytes in not increased [61].

Melasma is notoriously difficult to treat and even when treated successfully, recurrence rates are high, especially during the summer. Laser therapy should not be regarded as the treatment of choice, but should be tried if conservative therapies with available topical depigmenting agents, with associated strict sun avoidance, have failed to show any success. In general, epidermal melasma will respond better to laser treatment, while mixed epidermal/dermal melasma shows a poor response [62]. Even if initial lightening can be achieved, later darkening is often noted. Darker-skinned patients are even at an increased risk of darkening of the melasma after laser treatment. After rather disappointing results with the QS lasers (Nd:YAG, alexandrite, ruby), IPL, and ablative lasers (CO_2 and Er:YAG), non-ablative fractional photothermolysis was thought to be the solution to treating melasma with lasers. This was based on the fact that besides epidermal components, melanin has been observed within the micro-epidermal necrotic debris, which after fractional photothermolysis gets extruded. For more details, see the chapter on fractional photothermolysis [Bogdan Allemann and Kaufman, pp. 56–66]. There are reports on successful treatment of melasma with non-ablative fractionated lasers [63]. However, when using non-ablative fractional photothermolysis, the energy and density chosen need to be low, and treatment intervals need to be spaced well apart in order to prevent post-inflammatory hyperpigmentation, which still seems to occur in up to 50% of treated patients.

Post-Inflammatory Hyperpigmentation

Post-inflammatory hyperpigmentation can be epidermal and/or dermal in nature, similar to what is seen in melasma. In general, its treatment is unsatisfactory with an unpredictable outcome. Treatment can be performed with the various QS lasers, and epidermal post-inflammatory hyperpigmentation bears a better chance of treatment success than the dermal variety. Even if an initial improvement can be achieved, recurrences or even a worsening of the post-inflammatory hyperpigmentation are not infrequent, as melanocytes are further stimulated by the trauma induced by the laser.

In addition, individual treatment spots may develop central lightening together with peripheral post-inflammatory hyperpigmentation. Test sites are suggested, and patients should be informed about the possibility of a poor treatment outcome. Based on the same concept of melasma clearing, non-ablative fractional photothermolysis may offer a potential treatment option for post-inflammatory hyperpigmentation.

Dermal Pigmented Lesions

Nevus Ota – also called nevus fuscoceruleus ophthalmomaxillaris – presents as grayish blue-to-black macules of various shapes and sizes, usually located unilaterally within the distribution of the first two branches of the trigeminal nerve, and predominantly seen in Asian and black women. Occasionally, the pigmentation is bilateral and then the condition is called Hori's nevus or acquired nevus of Ota-like macules. Nevus of Ito (nevus fuscoceruleus acromiodeltoideus) is basically the same as nevus of Ota, only located within the areas innervated by the supraclavicular and cutaneous brachii lateralis nerves. The majority of these nevi (except Hori's nevus) are present at birth; the rest develop around puberty. Rare cases of melanoma arising in nevi of Ota have been reported. Histologically, there are increased numbers of pigmented elongated dendritic melanin-producing melanocytes within the upper dermis. Hyperpigmentation of the basal epidermis may be observed. As there are scattered melanocytes within the dermis, the pigment-specific QS lasers with adequate penetration into the dermis are best able to treat these conditions.

Successful clearance rates have been observed with the QS ruby, QS alexandrite, and QS Nd:YAG lasers [64–67]. Since the QS Nd:YAG laser has less melanin absorption than the other QS lasers, and thus carries less risk of creating the unwanted epidermal pigmentary changes usually seen in darker-skinned patients, this is the laser of choice for such patients.

Usually, gradual clearing can be observed and repetitive treatments (at least 4–8) are needed, ideally in 8- to 12-week intervals, in order for the macrophages to have time to remove the pigment. Side effects are hypo- and hyperpigmentation, which are mostly transient. Rare scarring can be observed. This is more frequently seen with the lower-wavelength QS lasers (ruby and alexandrite). Recurrence, although rare, is possible. As the treatments may be performed close to the eye, optimal eye protection with ocular shields may be required. There are reports on the successful clearing of nevus of Ota with non-ablative fractional photothermolysis [68]. However, it is unlikely that these devices will be as successful as QS lasers.

Patient Selection, Pre- and Post-Treatment Care

Before treating any pigmented lesion with a laser, a firm diagnosis must be obtained. If malignancy is suspected, a biopsy is crucial. Patients with a tendency to form keloids should only be treated with caution, and perhaps not at all if on oral retinoid therapy. Patients undergoing laser treatments should not have a suntan at any time and should consequently use sunscreen during the entire treatment period. Furthermore, the patient's Fitzpatrick skin type and tendency towards post-inflammatory hyperpigmentation must be considered, and treatments and settings chosen accordingly.

Patients need to be informed in detail of the treatment course and the expected outcome, as well as possible side effects and recurrence rates. An informed consent should be signed by the patient. A picture before, during, and at the end of the treatment can be helpful to monitor progress or therapy failure. Test spots are sometimes recommended before treatment. A combination and successive use of various pigment-specific lasers may also help to achieve clearing.

Due to their specificity for melanin, most pigmented lesions are treated with the QS lasers. Those lasers, with their highly-energy ultra-short pulses, are the most dangerous laser systems for the retina. Hence, during the treatment, appropriate eye protection should be worn by everyone in the room, and not be removed until the laser is switched to standby mode.

In the treatment of pigmented lesions, a topical anesthetic cream or injected local anesthesia may be utilized prior to treatment.

As mentioned above, immediately after the treatment with QS lasers a white discoloration or frost within the treated area is observed, which actually represents the desired tissue response. The white coloration is followed by erythema and mild swelling and then crusting. The crusting can last up to 7–10 days. It is crucial to inform the patient about the course of healing and that they should not to manipulate the crusts. Wound healing cream and sunscreen should be applied.

Conclusion and Future Trends

When treating pigmented lesions with lasers, the first and foremost step is to verify that the lesion intended for treatment is of benign nature. Should there be any doubt, a histological verification is essential.

When treating benign pigmented lesions, choosing the right laser device is based on various factors, including the presumed absorption spectra of the target, the depth of penetration aimed for, the size of the target structure, as well as the skin type of the patient. To efficiently and safely treat pigmented lesions, without harming the surrounding tissue, the pulse duration should ideally be in the nanosecond range, as is the case for the various QS lasers. Those QS lasers have been demonstrated in numerous studies to be effective and safe in the treatment of various pigmented lesions.

The QS ruby (694 nm), the QS alexandrite (755 nm), and the frequency-doubled Nd:YAG (532 nm) lasers all represent excellent treatment options for epidermal pigmented lesions. The QS Nd:YAG (1,064 nm), the QS ruby laser (694 nm), the QS alexandrite (755 nm) lasers are the best laser choices for dermal pigmented lesions, with the QS Nd:YAG (1,064 nm) laser the preferred choice when treating darker-skinned patients.

References

1 Anderson RR, Parrish JA: Selective photothermolysis: precise microsurgery by selective absorption of pulsed radiation. Science 1983;220:524–527.
2 Murphy GF, Shepard RS, Paul BS, Menkes A, Anderson RR, Parrish JA: Organelle-specific injury to melanin-containing cells in human skin by pulsed laser irradiation. Lab Invest 1983;49:680–685.
3 Kurban AK, Morrison PR, Trainor SW, Tan OT: Pulse duration effects on cutaneous pigment. Lasers Surg Med 1992;12:282–287.
4 Sherwood KA, Murray S, Kurban AK, Tan OT: Effect of wavelength on cutaneous pigment using pulsed irradiation. J Invest Dermatol 1989;92:717–720.
5 Polla LL, Margolis RJ, Dover JS, Whitaker D, Murphy GF, Jacques SL, Anderson RR: Melanosomes are a primary target of Q-switched ruby laser irradiation in guinea pig skin. J Invest Dermatol 1987;89:281–286.
6 Dover JS, Margolis RJ, Polla LL, Watanabe S, Hruza GJ, Parrish JA, Anderson RR: Pigmented guinea pig skin irradiated with Q-switched ruby laser pulses. Morphologic and histologic findings. Arch Dermatol 1989;125:43–49.
7 Anderson RR, Parrish JA: The optics of human skin. J Invest Dermatol 77:13–19.
8 Margolis RJ, Dover JS, Polla LL, Watanabe S, Shea CR, Hruza GJ: Visible action spectrum for melanin-specific selective photothermolysis. Lasers Surg Med 1989;9:389–397.
9 Sherwood KA, Murray S, Kurban AK, Tan OT: Effect of wavelength on cutaneous pigment using pulsed irradiation. J Invest Dermatol 1989;92:717–720.
10 Anderson RR, Margolis RJ, Watenabe S, Flotte J, Hruza GJ, Dover JS: Selective photothermolysis of cutaneous pigmentation by Q-switched Nd:YAG laser pulses at 1,064, 532 and 355 nm wavelength. J Invest Dermatol 1989;93:28–32.
11 Alexiades-Armenakas MR, Dover JS, Arndt KA: Laser therapy (chapter 137); in Bolognia JL, Jorizzo JL, Rapini MD: Bologna Dermatology. New York, Mosby, 2007.
12 Habif T: Dermatologic surgical procedures; in Habif T (ed): Clinical Dermatology. Maryland Heights, Mosby, 2004, pp 937–938.
13 Watanabe S: Basics of laser application to dermatology. Arch Dermatol Res 2008;300(suppl 1):S21–S30.
14 Laubach HJ, Tannous Z, Anderson RR, Manstein D: Skin responses to fractional photothermolysis. Lasers Surg Med 2006;38:142–149.
15 Goldberg DJ: Benign pigmented lesions of the skin: treatment with the Q-switched ruby laser. J Dermatol Surg Oncol 1993;19:376–379.
16 Nanni CA, Alster TS: Successful treatment of a Becker's nevus using 694 nm long-pulsed ruby laser energy. Dermatol Surg 1998;24:1032–1034.
17 Raulin C, Schönermark MP, Greve B, Werner S: Q-switched ruby laser treatment of tattoos and benign pigmented skin lesions: a critical review. Ann Plast Surg 1998;41:555–565.
18 Chan HH, Kono T: Nevus of Ota: clinical aspects and management. Skinmed 2003;2:89–96, quiz 97–98.

19 Kono T, Nozaki M, Chan HH, Mikashima Y: A retrospective study looking at the long-term complications of Q-switched ruby laser in the treatment of nevus of Ota. Lasers Surg Med 2001;29:156–159.
20 Tanzi E, Lupton J, Alster T: Lasers in dermatology: four decades of progress. J Am Acad Dermatol 2003;49:1–31.
21 Chan HH, King WW, Chan ES, Mok CO, Ho WS, Van Krevel C, Lau WY: In vivo trial comparing patients' tolerance of Q-switched alexandrite (QS Alex) and Q-switched neodymium:yttrium-aluminum-garnet (QS Nd:YAG) lasers in the treatment of nevus of Ota. Lasers Surg Med 1999;24:24–28.
22 Goldberg DJ: Laser treatment of pigmented lesions; in Alster TS, Apfelberg DB (eds): Cosmetic Laser Surgery, ed 2. New York, Wiley-Liss, 1999, pp 279–288.
23 Kilmer SL, Garden JM: Laser treatment of pigmented lesions and tattoos (review). Semin Cutan Med Surg 2000;19:232–244.
24 Alster TS: Complications of dermatologic laser surgery: prevention and management. J Am Acad Dermatol, 62nd Ann Meet, February 2004.
25 Grevelink JM, Duke D, Van Leeuwen RL, Gonzalez E, DeCoste SD, Anderson RR: Laser treatment of tattoos in darkly pigmented patients: efficacy and side effects. J Am Acad Dermatol 1996;34:653.
26 Kilmer SL, Anderson RR: Clinical use of the Q-switched ruby and the Q-switched Nd:YAG (1,064 and 532 nm) lasers for treatment of tattoos. J Dermatol Surg Oncol 1993;19:330–338.
27 Rashid T, Hussain I, Haider M, Haroon TS: Laser therapy of freckles and lentigines with quasi-continuous, frequency-doubled, Nd:YAG (532 nm) laser in Fitzpatrick skin type IV: a 24-month follow-up. J Cosmet Laser Ther 2002;4:81–85.
28 Ortonne JP, Pandya AG, Lui H, Hexsel D: Treatment of solar lentigines. J Am Acad Dermatol. 2006;54(suppl 2):S262–S271.
29 Polla LL, Margolis RJ, Dover JS, Whitaker D, Murphy GF, Jacques SL, Anderson RR: Melanosomes are a primary target of Q-switched ruby laser irradiation in guinea pig skin. J Invest Dermatol 1987;89:281–286.
30 Dover JS, Margolis RJ, Polla LL, Watanabe S, Hruza GJ, Parrish JA, Anderson RR: Pigmented guinea pig skin irradiated with Q-switched ruby laser pulses: morphologic and histologic findings. Arch Dermatol 1989;125:43–49.
31 Ara G, Anderson RR, Mandel KG, Ottesen M, Oseroff AR: Irradiation of pigmented melanoma cells with high intensity pulsed radiation generates acoustic waves and kills cells. Laser Surg Med 1990;10:52–59.
32 Dover JS, Margolis RJ, Polla LL: Pigmented guinea pig skin irradiated with Q-switched ruby lasers. Arch Dermatol 1989;125:43–49.
33 Watanabe S: Basics of laser application to dermatology. Arch Dermatol Res 2008;300(suppl 1):S21–S30.
34 Handley J: Adverse events associated with nonablative cutaneous visible and infrared laser treatment. J Am Acad Dermatol 2006;55:482–489.
35 Kim KH, Geronemus RG: Tattoo removal. Med Lett 2003;45:95–96.
36 Goldberg DJ: Laser Dermatology, Pearls and Problems. Oxford, Blackwell Publishing, 2008.
37 Todd MM, Rallis TM, Gerwels JW, Hata TR: A comparison of 3 lasers and liquid nitrogen in the treatment of solar lentigines: a randomized, controlled, comparative trial. Arch Dermatol 2000;136:841–846.
38 Kopera D, Hohenleutner U, Landthaler M: Q-switched ruby laser application is safe and effective for the management of actinic lentigo (topical glycolic acid is not). Acta Derm Venereol 1996;76:461–463.
39 Li YT, Yang KC: Comparison of the frequency-doubled Q-switched Nd:YAG laser and 35% trichloroacetic acid for the treatment of face lentigines. Dermatol Surg 1999;25:202–204.
40 Kopera D, Hohenleutner U, Landthaler M: Quality-switched ruby laser treatment of solar lentigines and Becker's nevus: a histopathological and immunohistochemical study. Dermatology 1997;194:338–343.
41 Trafeli JP, Kwan JM, Meehan KJ, Domankevitz Y, Gilbert S, Malomo K, Ross EV: Use of a long-pulse alexandrite laser in the treatment of superficial pigmented lesions. Dermatol Surg 2007;33:1477–1482.
42 Rosenbach A: Treatment of medium-brown solar lentigines using an alexandrite laser designed for hair reduction. Arch Dermatol 2002;138:547–548.
43 Rashid T, Hussain I, Haider M, Haroon TS: Laser therapy of freckles and lentigines with quasi-continuous frequency-doubled, Nd:YAG (532 nm) laser in Fitzpatrick skin type IV: a 24-month follow-up. J Cosmet Laser Ther 2002;4:81–85.
44 Sadighha A, Saatee S, Muhaghegh-Zahed G: Efficacy and adverse effects of Q-switched ruby laser on solar lentigines: a prospective study of 91 patients with Fitzpatrick skin type II, III, and IV. Dermatol Surg 2008;34:1465–1468.
45 Kawada A, Shiraishi H, Asai M, Kameyama H, Sangen Y, Aragane Y, Tezuka T: Clinical improvement of solar lentigines and ephelides with an intense pulsed light source. Dermatol Surg 2002;28:504–508.
46 Goldberg DJ: Benign pigmented lesions of the skin: treatment with the Q-switched ruby laser. J Dermatol Surg Oncol 1993;19:376–379.
47 Kawada A, Shiraishi H, Asai M, Kameyama H, Sangen Y, Aragane Y, Tezuka T: Clinical improvement of solar lentigines and ephelides with an intense pulsed light source. Dermatol Surg 2002;28:504–508.
48 Shimbashi T, Kamide R, Hashimoto T: Long-term follow-up in treatment of solar lentigo and cafe-au-lait macules with Q-switched ruby laser. Aesthetic Plast Surg 1997;21:445–448.
49 Grossman MC, Anderson RR, Farinelli W, Flotte TJ, Grevelink JM: Treatment of cafe au lait macules with lasers: a clinicopathologic correlation. Arch Dermatol 1995;131:1416–1420.
50 Kilmer SL, Wheeland RG, Goldberg DJ, Anderson RR: Treatment of epidermal pigmented lesions with the frequency-doubled Q-switched Nd:YAG laser: a controlled, single-impact, dose-response, multicenter trial. Arch Dermatol 1994;130:1515–1519.
51 Alster TS: Complete elimination of large café-au-lait birthmarks by the 510 nm pulsed dye laser. Plast Reconstr Surg 1995;96:1660–1664.
52 Kopera D, Hohenleutner U, Landthaler M: Quality-switched ruby laser treatment of solar lentigines and Becker's nevus: a histopathological and immunohistochemical study. Dermatology 1997;194:338–343.

53 Tse Y, Levine VJ, McClain SA, Ashinoff R: The removal of cutaneous pigmented lesions with the Q-switched ruby laser and the Q-switched Nd:YAG laser: a comparative study. J Dermatol Surg Oncol 1994;20:795–800.
54 Glaich AS, Goldberg LH, Dai T, Kunishige JH, Friedman PM: Fractional resurfacing: a new therapeutic modality for Becker's nevus. Arch Dermatol 2007;143:1488–1490.
55 Abecassis S, Spatz A, Cazeneuve C, Martin-Villepou A, Clerici T, Lacour JP, Avril MF: Melanoma within naevus spilus: 5 cases. Ann Dermatol Venereol 2006;133:323–328.
56 Grevelink JM, Gonzalez S, Bonoan R, Vibhagool C, Gonzalez E: Treatment of nevus spilus with the Q-switched ruby laser. Dermatol Surg 1997;23:365–369.
57 Moreno-Arias GA, Bulla F, Vilata-Corell JJ, Camps-Fresneda A: Treatment of widespread segmental nevus spilus by Q-switched alexandrite laser (755 nm, 100 nsec). Dermatol Surg 2001;27:841–843.
58 Gold MH, Foster TD, Bell MW: Nevus spilus successfully treated with an intense pulsed light source. Dermatol Surg 1999;25:254–255.
59 Schaffer JV, Orlow SJ, Lazova R, Bolognia JL: Speckled lentiginous nevus: within the spectrum of congenital melanocytic nevi. Arch Dermatol 2001;137:1655.
60 Grimes PE: Melasma: etiologic and therapeutic considerations. Arch Dermatol 1995;131:1453–1457.
61 Grimes PE, Yamada N, Bhawan J: Light microscopic, immunohistochemical, and ultrastructural alterations in patients with melasma. Am J Dermatopathol 2005;27:96–101.
62 Taylor CR, Anderson RR: Ineffective treatment of refractory melasma and post-inflammatory hyperpigmentation by Q-switched laser. J Dermatol Surg Oncol 1994;20:592–597.
63 Kroon MW, Wind BS, Beek JF, Wietze van der Veen JP, Nieuweboer-Krobotová L, Bos JD, Wolkerstorfer A: Nonablative 1550-nm fractional laser therapy versus triple topical therapy for the treatment of melasma: a randomized controlled pilot study. J Am Acad Dermatol 2011;64:516–523.
64 Lowe NJ, Wieder JM, Sawcer D, Burrows P, Chalet M: Nevus of Ota: treatment with high energy fluences of the Q-switched ruby laser. J Am Acad Dermatol 1993;29:997–1001.
65 Moreno-Arias GA, Camps-Fresneda A: Treatment of nevus of Ota with the Q-switched alexandrite laser. Laser Surg Med 2001;28:451–455.
66 Chan HH, Leung RS, Ying SY, Lai CF, Kono T, Chua JK, Ho WS: A retrospective analysis of complications in the treatment of nevus of Ota with the Q-switched alexandrite and Q-switched Nd:YAG lasers. Dermatol Surg 2000;26:1000–1006.
67 Watanabe S, Takahashi H: Treatment of nevus of Ota with the Q-switched ruby laser. N Engl J Med 1994;331:1745–1750.
68 Kouba DJ, Fincher EF, Moy RL: Nevus of Ota successfully treated by fractional photothermolysis using a fractionated 1440-nm Nd:YAG laser. Arch Dermatol 2008;144:156–158.

I. Bogdan Allemann, MD
Klinik für Plastische Chirurgie und Handchirurgie, Universitätsspital Zürich
Rämistrasse 100
CH–8091 Zürich (Switzerland)
Tel. +41 44 255 1111, E-Mail Inja.bogdan@usz.ch

Bogdan Allemann I, Goldberg DJ (eds): Basics in Dermatological Laser Applications.
Curr Probl Dermatol. Basel, Karger, 2011, vol 42, pp 97–110

Tattoo Removal

Maurice A. Adatto[a,b] · Shlomit Halachmi[c] · Moshe Lapidoth[c,d]

[a]Skinpulse Dermatology, Laser and Beauty Centers, Geneva, and [b]Dermatology Clinic, Inselspital, Bern, Switzerland;
[c]Laser Unit, Department of Dermatology, Rabin Medical Center, Petah Tiqwa, and [d]Sackler Faculty of Medicine,
Tel Aviv University, Tel Aviv, Israel

Abstract

Over 50,000 new tattoos are placed each year in the United States. Studies estimate that 24% of American college students have tattoos and 10% of male American adults have a tattoo. The rising popularity of tattoos has spurred a corresponding increase in tattoo removal. Not all tattoos are placed intentionally or for aesthetic reasons though. Traumatic tattoos due to unintentional penetration of exogenous pigments can also occur, as well as the placement of medical tattoos to mark treatment boundaries, for example in radiation therapy. Protocols for tattoo removal have evolved over history. The first evidence of tattoo removal attempts was found in Egyptian mummies, dated to have lived 4,000 years BC. Ancient Greek writings describe tattoo removal with salt abrasion or with a paste containing cloves of white garlic mixed with Alexandrian cantharidin. With the advent of Q-switched lasers in the late 1960s, the outcomes of tattoo removal changed radically. In addition to their selective absorption by the pigment, the extremely short pulse duration of Q-switched lasers has made them the gold standard for tattoo removal.

Over 50,000 new tattoos are placed each year in the United States. Studies estimate that 24% of American college students have tattoos and 10% of male American adults have a tattoo [1, 2]. Its initial popularity in certain social groups suggests that it served a role of non-conformist identity in particular, but it has become mainstream and is now frequently ornamental and artistic in primary intent. Many cite the main motivation for tattoos as a form of expressing personal identity. The rising popularity of tattoos has spurred a corresponding increase in tattoo removal. Tastes, relationships, and social status change, and the perceived non-conformism associated with tattoos leads to a certain degree of taboo in employment settings and certain social contexts. As a result, the rate of regret is as high as 50% [3]. The most commonly cited reasons for tattoo removal are change in relationships, enhancement of self-esteem, a desire to improve social acceptance, and work or family reasons [4].

Not all tattoos are placed intentionally or for aesthetic reasons. Traumatic tattoos due to unintentional penetration of exogenous pigments can occur from dirt in a road accident, from pencil tip graphite, or from gunpowder. Medical tattoos are placed to mark treatment boundaries, for example in radiation therapy. Tattoos can also be used in medicine as part of reconstructive therapies, as in areola reconstruction after mastectomy. Permanent and semi-permanent makeup are also tattoos; clients undergoing these procedures may not be aware that they are having a tattoo placed.

History of the Tattoo

The etymology of the word tattoo is derived directly from the Tahitian *tattow* (rhymes with cow). The root *ta* means 'drawing', and the word in its entirety means 'a drawing in the skin'. The word *tattow* first appeared in *Captain Cook's Journal* for July 1769. This word has since passed virtually unchanged into the European languages. It appeared for the first time in the French dictionary *Littré* in its 1858 edition, with the definition: 'the system of methods by which vegetable or mineral dyes are introduced beneath the epidermis to varying depths to produce colorations or designs which are visible, though absolutely indelible'. Note the carefully qualified reference to the longevity of a tattoo.

While the history of the term tattoo can be traced in writings, physical evidence of tattoos on humans predates written history. Rests of red ochre pigment used for preparing dyes were found in the grotto of Arcy-sur-Cure (Burgundy, France) and date back approximately 35,000 years [5]. The oldest human tattooed body discovered is a Bronze Age hunter, nicknamed Ötzi, who dates back 5,000 years. Found in 1991 at the border of Austria and Italy, frozen into a glacier, his body was covered with many black tattoos. Many of the tattoos were over joints, and it has been suggested that this may represent a therapeutic modality used at the time to treat arthritis, a practice still used in parts of the world.

Reasons to get tattoos are ever evolving. When humans were nomadic, tattooing carried tribes' laws, but with the advent of writing this function was no longer needed. Although forbidden by mainstream religions as far back as the days of Moses [6], tattooing was utilized over the ages to signal religious affiliations, strength, or social status. Tattoos were imported to Western culture by sailors returning from the East. At the beginning of the 20th century, tattoos were mainly used in Western Europe among sailors and other working class members. Later, tattoos were used to mark affiliations with certain groups, and gained popularity as 'clan' loyalty markings in motorcyclists or inmates. In the 1980s, the punk and gay movement adopted body art as a protest against the conservative middle class norms of society. Until the 1990s, body art remained a provocative part of various cultures.

In the past two decades, the popularity of tattoos has increased tremendously, with the number of tattoos both rising in the existing groups and expanding to mainstream society [7]. Studies have identified ten main reasons people ascribe as motivations to get tattoos, from spirituality to physical endurance, with the main ones being beauty, art, and fashion [8].

Techniques, Pigments, and Tattoo Biology

Tattoos were initially placed by perforating the skin with a sharp-tipped instrument of any material, after which a pigment was applied and rubbed into the skin. Amateur tattoos are still performed by piercing the skin with any sharp object (needle, pointed stick, sharpened bone) dipped in a colored pigment (India ink, soil, shoe polish etc.).

Mechanical tattooing came about through an innovative machine invented by Thomas Edison in 1876 (US patent 196,747), called Stencil-Pens. It was designed for use as an engraving device, but in 1891 Samuel O'Reilly modified the device to inject ink into the skin. The device consists of an electric motor driving a crankshaft similar to that of a sewing machine, which operates a metal stylus with a single needle or array of needles at its tip.

Professional tattooing technique depends on the country where the tattoo is performed. The Western technique with the electric machine developed by O'Riley is the most popular worldwide. In Japan, some tattoo artists still work with the traditional procedure, using needles put together into a wooden handle, called a *hari*, which can be maneuvered manually back and forth. The traditional Polynesian technique uses a sharp rake held by a handle that the artist taps using a rigid piece of bamboo.

Technique is critical in tattooing. As a rough rule, professional tattoos are placed more deeply in the skin and with greater precision in depth, whereas amateur tattoos tend to be placed more superficially and with greater variability in depth of pigment placement. These differences are key factors in the stability and durability of a tattoo, as well as in the ease of subsequent removal. For good tattooing results, the pigment should be injected just below the epidermis in a thin and uniform layer. Too much pigment will give a thick draw. If the tattoo is injected too deeply, pigment can be found in the subcutaneous fat or tendon sheath, where it will be inaccessible to laser removal.

The fate of tattoo pigment in the skin has been examined in few studies. Serial biopsies obtained after tattooing show that the dye is initially present in focal accumulations, corresponding to the injection sites, but disperses homogeneously in the epidermis and superficial dermis within 1–2 weeks [9]. After 1 month, the pigment is localized to basal cell cytoplasm in the epidermis and to macrophages and fibroblasts in the dermis, with a predominance of phagocytosed pigment appearing below the dermal-epidermal junction. Granulomatous reactions may be found at this stage. It is thought that during this first month, the damage to the basement membrane allows a certain degree of transepidermal pigment elimination of free pigment, before the pigment has been phagocytosed. Over the following months, the pigment is found deeper in the dermis, within fibroblasts and often within perivascular fibrotic tissue. This localization has been used to explain the durable nature of the pigment. Furthermore, a dense connective tissue matrix entraps these fibroblasts, further reducing their mobility and maintaining the phagocytosed pigment in place for years. It is thought that the uptake and long-term storage of ink particles and latex beads by the dermal and subcutaneous fibroblasts represent a specific non-inflammatory defense mechanism. The histiocyte, a term proposed by Kiyono in 1914 for a fixed macrophage on the basis of his studies using vital dye staining, is considered to include, in addition to true macrophages, fibroblasts showing endocytotic activities for small foreign bodies such as acid dyes, ink particles, and latex beads [10].

The size of the tattoo granules is critical to the appearance, as well as to their biology and subsequent removal. Smaller granules provide greater homogeneity in the tattoo appearance, but they are less visible and more prone to migration and elimination. Larger granules provide more visible color and greater longevity. Tattoo composition is quite variable, with granule diameters of 0.5–5 μm [11]. Black tattoo granules are relatively small and therefore diffuse more quickly after tattooing. Red and turquoise tattoos are comprised of larger granules; this may explain the reduced smudging seen with these pigments.

It is important to emphasize that the pigments used for tattooing are detoured from their normal intended use. Amateur tattoos are often carbon-rich mixtures handmade as secondary products, such as boiled ink or burned wood or vegetable matter, though some tattoo artists report spiking the mix with chipped automobile paint and other industrial sources to boost the color. This results in non-homogeneous granule size as well as in unknown purity and contaminants. Professional pigments injected into the skin are dyes borrowed from the chemical industry for other applications, such as dying textiles or painting cars or road signs. They are not intended for contact with human skin. Some of them are toxic, and most Western countries have pending legislation to prohibit their use. In Switzerland, a federal law regulates and prohibits certain restricted substances whose use is known to be toxic or potentially carcinogenic [12]. It is of the utmost importance to also note that the pigments used in standard tattoo practice have a purity of 80%, meaning that 20% of the product is comprised of impurities. Since there is no other regulatory body that mandates a listing of impurities, let alone their concentration, biodistribution, or toxicology, the body art consumer consumes a product with significant uncertainty of its composition and safety.

The only ink that has been developed for tattooing purposes is an ink composed of a biodegradable pigment surrounded by a membrane that can be easily broken by standard Q-switched lasers [13]. The tattoo particles behave exactly as other pigments. The main difference is that when the envelope is broken by the laser impact, it liberates the dye within the dermis, allowing the pigment to be 'washed out' by the body. The pigment chosen is non-toxic and non-carcinogenic. This ink can be removed in approximately one third of the number of session required with traditional tattoo pigments.

Complications of Tattooing

They are many complications of tattooing. The most common adverse event is infections, due to bacteria (mainly *Staphylococcus* and *Streptococcus*), viruses (hepatitis B and C or HIV), or fungi [14]. The second most common complication is an inflammatory response to the pigment, usually manifesting as a granulomatous reaction [15]. In rare cases, a granuloma found within a tattoo can reveal an existing sarcoidosis [16]. Mercury (red) and cadmium (yellow) pigments are particularly likely to induce allergic reactions, though reactions to chromium (green), cobalt (blue), and beige pigments have also been reported [17]. In the case of cadmium, the trigger is often sun exposure, which appears to trigger a photoallergic reaction. Since red tattoos often contain traces of yellow, they may also induce phototoxic reactions.

Keratoacanthomas and other skin tumors, including melanoma, have also been reported after tattooing [18]. Due to the million of tattoos annually performed and the very small number of tumors reported, a causal relationship is difficult to ascertain. However, particular care is needed when a melanoma arises within or near a tattoo, as the black tattoo pigment can mask a developing melanoma [19]. In addition, the common finding of tattoo pigment in the lymph nodes draining a tattoo site can complicate the staging of melanoma during nodal dissection, or even mimic melanoma [20, 21].

Tattoo toxicity was recently studied by Bäumler et al. [22] as a function of pigment concentration in the tissue. The group introduced red pigment PR22 at 98% purity into human or porcine skin and measured the concentration of the pigment in various locations. Based on the distribution profile of this marker ink, they concluded that dyes which are potential carcinogens may reach levels in the skin that could lead to a potential health problem.

Tattoo Removal

Protocols for tattoo removal have evolved over history. The first evidence of tattoo removal attempts was found in Egyptian mummies, dated to have lived 4,000 years BC. Ancient Greek writings describe tattoo removal with salt abrasion or with a paste containing cloves of white garlic mixed with Alexandrian cantharidin. Tattoo removal techniques can be broadly divided into four approaches: mechanical, chemical, ablative, and selective [23, 24]. Mechanical methods involve physical elimination of the pigment from the tissue, by methods such as dermabrasion, excision, or grafting. Excisional techniques may be appropriate for small tattoos when other methods would be ineffective and when the scar would be acceptable. The benefits of excision are full tattoo removal, a single visit, and perhaps its lower cost (fig. 1). A new approach in grafting, involving an enzymatically modified epidermis, has been proposed [25]. A number of chemical methods have been used, either in conjunction with dermabrasion or as monotherapy. The

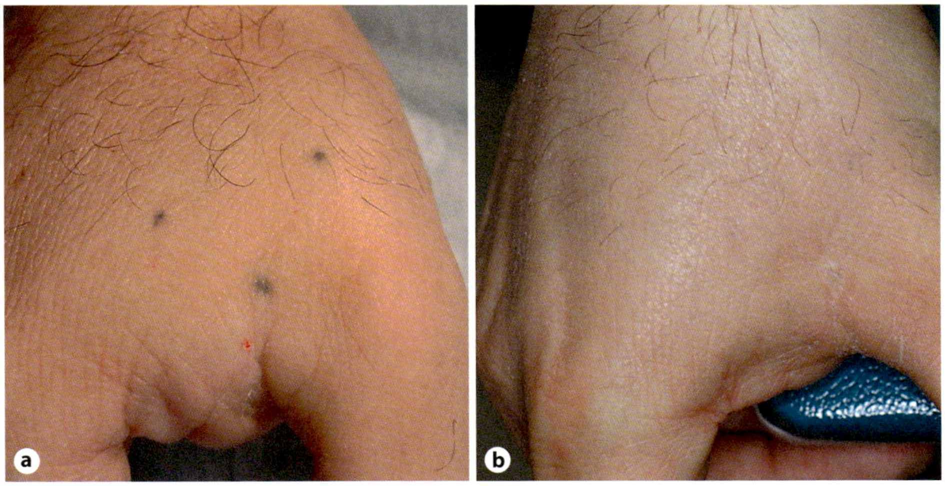

Fig. 1. Before (**a**) and after (**b**) excision. Pigment has been completely removed.

protocol involves tannic acid and silver nitrate, as first described by Varriot in 1988, and is still used by some professional tattoo artists [26, 27]. Phenol and trichloroacetic acid peels are also still in use [28]. While strictly speaking a pharmacological approach rather than a chemical one, the immune modifier imiquimod has been applied to tattoos, with or without tretinoin to enhance epidermal penetration, with mixed results [29]. While effective to varying degrees, these techniques too often result in hypopigmentation, hyperpigmentation, or scarring (fig. 2). Ablative methods involve the use of non-selective energy sources, such as cryotherapy, electrodessication, or non-selective lasers, such as the long-pulse Nd:YAG or continuous-wave CO_2 laser to remove the epidermis and damage the basement membrane, allowing transepidermal elimination of the pigment from the dermis. With the advent of Q-switched lasers in the late 1960s, the outcomes of tattoo removal changed radically. In addition to their selective absorption by the pigment, the extremely short pulse duration of the Q-switched lasers has made them the gold standard for tattoo removal.

The First Lasers

The first laser used on human skin was a Q-switched ruby by Leon Goldman, who tested it first on normal skin [30]. He later published the use of the same laser on a tattoo [31]. Encouraged by the good results, he then published a study comparing the effects of three lasers on tattoos: Q-switched ruby, Q-switched Nd:YAG, and argon. The results were remarkable, particularly with the Q-switched ruby, with hypertrophic scarring in only 3 of 116 patients (1.85%). Unfortunately, this study makes no mention of the cosmetic results with respect to the efficacy of pigment clearance [32]. Though the results of the early work of Goldman with the Q-switched ruby laser were encouraging, the technique was abandoned for almost 20 years. The reasons were mainly technical and commercial. At the time the Q-switched ruby laser was expensive to purchase and to maintain. It was more a laboratory system than a laser that could be used in everyday dermatological clinics.

During the 1970s and 1980s, continuous wave (CW) CO_2 and CW argon lasers became popular

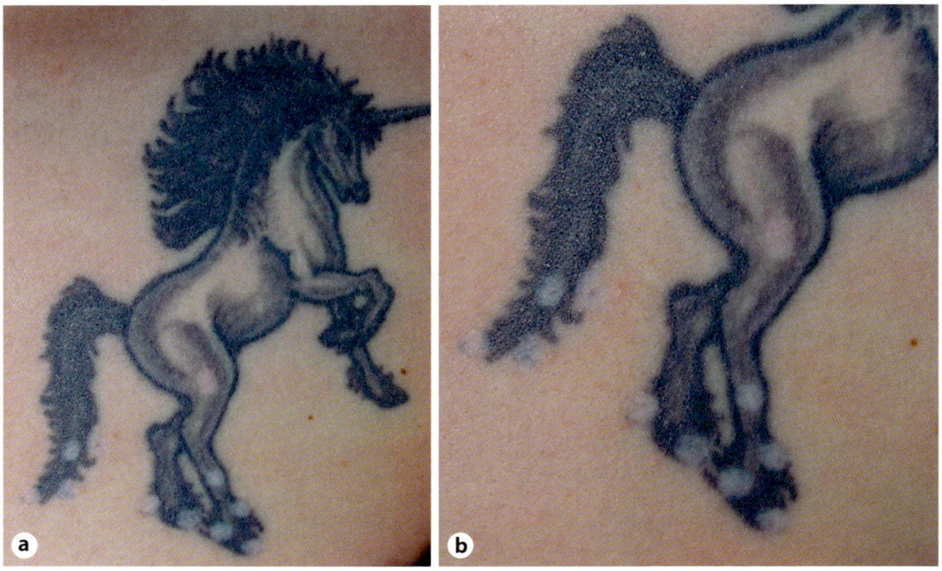

Fig. 2. Scars after tannic acid injections. **a** Full tattoo. **b** Close up.

among independent practitioners. They were small, affordable, and could be used for multiple dermatological indications. The CW CO_2 was used as an ablative tool for various indications, and the CW argon was the first laser used to treat vascular conditions such as port-wine stains [33, 34]. One study summarized the findings of many regarding the cosmetic outcome of tattoo removal with CW CO_2 laser: although the laser delivered better results than chemical or mechanical methods, there was still unacceptable scarring [35]. CW lasers were capable of removing pigmented lesions by ablating or coagulating superficial portions of the skin, but their non-specific tissue effect often led to unpredictable textural or pigmentary changes [36].

The Modern Era of Laser Tattoo Removal

A revolution in laser tattoo removal was inaugurated when the concepts of selective photothermolysis and thermal relaxation time were presented by Anderson and Parrish [37] in 1983. The now-actualized theory proposed that laser-induced injury could be confined to microscopic targets if there were selective light absorption by the target and not by surrounding structures and that if the light was delivered and absorbed within a period of time shorter than or equal to the time needed for the target to transfer the absorbed energy to surrounding structures. To a good approximation, tattoo pigments that have been phagocytosed by dermal stromal cells behave similarly to melanosomes.

The absorption of light pulses by the tattoo pigments is the first and most important step. Two elements are important for optimal light pulse selection. First, it is critical that the laser wavelength used for tattoo removal match the absorption spectrum of the tattoo pigment [38]. If there is no absorption, there can be no reaction. Each color within a tattoo requires treatment with a laser wavelength that is selectively absorbed, with minimal absorption by primary endogenous chromophores like hemoglobin or melanin. Second, the laser pulse duration should be shorter than the thermal relaxation time of the target. Pulses in the nanosecond range are typically used. Q-switching makes it possible to deliver very high energies, up to gigawatts, in extremely short pulses.

Three effects come into play upon the absorption of Q-switched laser by the tattoo pigment. The absorption of this highly intense light by the tattoo pigments leads to rapid heating, with temperatures reaching as high as 900°C. The heating induced by light absorption has been termed a photothermal effect. As the particle is heated, it undergoes thermal expansion. Since Q-switched laser pulses are delivered over nanoseconds, the particle is heated extremely rapidly. The snap heating results in violent expansion, generating shock waves, audible as a popping sound with each laser pulse. The generation of acoustic shock waves by the rapid heating induced by light absorption has been termed the photoacoustic effect. These strong acoustic waves generate forces whose tensile strengths exceed the fracture thresholds, leading to fragmentation of the tattoo pigment particles. The high temperatures also cause rapid vaporization of water, leading to the formation of gas bubbles. When formed at this fast rate and under unstable pressure conditions, the gas bubbles can undergo cavitation, further fracturing the pigment particle. The shear forces induced by rapid motion of the particulate matter, gas bubbles, and cavitation also cause rupture of cells housing the pigment, freeing the fragmented pigment from its sequestering phagocyte and rendering it more accessible for clearance. The photoacoustic effect depends in part on the particle size, with larger particles being significantly more susceptible; smaller particles, while less susceptible, are also less visible and therefore not as critical a target in laser tattoo removal [39].

The rapid absorption of energy can also directly disrupt the chemical bonds of the pigment; a phenomenon termed the photochemical effect.

Immediately after the laser pulse, the epidermis appears white. This is due to the formation of gas as water is vaporized by heating. Gas is also generated within the pigment particles, contributing to their fracture. In the days and weeks following treatment, an acute inflammatory infiltrate surrounds the pigment and debris from the fragmented cells. This process of inflammation and phagocytosis clears a portion of the tattoo pigment from the dermis. The tattoo particles can be found in the regional lymph nodes. A small amount of pigment can be eliminated transepidermally during the mild desquamation that can occur after treatment [40]. However, it is likely that some of the fragmented particles stay in place within the dermis, but as they are fragmented into smaller granules, their optical scattering coefficient changes and they are less visible.

The immediate lightening of epidermis is due to the gas bubble formation and edema as well as to the fragmentation of the pigment into smaller particles [41]. The fading that is observed over the weeks following treatment is due to phagocytosis and clearance of the fragmented pigment particles. Free pigment is intracellular again within approximately 4 weeks after treatment [42]. Due to this timeline, subsequent treatment should be performed after at least 4 weeks. Longer intervals may reduce the risk of pigmentary changes in the skin (fig. 3).

Benefits of the Use of Q-Switched Lasers

It is well established that the Q-switched lasers are the gold standard in lasers to remove tattoos. Four wavelengths are now available with Q-switching: frequency-doubled Nd:YAG at 532 nm (5–10 ns), ruby at 694 nm (25–40 ns), alexandrite at 755 nm (50–100 ns), and Nd:YAG at 1,064 nm (5–10 ns) (table 1). Comparative studies of the four Q-switched lasers have been reported. Though ruby, alexandrite, and Nd:YAG

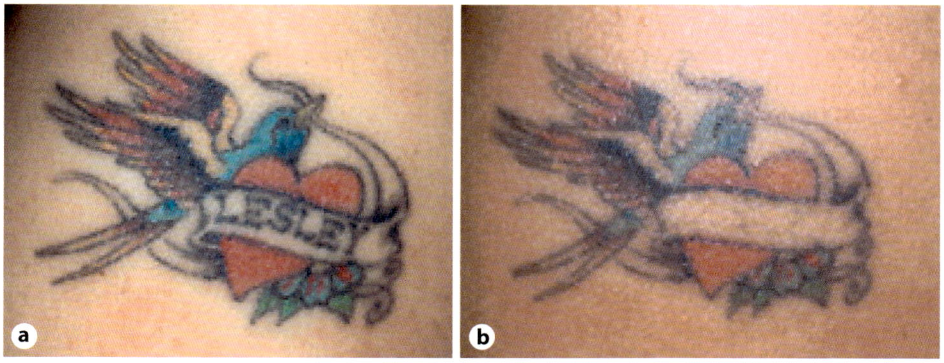

Fig. 3. Before (**a**) and after (**b**) 6 laser treatments with QS-YAG 1,064 nm.

Table 1. Four wavelengths available with Q-switching

Tattoo colors	Pulse duration ns	Wavelength nm	Q-switched lasers
Red	5–10	532	Nd:YAG
Black, blue, green	25–40	694	ruby
Black, blue, green	50–100	755	alexandrite
Black, blue	5–10	1,064	frequency-doubled Nd:YAG

at 1,064 nm are all effective in dark tattoos, the ruby and the alexandrite have been reported to be somewhat more efficient than Nd:YAG [43–47]. It is generally thought that Nd:YAG at 1,064 nm may be preferable in darker skin types due to its lower absorption by melanin and its greater tissue penetration. However, a study of Q-switched ruby and Nd:YAG lasers used in treating carbon-based tattoos in Fitzpatrick V–VI patients showed mild transient hyperpigmentation with both lasers and mild textural changes with the Q-switched Nd:YAG laser [48]. The alexandrite is most effective in treating blue and green tattoos, and the frequency-doubled Nd:YAG may be effective in red tattoos, though the removal of red tattoos in general is challenging due to lack of appropriately selective wavelengths that are not absorbed by endogenous chromophores [45].

Blue-black carbon-based amateur tattoos respond best, requiring on the order of 5 sessions of Q-switched ruby, alexandrite, or Nd:YAG at 1,064 nm. In contrast, professional tattoos containing greater pigment density could require 10–15 sessions [46]. Traumatic tattoos, such as gravel particles can also be treated, though their response is less predictable due to the heterogeneous nature of the impacted particles [49]. Caution must be taken with traumatic tattoos induced by gunpowder or fireworks, as the laser impact can cause microexplosions and provoke pox-like scars [50]. Amalgam tattoos, due to the deposition of metallic particles in the oral mucosa after dental procedures, can be treated with Q-switched alexandrite, ruby, or Nd:YAG [51].

Brown dyschromia secondary to parenteral iron injections can be treated as well [52]. Other forms of hyperpigmentation have been treated with variable success. Good results are reported by some in treating post-sclerotherapy pigmentation and gray pigmentation induced by antimalarials, imipramine, and minocycline [53–56].

Caveats with the Use of Q-Switched Lasers

The concerns surrounding laser tattoo removal fall into two groups: efficacy of pigment elimination and safety of treatment. The concern of safety can also be discussed in two categories: risks related to the effect of the laser on tissue, and dangers due to the properties of the tattoo pigments. The second component poses a true risk, as neither the client nor the tattoo removal provider can predict whether there may be unwanted reactions to tattoo components when tattoos are fragmented by lasers.

Adverse Events Associated with Q-Switched Lasers

The first adverse event published with Q-switched lasers was pigment darkening [57]. This phenomenon has since been reported by others and recognized as a true risk in treatment of certain colored tattoos, primarily white, skin-colored, pink, or red, and frequently in permanent makeup due to its use of these shades [58, 59]. The darkening is explained by conversion of ferric oxide (Fe_2O_3), which has a reddish hue, to the black ferrous oxide (FeO). A similar reaction occurs when titanium dioxide, used in white dyes or mixed in bright colors to enhance the brilliance of tattoos, is exposed to laser light [60]. This reaction is a pure chemical transformation induced by absorption of the laser by the ferric oxide. In rare instances, the darkened pigment may be more desirable to the client than the original tattoo [61]. However, in most cases the black pigment is undesirable and is resistant to further treatment by Q-switched lasers. It is therefore highly advisable to perform a test spot when a dye suspected of containing ferric oxide is present within a tattoo, and particularly when treating permanent makeup. If the pigment to be removed is on the face or another cosmetically sensitive area, the client may wish to have the test spot performed in an inconspicuous location. If such an area is not available within the original tattoo to be removed, the client may return to the tattoo artist to request that a small tattoo, using the same pigment used in the initial tattoo (if available), be placed in the axilla. After a month, a Q-switched laser test spot is performed on this limited area. If this is not possible and pigment darkening does occur, the test spot can be excised with a small punch biopsy. The darkened tattoo can also be removed, albeit with risk of scarring, using ablative CO_2 or erbium YAG lasers [62, 63].

Bäumler and colleagues [64] looked at the decomposition products of tattoo pigment, two widely used azo compounds, after a Q-switched laser exposure. The high laser intensities cleaved the azo compounds, leading to an increase in decomposition products such as 2-methyl-5-nitroaniline, 2- to 5-dichloraniline, and 4-nitro-toluene, which are toxic or even carcinogenic compounds. Chemical analysis shows that the tattoo colorants already contain such compounds before laser irradiation. These in vitro findings, in the context of the high number of tattoos placed and removed annually, call for formal studies to assess the risk of laser-induced decomposition products in humans.

Since Q-switched lasers used for tattoo removal are also absorbed to an extent by melanin, pigmentary change is a potential adverse event with the use of any of the lasers, but the concern is higher with ruby and alexandrite. Hyperpigmentation is generally transient, though weeks or months may elapse before complete resolution. Hyperpigmentation that results from laser tattoo removal should not be treated with pigmented lesion lasers. Hydroquinone, and of course strict sun avoidance, may be of help. Hypopigmentation can also occur and is slower to resolve. The areas can be masked during recovery with a dihydroxyacetone-containing self-tanning product. Other potential adverse events include textural changes or scarring. To reduce the risk, areas that are known to be prone to hypertrophic scarring, such as the chest, neck, and upper back and arms, should be treated with lower fluence.

Another frequent side effect is scarring following treatment of multi-layered tattoos [65]. Many people apply a second camouflage tattoo ('retattoo') over an undesired first tattoo [66]. However, when the final tattoo outcome is not desirable, and the patient seeks removal of the entire tattoo composition, they may forget to report a prior underlying tattoo. For this reason, patients should always be asked if they have a 'retattooed' motif. The scarring phenomenon is likely due to the high density of pigment in these layered tattoos and the very strong absorption of energy by them. This produces heat so intense that it causes damage even to the surrounding dermis. 'Retattooing' can also be a possible cause of resistance to treatment as different pigments were used in the two tattoo layers, increasing the difficulty of removal by Q-switched lasers.

Allergy to pigments has already been briefly discussed. In some cases, allergy to red pigment can be successfully treated by removing the pigment by Q-switched laser at 532 nm [67]. This will reduce the tenderness as well as the itching due to the allergy. However, there is also potential for a rare hypersensitivity reaction after using the Q-switched laser, which may be immediate or delayed [68, 69]. A possible mechanism for these allergies may be the liberation of fragmented tattoo particles from the intracellular compartment, presenting them as new allergens or as foreign bodies. These reactions require treatment as would any allergic or foreign body reaction.

A very rare complication of laser treatment is chrysiasis. Classic chrysiasis is a blue-gray skin discoloration that can occur in sun-exposed areas of patients treated with gold salt. Localized chrysiasis has been reported after the use of Q-switched laser for treatment of post-inflammatory hyperpigmentation [70]. The Q-switched laser can induce structural alteration of the gold deposits similar to that induced by ultraviolet light exposure, leading to blue-gray discoloration within the treated area. This discoloration can be improved with long-pulsed laser (normal mode ruby laser) [71].

Resistance of Pigments to Q-Switched Lasers

Resistance of the pigment to available lasers is a key problem. As discussed, four wavelengths are available for Q-switched lasers: 532, 694, 755, and 1,064 nm. The laser wavelength must be absorbed by the pigment to interact with it and promote the photoacoustic shock. If there is no absorption, the laser will be ineffective. A study by Bäumler et al. [72] on composition of tattoo pigments showed that some colors, like yellow, have their absorption peak in a range where no laser exists, at 480 nm. Hence yellow pigments respond poorly to treatment with the existing Q-switched lasers. Of the 41 colored substances analyzed, 16 represented individual chemicals of well-known organic structure pigments.

A tremendous challenge in tattoo removal arises from the artistic desire of the body artist to tailor-make the color. As discussed earlier, effectiveness of tattoo removal by Q-switched ruby depends on both the chemical makeup of the pigment and the particle size. Most tattoo dyes applied to the skin are combinations of pigments, often with titanium dioxide as a brightening agent. Two tattoos that look alike in color may in fact present two very different

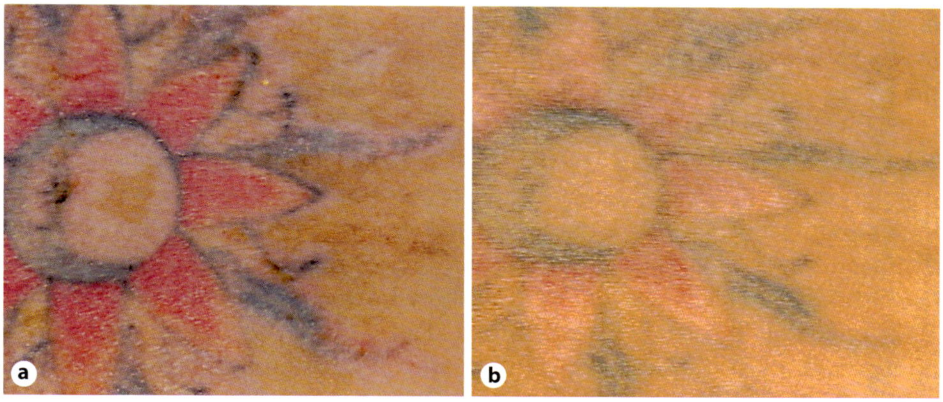

Fig. 4. Before (**a**) and after (**b**) 3 treatments with the fractional Lux 1540. Photo courtesy of David B. Vasily, MD.

mixtures of pigments. Furthermore, transmission electron microscopy of the same dye color can show particles of vastly different sizes. The variability in composition that is inherent to the unregulated tattoo pigment industry renders it a great challenge to guess which tattoo will respond in a favorable manner to Q-switched laser based on clinical appearance alone. The mixture of different compounds and the addition of titanium dioxide make modern recent decorative tattoos more resistant to laser removal. The management of these resistant pigments should be explained very clearly to the patient, including specific mention that some colors, especially yellow, green, and blue, may resist removal, and that some could darken and then resist removal. In the case of a multi-colored tattoos, it is best to first perform a test spot [73].

Non-ablative fractional lasers may provide improved methods to remove resistant pigments (fig. 4). This technology creates columns of coagulated tissue within the dermis at various depths depending on the laser fluence used. The columns of coagulated tissue are spaced, leaving intact tissue in between. The exact mechanism involved in enhanced pigment elimination is not yet clear, but good results can be achieved.

Future Directions

Laser tattoo removal is effective for some pigments, but the state of the art should be improved. A better understanding of the mechanisms by which the pigment particles are cleared will provide enhanced approaches to treatment. While it is known that pigment fragmentation plays a major role in the initial reduction of color intensity, this pigment is rephagocytosed by dermal cells within a few days [42]. It may be that the smaller fragments are not truly eliminated from the tissue but are simply rendered less visible due to their altered light-reflecting properties and their deeper location within the dermis. Transepidermal elimination and lymphatic transport, which contribute to pigment clearance, may be targets for improved tattoo removal treatments. Non-ablative fractional lasers may contribute to such mechanisms.

A key factor in successful tattoo removal is the makeup of the tattoo pigment itself. It is of the utmost importance that body artists use safer well-defined pigments, which have been specifically designed for tattoos. The currently used inks are poorly defined in chemical composition and bear impurities at levels as high as 20% [22]. These can induce hypersensitivity, toxicity, or carcinogenicity. The pigment is also transported to other organs, e.g. lymph nodes [74]. Moreover, the energy delivered by lasers can heat or breaks chemical bonds within the molecule. At temperatures above 280°C, some azo dyes form 3,3′-dichlorobenzidine, a known genotoxin [75]. Azo dyes are frequently used for tattooing because of their longevity and color intensity [74]. Therefore, the health risks of tattooing in humans as well those of laser tattoo removal should be further investigated.

In 2006, Rox Anderson developed the concept of a specific ink for tattooing which could be eliminated more quickly by available Q-switched laser systems. The ink was composed of a biodegradable pigment surrounded by a membrane. The Q-switched laser impact broke the membrane, and thus liberated this specific pigment. As it was biodegradable and non-toxic for the body, the risks of potential toxicity or carcinogenicity after laser removal were highly reduced. Unfortunately, this concept was abandoned, one of the reasons being that tattooists did not want to use an ink on people that already had it in mind to remove their tattoo.

Conclusion

'Getting a tattoo, owning a tattoo and removing a tattoo are three risky acts' [76]. As long as tattoos exist, tattoos will be sought out, and tattoos will be regretted. Within the last 10 years, Q-switched laser systems have enormously improved the aesthetic result of tattoo removal. During this time, our understanding of the mechanisms of pigment elimination has also improved. However, the challenges have also evolved. 'Improved' brighter inks of unknown composition continue to enter the market. As a result, the use of lasers has become more difficult, with more potential complications (e.g. darkening) and pigment resistance. These difficulties notwithstanding, the population seeking tattoo removal perceives lasers as magic wands, which can eliminate tattoos completely, with few treatments, and without risk. We will not change human nature. However, tattoos can be changed, with emphasis on safer injection techniques, less toxic chemicals, and more easily reversed pigments. This will not be well met by some body artists, who believe that tattoos are meant to last a lifetime. For them, the answer must be that for the lifetime to remain healthy, tattoos must adhere to public health requirements, and as artists of the body itself, they must strive always to enhance and not risk harm to their medium.

References

1 Makkai T, McAllister I: Prevalence of tattooing and body piercing in the Australian community. Commun Dis Intell 2001;25:67–72.
2 Armstrong ML, Roberts AE, Koch JR, Saunders JC, Owen DC, Anderson RR: Motivation for contemporary tattoo removal: a shift in identity. Arch Dermatol 2008;144:879–884.
3 Antoszewski B, Sitek A, Fijałkowska M, Kasielska A, Kruk-Jeromin J: Tattooing and body piercing – what motivates you to do it? Int J Soc Psychiatry 2010;56:471–479.
4 Varma S, Lanigan SW: Reasons for requesting laser removal of unwanted tattoos. Br J Dermatol 1999;140:483–485.
5 Adatto M: Living Skin – Decorative Tattooing and Dermatology. Basel, Editiones Roche, 1993, p. 18.
6 Leviticus 19:28.
7 DeMello M: Bodies of inscription: a cultural history of the modern tattoo community. Durham, Duke University Press, 2000.
8 Wohlrab S, Stahl J, Kappeler P: Modifying the body: motivations for getting tattooed and pierced. Body Image 2007;4:87–95.

9 Taylor CR, Anderson RR, Gange RW, Michaud NA, Flotte TJ: Light and electron microscopic analysis of tattoos treated by Q-switched ruby laser. J Invest Dermatol 1991;97:131–136.
10 Fujita H, Nishii Y, Yamashita K, Kawamata S, Yoshikawa K: The uptake and long-term storage of India ink particles and latex beads by fibroblasts in the dermis and subcutis of mice, with special regard to the non-inflammatory defense reaction by fibroblasts. Arch Histol Cytol 1988;51:285–294.
11 Lea PJ, Pawlowski A: Human tattoo: electron microscopic assessment of epidermis, epidermal-dermal junction, and dermis. Int J Dermatol 1987;26:453–458.
12 Swiss Ministry for Health (Bundesamt für Gesundheit). www.bag.admin.ch.
13 U.S. patent number: 6, 122 (Issued 1/11/00).
14 Greif J, Hewitt W: The living canvas. Adv Nurse Pract 1998;6:26–31, 82.
15 Bagley MP, Schwartz RA, Lambert WC: Hyperplastic reaction developing within a tattoo: granulomatous tattoo reaction, probably to mercuric sulfide (cinnabar). Arch Dermatol 1987;123:1557, 1560–1561.
16 Tran D, Ashton RE, Cotterill JA: Sarcoidosis presenting as tattoo granuloma inadvertently treated with laser therapy. J Cutan Laser Ther 2000;2:41–43.
17 Kaur RR, Kirby W, Maibach H: Cutaneous allergic reactions to tattoo ink. J Cosmet Dermatol 2009;8:295–300.
18 Goldenberg G, Patel S, Patel MJ, Williford P, Sangueza O: Eruptive squamous cell carcinomas, keratoacanthoma type, arising in a multicolor tattoo. J Cutan Pathol 2008;35:62–64.
19 Khan IU, Moiemen NS, Firth J, Frame JD: Malignant melanoma disguised by a tattoo. Br J Plast Surg 1999;52:598.
20 Friedman T, Westreich M, Mozes SN, Dorenbaum A, Herman O: Tattoo pigment in lymph nodes mimicking metastatic malignant melanoma. Plast Reconstr Surg 2003;111:2120–2122.
21 Moehrle M, Blaheta HJ, Ruck P: Tattoo pigment mimics positive sentinel lymph node in melanoma. Dermatology 2001;203:342–344.
22 Engel E, Santarelli F, Vasold R, Maisch T, Ulrich H, Prantl L, König B, Landthaler M, Bäumler W: Modern tattoos cause high concentrations of hazardous pigments in skin. Contact Dermatitis 2008;58:228–233.
23 Peris Z: Removal of traumatic and decorative tattoos by dermabrasion. Acta Dermatovenerol Croat 2002;10:15–19.
24 Goldstein N, Penoff J, Price N, Ceilley RI, Goldman L, Hay-Roe V, Miller TA: VIII. Techniques of removal of tattoos. J Dermatol Surg Oncol 1979;5:901–910.
25 Takushima A, Okazaki M, Harii K: Tattoo removal by grafting enzymatically separated epidermis. Ann Plast Surg 2001;47:398–403.
26 Variot G: Nouveau procédé de destruction des tatouages. C R Soc Biol 1888;11:836–838.
27 van der Velden EM, van der Walle HB, Groote AD: Tattoo removal: tannic acid method of Variot. Int J Dermatol 1993;32:376–380.
28 Piggot TA, Norris RW: The treatment of tattoos with trichloracetic acid: experience with 670 patients. Br J Plast Surg 1988;41:112–117.
29 Solis RR, Diven DG, Colome-Grimmer MI, Snyder N 4th, Wagner RF Jr: Experimental nonsurgical tattoo removal in a guinea pig model with topical imiquimod and tretinoin. Dermatol Surg 2002;28:83–86, discussion 86–87.
30 Goldman L, Blaney DJ, Kindel DJ Jr, Franke EK: Effect of the laser beam on the skin: preliminary report. J Invest Dermatol 1963;40:121–122.
31 Goldman L, Wilson RG, Hornby P, Meyer RG: Radiation from a Q-switched ruby laser: effect of repeated impacts of power output of 10 megawatts on a tattoo of man. J Invest Dermatol 1965;44:69–71.
32 Goldman L, Rockwell RJ, Meyer R, Otten R, Wilson RG, Kitzmiller KW: Laser treatment of tattoos: a preliminary survey of three year's clinical experience. JAMA 1967;201:841–844.
33 McBurney EI, Rosen DA: Carbon dioxide laser treatment of verrucae vulgares. J Dermatol Surg Oncol 1984;10:45–48.
34 Apfelberg DB, Maser MR, Lash H: Extended clinical use of the argon laser for cutaneous lesions. Arch Dermatol 1979;115:719–721.
35 Arellano CR, Leopold DA, Shafiroff BB: Tattoo removal: comparative study of six methods in the pig. Plast Reconstr Surg 1982;70:699–703.
36 Apfelberg DB, Laub DR, Maser MR, Lash H: Pathophysiology and treatment of decorative tattoos with reference to argon laser treatment. Clin Plast Surg 1980;7:369–377.
37 Anderson RR, Parrish JA: Selective photothermolysis: precise microsurgery by selective absorption of pulsed radiation. Science 1983;220:524–527.
38 Baumler W, Eibler ET, Hohenleutner U, Sens B, Sauer J, Landthaler M: Q-switch laser and tattoo pigments: first results of the chemical and photophysical analysis of 41 compounds. Lasers Surg Med 2000;26:13–21.
39 Ho DD, London R, Zimmerman GB, Young DA: Laser-tattoo removal – a study of the mechanism and the optimal treatment strategy via computer simulations. Lasers Surg Med 2002;30:389–397.
40 Ferguson JE, Andrew SM, Jones CJ, August PJ: The Q-switched neodymium:YAG laser and tattoos: a microscopic analysis of laser-tattoo interactions. Br J Dermatol 1997;137:405–410.
41 Dozier SE, Diven DG, Jones D, Brysk M, Sanchez RL, Motamedi M: The Q-switched Alexandrite laser's effects on tattoos in guinea pigs and harvested human skin. Dermatol Surg 1995;21:237–240.
42 Pfirrmann G, Karsai S, Roos S, Hammes S, Raulin C: Tattoo removal – state of the art. J Dtsch Dermatol Ges 2007;5:889–897.
43 Lin T, Jia G, Rong H, Li J, Zhou Z: Comparison of a single treatment with Q-switched ruby laser and Q-switched Nd:YAG laser in removing black-blue Chinese tattoos. J Cosmet Laser Ther 2009;11:236–239.
44 Levine VJ, Geronemus RG: Tattoo removal with the Q-switched ruby laser and the Q-switched Nd:YAGlaser: a comparative study. Cutis 1995;55:291–296.
45 Kilmer SL, Anderson RR: Clinical use of the Q-switched ruby and the Q-switched Nd:YAG (1064 nm and 532 nm) lasers for treatment of tattoos. J Dermatol Surg Oncol 1993;19:330–338.
46 Leuenberger ML, Mulas MW, Hata TR, Goldman MP, Fitzpatrick RE, Grevelink JM: Comparison of the Q-switched alexandrite, Nd:YAG, and ruby lasers in treating blue-black tattoos. Dermatol Surg 1999;25:10–14.
47 Zelickson BD, Mehregan DA, Zarrin AA, Coles C, Hartwig P, Olson S, Leaf-Davis J: Clinical, histologic, and ultrastructural evaluation of tattoos treated with three laser systems. Lasers Surg Med 1994;15:364–372.

48 Lapidoth M, Aharonowitz G: Tattoo removal among Ethiopian Jews in Israel: tradition faces technology. J Am Acad Dermatol 2004;51:906–909.
49 Trcilius AM: Effective treatment of traumatic tattoos with a Q-switched Nd:YAG laser. Lasers Surg Med 1998;22:103–108.
50 Taylor CR: Laser ignition of traumatically embedded firework debris. Lasers Surg Med 1998;22:157–158.
51 Shah G, Alster TS: Treatment of an amalgam tattoo with a Q-switched alexandrite (755 nm) laser. Dermatol Surg 2002;28:1180–1181.
52 Raulin C, Werner S, Greve B: Circumscripted pigmentations after iron injections-treatment with Q-switched laser systems. Lasers Surg Med 2001;28:456–460.
53 Tafazzoli A, Rostan EF, Goldman MP: Q-switched ruby laser treatment for postsclerotherapy hyperpigmentation. Dermatol Surg 2000;26:653–656.
54 Becker-Wegerich PM, Kuhn A, Malek L, Lehmann P, Megahed M, Ruzicka T: Treatment of nonmelanotic hyperpigmentation with the Q-switched ruby laser. J Am Acad Dermatol 2000;43(part 1):272–274.
55 Atkin DH, Fitzpatrick RE: Laser treatment of imipramine-induced hyperpigmentation. J Am Acad Dermatol 2000 43(part 1):77–80.
56 Friedman IS, Shelton RM, Phelps RG: Minocycline-induced hyperpigmentation of the tongue: successful treatment with the Q-switched ruby laser. Dermatol Surg 2002;28:205–209.
57 Anderson RR, Geronemus R, Kilmer SL, Farinelli W, Fitzpatrick RE: Cosmetic tattoo ink darkening: a complication of Q-switched and pulsed-laser treatment. Arch Dermatol 1993;129:1010–1014.
58 Varma S, Swanson NA, Lee KK: Tattoo ink darkening of a yellow tattoo after Q-switched laser treatment. Clin Exp Dermatol 2002;27:461–463.
59 Chang SE, Kim KJ, Choi JH, Sung KJ, Moon KC, Koh JK: Areolar cosmetic tattoo ink darkening: a complication of Q-switched alexandrite laser treatment. Dermatol Surg 2002;28:95–96.
60 Ross EV, Yashar S, Michaud N, Fitzpatrick R, Geronemus R, Tope WD, Anderson RR: Tattoo darkening and nonresponse after laser treatment: a possible role for titanium dioxide. Arch Dermatol 2001;137:33–37.
61 Casparian JM, Krell J: Using a side effect to therapeutic advantage: the darkening of red eyebrow tattoo pigment following Q-switched laser treatment. Dermatol Surg 2000;26:255–258.
62 Fitzpatrick RE, Lupton JR: Successful treatment of treatment-resistant laser-induced pigment darkening of a cosmetic tattoo. Lasers Surg Med 2000;27:358–361.
63 Rudlinger R: Successful removal by ruby laser of darkened ink after ruby laser treatment of mismatched tattoos for acne scars. J Cutan Laser Ther 2000;2:37–39.
64 Vasold, R., Naarmann N, Ulrich H, Fischer D, König B, Landthaler M, Bäumler W: Tattoo pigments are cleaved by laser light-the chemical analysis in vitro provide evidence for hazardous compounds. Photochem Photobiol 2004;80:185–190.
65 Alora MBT, Arndt KA, Taylor CR: Scarring following Q-switched laser treatment of 'double tattoos'. Arch Dermatol 2000;136:269–270.
66 Adatto MA: In: Living Skin – Decorative Tattooing and Dermatology. Basel, Editiones Roche, 1993, p 84–86.
67 Dave R, Mahaffey PJ: Successful treatment of an allergic reaction in a red tattoo with the Nd-YAG laser. Br J Plast Surg 2002;55:456.
68 England RW, Vogel P, Hagan L: Immediate cutaneous hypersensitivity after treatment of tattoo with Nd:YAG laser: a case report and review of the literature. Ann Allergy Asthma Immunol 2002;89:215–217.
69 Ashinoff R, Levine VJ, Soter NA: Allergic reactions to tattoo pigment after laser treatment. Dermatol Surg 1995;21:291–294.
70 Trotter MJ, Tron VA, Hollingdale J, Rivers JK: Localized chrysiasis induced by laser therapy. Arch Dermatol 1995;131:1411–1414.
71 Yun PL, Arndt KA, Anderson RR: Q-switched laser-induced chrysiasis treated with long-pulsed laser. Arch Dermatol 2002;138:1012–1014.
72 Bäumler W, Eibler ET, Hohenleuter U, Sens B, Sauer J, Landthaler M: Q-switch laser and tattoo pigments: first results of the chemical and photophysical analysis of 41 compounds. Lasers Surg Med 2000;26:13–21.
73 Timko AL, Miller CH, Johnson FB, Ross EV: In vitro quantitative chemical analysis of tattoo pigments. Arch Dermatol. 2001;137:143–147.
74 Engel E, Vasold R, Santarelli F, Maisch T, Gopee MV, Howard PC, Landthaler M, Bäumler W: Tattooing of skin results in transportation and light-induced decomposition of tattoo pigments – a first quantification in vivo using a mouse model. Exp Dermatol 2010;19:54–60.
75 Chen S, Kao CM, Huang MH: Assessment of genotoxicity of benzidine and its structural analogues to human lymphocytes using comet assay. Toxicol Sci 2003;72:283–288.
76 Anderson RR: Shedding some lights on tattoos? Photochem Photobiol 2004;80:155–156.

Maurice A. Adatto, MD
Skinpulse Dermatology, Laser and Beauty Centers
5, Rd-Pt Plainpalais
CH–1205 Geneva (Switzerland)
Tel. +41 22 307 2777, E-Mail madatto@skinpulse.ch

Bogdan Allemann I, Goldberg DJ (eds): Basics in Dermatological Laser Applications.
Curr Probl Dermatol. Basel, Karger, 2011, vol 42, pp 111–121

Hair Removal

Merete Haedersdal · Christina S. Haak

Department of Dermatology, University of Copenhagen, Bispebjerg Hospital, Copenhagen, Denmark

Abstract

Hair removal with optical devices has become a popular mainstream treatment that today is considered the most efficient method for the reduction of unwanted hair. Photothermal destruction of hair follicles constitutes the fundamental concept of hair removal with red and near-infrared wavelengths suitable for targeting follicular and hair shaft melanin: normal mode ruby laser (694 nm), normal mode alexandrite laser (755 nm), pulsed diode lasers (800, 810 nm), long-pulse Nd:YAG laser (1,064 nm), and intense pulsed light (IPL) sources (590–1,200 nm). The ideal patient has thick dark terminal hair, white skin, and a normal hormonal status. Currently, no method of lifelong permanent hair eradication is available, and it is important that patients have realistic expectations. Substantial evidence has been found for short-term hair removal efficacy of up to 6 months after treatment with the available systems. Evidence has been found for long-term hair removal efficacy beyond 6 months after repetitive treatments with alexandrite, diode, and long-pulse Nd:YAG lasers, whereas the current long-term evidence is sparse for IPL devices. Treatment parameters must be adjusted to patient skin type and chromophore. Longer wavelengths and cooling are safer for patients with darker skin types. Hair removal with lasers and IPL sources are generally safe treatment procedures when performed by properly educated operators. However, safety issues must be addressed since burns and adverse events do occur. New treatment procedures are evolving. Consumer-based treatments with portable home devices are rapidly evolving, and presently include low-level diode lasers and IPL devices.

Hair removal with optical devices has successfully been undertaken since 1996, and today photoepilation with lasers and intense pulsed light (IPL) sources is considered the most efficient method for the reduction of unwanted hair [1–4]. The market for hair removal is substantial, and millions of people all over the world remove unwanted hair at regular intervals with considerable amounts of time and expense being spent to achieve hair-free appearances. Traditional treatments include shaving, plucking, waxing, chemical depilatories, and electrolysis [5]. None of these methods are ideal due to limited and short-term efficacies. Electrolysis is a tedious procedure that may, however, be considered for smaller hair-bearing areas and for candidates not suitable for optical hair removal procedures.

Excess hair growth covers a broad range of severity and may present as hypertrichosis or hirsutism, although most individuals seek hair removal treatments primarily because of cosmetic concerns. Common cosmetic treatments include the axillae, bikini line, legs, and face in women, as well as the chest, back, and shoulders in men. Hirsutism is defined as an androgen-dependent male pattern of

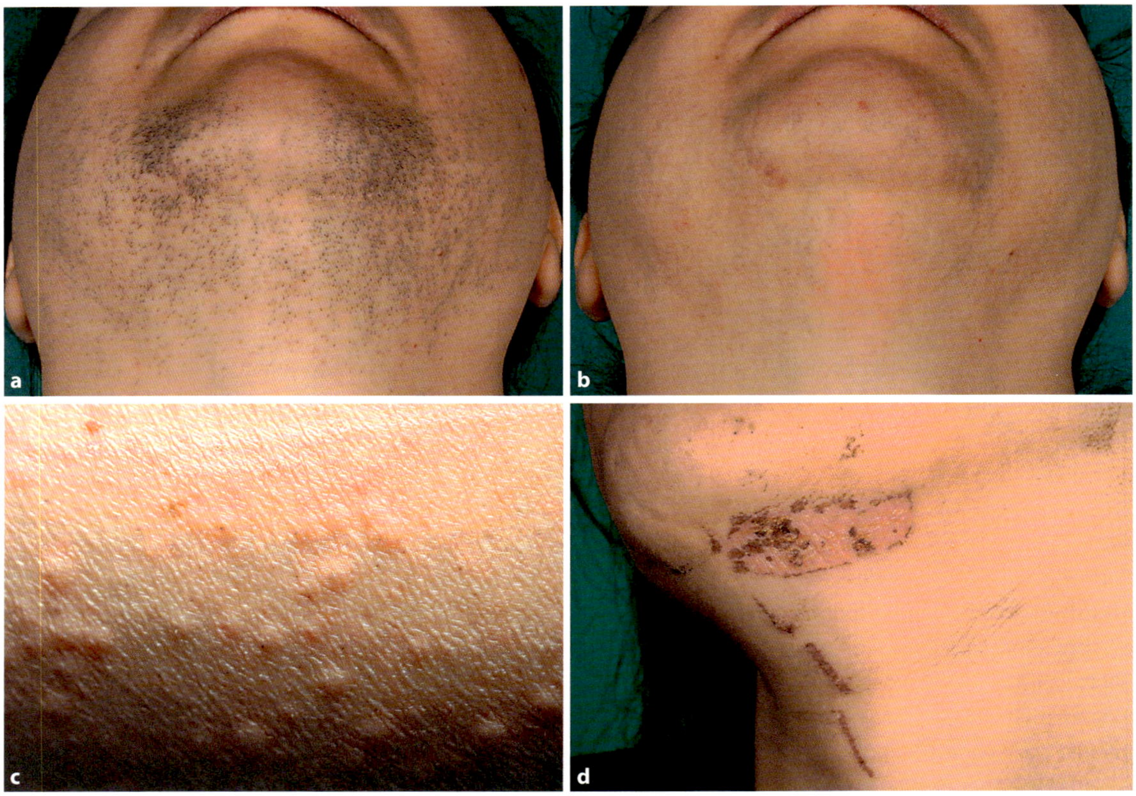

Fig. 1. Before (**a**) and 3-month follow-up (**b**) photos of a hirsute patient who received 6 consecutive diode laser treatments at intervals of 4–6 weeks. Intended immediate skin reactions, erythema, and perifollicular edema, after laser and IPL treatments (**c**). Severe burn damage with ulceration and crusting in a patient with skin type IV at 2 days after receiving IPL treatment with an excessive fluence (**d**).

hair distribution in women and affects between 2 and 10% of women of reproductive age (fig. 1a, b). Causes of hirsutism include excess androgen production by the ovaries or the adrenals, increased sensitivity of the skin to androgens, and an inherent functional difference in the androgen receptors. Hirsutism is most commonly due to polycystic ovary syndrome [6]. Hypertrichosis is defined as an increase in hair growth that is not androgen dependent, and which may present at any body site. Hypertrichosis occurs in all genders and may be acquired or congenital, it may present from genetic predisposition, from medications (e.g. ciclosporin, prednisolone, or phenytoin), or it may be part of a variety of diseases like porphyrias, thyroid disorders, internal malignancies, malnutrition, or anorexia nervosa [7]. Pseudofolliculitis barbae, excess hair in pilonidal sinus disease, and hair-bearing flaps used for reconstruction are other dermatological conditions that respond well to hair removal with lasers and light sources [8, 9].

Mechanisms of Laser Hair Removal

Light can potentially destroy hair follicles by 4 different mechanisms: (1) a photothermal reaction due to local heating; (2) a photomechanical reaction due to shock waves; (3) a photochemical reaction due to toxic mediators induced by the combination of a topical photosensitizer and light exposure (photodynamic therapy); (4) optical bipolar radiofrequency. Photothermal destruction of hair follicles today constitutes the fundamental concept for hair removal. However, all 4 mechanisms have been examined for hair removal [1, 2]. So far, photodynamic therapy has been examined in pilot studies, and shows potential for future hair removal with the advantage of treating non-pigmented hairs [2]. Hair reduction due to photomechanical destruction with nanosecond Q-switched lasers was examined at the early stage of laser hair removal with and without a topical carbon suspension, leading to temporary hair loss. This treatment approach is today considered of minor relevance due to insufficient efficacy.

Photothermal destruction of hair follicles is based on the concept of selective photothermolysis. Selective photothermolysis outlines that selective thermal damage will take place to a pigmented target structure when a specific wavelength is delivered at a sufficient fluence level during a time equal to or less than the thermal relaxation time of the target [10]. Applying the principle of selective photothermolysis to hair removal means that laser and IPL procedures are dependent on the presence of melanin in the hair shaft since melanin is the target chromophore that absorbs energy from specific wavelengths, resulting in spatially confined thermal damage to melanin-containing structures when pulse durations correspond to the thermal relaxation time. However, recent evidence suggests that the goal of laser hair removal is to damage stem cells in the bulge area of the outer root sheath, which requires heat diffusion from melanin in the hair shaft [11]. The concept of thermal damage time has therefore been introduced for hair removal where suggested pulse durations are longer than the thermal relaxation time, allowing dissipation of thermal damage to the follicular stem cells [11].

By adjusting wavelengths and pulse durations from lasers and IPL devices, it is possible to obtain deep selective heating of the hair shaft, the follicular epithelium, and the matrix region. Wavelengths in the red and near-infrared parts of the electromagnetic spectrum are suitable for targeting follicular and hair shaft melanin, and the 600 to 1,100 nm range covers an optical window that is ideal for hair removal due to: (1) sufficient energy absorption by melanin; (2) decreased absorption from the competitive skin chromophores, oxyhemoglobin, and water; (3) light penetration to deeper dermal structures [3]. Pulse durations are typically used in the range of 10–100 ms. Epidermal melanin provides a competitive chromophore and it is essential to protect the epidermis from unspecific damage, especially in darker-skinned individuals with types IV–VI. Selective cooling of the epidermis has been shown to minimize epidermal injury [12]. Cooling can be obtained before, during, and after laser treatment (pre-, parallel-, and post-cooling) as contact cooling (cooled sapphire, metal, or glass plates integrated into the handpiece; cooled gel layer), cold air ventilation, and dynamic cooling devices (where pulsed cryogen spray is used as the cooling agent) [13].

Selecting Appropriate Treatment Parameters

Optical devices for hair removal include the normal mode ruby laser (694 nm), normal mode alexandrite laser (755 nm), pulsed diode lasers (800, 810 nm), long-pulse Nd:YAG laser (1,064 nm), and IPL sources (590–1,200 nm) (table 1) [14–16]. These devices operate in the red or near-infrared wavelength regions corresponding to the recommended optical window in the wavelength range of 600–1,100 nm. Today, the alexandrite laser, diode laser, and IPL devices cover the majority of hair removal treatments, whereas the ruby laser has a minor role, mainly due to a substantial absorption in melanin, which implies a potential risk of adverse events, especially in tanned and darker skin types [13]. Regarding safety, the long-pulse Nd:YAG laser is attractive in individuals

Table 1. Choosing treatment parameters with lasers and IPL sources for hair removal

Wavelength	• Wavelengths used for hair removal have to be adjusted to patient skin type: – normal mode ruby laser, 694 nm: skin types I–III – normal mode alexandrite laser, 755 nm: skin types I–IV – pulsed diode laser, 800/810 nm: skin types I–IV – normal mode Nd:YAG laser, 1,064 nm: skin types I–VI – IPL, 590–1,200 nm: skin types I–IV (depending on filter used and output wavelength spectrum) • Shorter wavelengths are absorbed more strongly by melanin than longer wavelengths • Longer wavelengths penetrate deeper and are safer to use in individuals with darker skin types
Pulse duration	• Pulse durations should be adjusted to the size of the hair structure and typically operate in the long millisecond range between 10 and 100 ms • Longer pulse durations are safer to use in individuals with darker skin types
Spot size	• Larger spot sizes penetrate deeper than smaller spot sizes due to reduced scattering • Larger spot sizes speed up treatment procedures for the convenience of patients and laser operators
Fluence	• Efficacy correlates with the delivered fluence • The fluence level has to be adjusted to the actual amount of target chromophore in terms of hair color and hair density • The immediate treatment response should be vaporization of the hair shaft that in a few seconds may be followed by mild erythema and possibly perifollicular edema (fig. 1c), typically resolving within 60 min • The operator should be experienced in evaluating immediate treatment endpoints in order to adjust the optimal fluence level without any initial signs of adverse events (intense erythema, confluent edema, or gray or white discoloration)

with darker skin types IV–VI. The physical parameters within the specific devices vary considerably in terms of wavelength, pulse duration, spot size, and fluence. Treatment parameters in terms of wavelength, pulse duration, spot size, and fluence must be individually selected and adjusted to the clinical situation before commencing treatment sessions [13] (table 1).

Expected Benefits

The expected benefit from optical hair removal procedures depends on individual factors like skin type, hair, and hormonal status. The ideal patient has thick dark terminal hairs, white skin, and a normal hormonal status without any signs of polycystic ovary syndrome or other hormonal imbalances. Candidates with thin vellus hairs or white, grey, or red hairs on the other hand do not respond sufficiently well to photoepilation with currently available lasers and IPL systems. It is regarded as insignificant for the treatment outcome whether hair follicles are targeted during specific growth phases, since the human hair follicles remain pigmented during the entire hair cycle, thus providing a sufficient target [4]. Dark skin types and sun exposure influences the treatment outcome with an increased risk of inducing side effects. Therefore, suntanned patients are advised to avoid treatment until the tan has faded, and patients with darker skin types IV–VI must

Table 2. Practical aspects of optical hair removal procedures

Patient selection and pretreatment care	• The ideal candidate for optical hair removal has pale skin, dark, thick hair, normal hormone status, and realistic expectations towards treatment procedures, efficacy, and risk of adverse effects • Patients with darker skin types are at risk of side effects and should be treated with caution • Patients with underlying medical conditions should be identified and treated accordingly • A suntan should be recognized and treatments postponed until it has faded due to increased risk of adverse effects • Patients on oral retinoids should not undergo laser hair removal for at least 6 months after discontinuation of the medication due to increased risk of keloids • Patients at risk of developing paradoxical hair growth should be identified; typically young females with skin types III–IV and ill-defined frontal hairlines • Patients should be advised to refrain from plucking or waxing prior to treatment since these procedures result in loss of tissue target • Shaving, bleaching, or chemical depilatories are acceptable alternatives • Patients should shave the treatment area the day before laser treatment; alternatively shaving may be performed in the clinic with subsequent removal of hairs with adhesive tape to avoid burns from loose hairs • Topical anesthetics may be applied under occlusion 1 h before treatment to avoid pain and discomfort
Treatment procedures	• Equipment must be examined for any defects or damage • Cooling systems must be turned on and tested • Everyone in the treatment room must wear goggles for eye protection • When treating the face area, the patient's eyes must be covered with gauze and goggles since goggles alone may permit some light to enter from the sides • When treating near the eyes, the cornea must be covered with metal shields • Do not treat inside the orbital rim because of risk of eye injury • Pigmented lesions in the treatment area, such as ephelides and nevi, should be marked in white to avoid light absorption • Treatment parameters must be adjusted to patient skin type and chromophore
Post-treatment care	• Cooling with ice packs may reduce erythema and swelling • Topical steroids may be used in patients with intense immediate skin reactions, such as pronounced erythema, confluent edema, or gray or white discoloration that may develop into burns and adverse effects • Patients should be instructed to avoid sun exposure and to use proper sun protection during the entire treatment period to avoid a suntan • Patients should be instructed on possible hair shedding in the first weeks after treatment, not to be mistaken as hair regrowth

be treated with caution. Patient selection thus remains mandatory for obtaining successful treatment outcomes (table 2).

To date, no method of lifelong permanent hair eradication has become available, and it is important that patients have realistic expectations before treatments are commenced. Concerning *permanent hair reduction*, the FDA has given the definition: 'Permanent hair reduction refers to a long-term, stable reduction in the number of hairs re-growing after a treatment regime, which may include several treatment sessions. The number of re-growing hairs must be stable for a period of time longer than the complete growth cycle of hair follicles, which varies from four to twelve months according to body location' [17]. Unfortunately, this definition may lead to unrealistic patient expectations since 'permanency' from the patients' points of view typically means that hairs do not regrow and that the hair-free appearance lasts persistently. *Temporary hair loss* is defined as a delay in hair growth, which usually lasts for 1–3 months, consistent with the induction of telogen [2]. Several factors contribute to the patients' overall subjective impression of hairiness and include the number and rate of regrowing hairs as well as hair thickness and the color of regrowing hairs. The most substantial data for hair removal effectiveness exist for a reduction in the number of regrowing hairs, whereas information is both sparse and anecdotal concerning hair thickness and color of regrowing hairs. It is considered essential to evaluate hair removal efficacy from an evidence-based point of view in order to provide patients with realistic preoperative expectations. Thus, each patient's specific expectations have to match the capabilities and facts of laser and IPL hair removal before commencing optical treatment sessions for hair reduction.

Substantial evidence has been found for *short-term hair removal efficacy* of up to 6 months after epilation with the normal mode ruby laser, normal mode alexandrite laser, pulsed diode lasers, long-pulse Nd:YAG laser, and IPL sources (590–1,200 nm) [18, 19]. The efficacy is improved when repetitive treatments are given [16, 20–23] and the short-term efficacy from photoepilation is superior to conventional treatments with shaving [16, 22, 24–27], wax epilation [26, 28], and electrolysis [28, 29]. Overall, the short-term efficacy is reported to be between 30 and 70% hair reductions up to 6 months after the final treatment, with treatment outcomes depending on treatment setting and a few studies reporting higher and lower efficacy outcomes. Evidence has been found for *long-term hair removal efficacy* beyond 6 months postoperatively after repetitive treatments with the alexandrite, diode, and the long-pulse Nd:YAG lasers [16, 18, 19, 22, 30, 31]. It may be possible that repetitive treatments with the ruby laser and IPL are capable of inducing long-term hair reduction as well; although the actual long-term evidence is sparse for IPL devices and varying results have been reported from ruby laser treatments. Further documentation is required, especially on the long-term efficacy of IPL treatments [32]. The overall results from long-term studies with the alexandrite, diode, and Nd:YAG lasers vary considerably in controlled trials, but show on average 50% hair reduction from repetitive treatments with these devices at 6–20 months after treatment [16, 22, 30, 31].

Normal Mode Ruby Laser: 694 nm

The ruby laser was the earliest laser used for hair removal but is now more seldom used, which may be explained by its limited clinical applications and relatively high incidence of adverse effects due to a high absorption rate of the epidermal melanin. In controlled trials, the short-term hair reduction has been reported to be significantly better after ruby laser treatments (38–49% hair reduction at 2–4 months) than after electrolysis, waxing, or shaving, or untreated control skin (no significant decrease) [26, 28], whereas complete hair regrowth was found at 3 months after shaving and 1–3 ruby laser treatments in another study [33]. No long-term hair reduction was obtained at 12 months after 3 ruby laser treatments in one controlled study [28], while Dierickx et al. [34] found obvious hair loss 1 and 2 years after 1 ruby laser treatment in 4/7 individuals. One randomized controlled trial has described that 3 repetitive treatments with the ruby laser resulted in 19% hair reduction at 5 months after the last treatment, in contrast to only 6% hair reduction following 2 treatments [20]. Compared to newer treatment modalities, one controlled study has shown the ruby laser to be less efficient than IPL at 6 months after 3 repetitive treatments with hair reductions of 49% in IPL-treated patients and 21% in patients treated with the ruby laser [14].

Regarding adverse events, hypopigmentation is the most frequently reported, especially in darker skin types [35].

Normal Mode Alexandrite Laser: 755 nm
Today, the best available evidence supporting optical hair removal is found for alexandrite lasers with a total of more than 20 controlled trials, of which a brief summary is given in the following. The efficacy of the alexandrite laser has on a short-term basis (up to 6 months postoperatively) been found to be superior to conventional treatments using shaving and electrolysis [15, 27]. The alexandrite laser has been compared with other photoepilation equipment in several studies. In comparison to the diode laser, most studies have reported similar patient tolerance and similar hair reductions at 3 months (59%) and 6 months postoperatively (37–72%, showing increasing efficacy with increased numbers of treatments) [23, 31, 36–38]. Compared with the Nd:YAG laser, both devices have been reported to be efficient on a short-term basis (alexandrite laser 59–70% hair reduction at 2–3 months postoperatively versus 32–48% for the Nd:YAG laser) [38, 39]. In terms of long-term hair reduction, substantial and similar efficacies were reported at 18 months after 4 treatments with both the alexandrite laser (76–84%) and the Nd:YAG laser (74%) [30]. Combining the alexandrite laser with either the diode or long-pulse Nd:YAG lasers did not improve treatment outcomes, but led to a higher frequency of adverse events [9, 38, 39]. In comparison to IPL, diverging results have been published, which probably reflect the diversity of the available IPL devices. In one study, 6 treatments with the alexandrite laser resulted a significantly larger decrease in hair counts than 6 treatments with the IPL (46 vs. 27% hair reduction 6 months postoperatively) [40], while 2 controlled trials demonstrated similar treatment efficacies at 6 months after 2–7 treatments [23, 36]. Finally, 2 randomized controlled trials (RCTs) evaluated the efficacy of the alexandrite laser in combination with either topical application of eflornithine or placebo cream [29, 41]. In one of the studies, the follow-up was, however, only 2 weeks. At this time point, the eflornithine-treated skin showed significantly fewer hairs upon examination (93 vs. 67.9% hair reduction) [29]. Similarly, Smith et al. [41] showed a significantly better short-term hair reduction in their group of patients treated with eflornithine before laser treatment, but no difference was found at 34 weeks after treatment.

Pulsed Diode Laser: 800 and 810 nm
By virtue of using semiconductors instead of flashlamps to produce laser light, diode lasers can vary widely in their emitted wavelengths. Diode lasers currently used for hair removal emit 800 or 810 nm light, and substantial evidence is available to support these devices [18, 19]. Their efficacy has been proven to be significantly better than shaving in both short- and long-term studies (13–36% hair reduction in laser-treated areas after 12 months vs. no effect in shaved areas) [16, 24]. Several studies have compared the diode laser with the alexandrite laser, and similar treatment outcomes were seen in all studies except one [23, 31, 35, 37–39]. Compared with the long-pulse Nd:YAG laser, the diode laser conveyed better results at 2–3 months after treatment (58.7–59.7 vs. 31.9–47.7%) while similar, almost complete, hair regrowth was seen for both lasers at 9 months postoperatively [38, 39, 42]. No difference in outcome measures was reported for the diode laser compared to IPL in 2 studies with follow-up times of 6 months, both resulting in approximately 50–70% hair reduction after 3–7 treatments [23, 36]. One randomized controlled trial showed that combining the diode laser with a liposomal melanin spray increased the hair reduction efficacy of the diode laser marginally at 6 months after 3 treatments (14 vs. 10% hair reduction) [43]. A new treatment approach is linear scanned hair removal with an 808 nm diode laser, which has the benefit of fast treatment times [44].

Long-Pulse Nd:YAG Laser: 1,064 nm
Light at 1,064 nm is significantly less absorbed by melanin compared to ruby, alexandrite, diode, and IPL systems and penetrates deeply into dermis, down to 4–6 mm. Originally it was used in Q-switched mode with or without topical carbon particles, but with limited success. The millisecond systems favor treatment for especially darker skin types. The efficacy was proven to be significantly better than shaving in both short-term (24% hair reduction versus 0% at 3 months) and long-term studies (>75% hair reduction in 11% of the patients versus 0% of shaved patients at 10 months) [22, 25]. Five repetitive treatments improved the long-term treatment outcome compared to a single treatment (>50% hair reduction in 40% of patients vs. <25% 12 months postoperatively) [22]. For comparisons with alexandrite and diode lasers, see the above sections. The short-term hair removal efficacy after 6 weeks was similar for 1 treatment with long-pulse Nd:YAG laser (73%) and IPL treatment (64%) [21]. Eflornithine increased the efficacy of the Nd:YAG laser on a short-term basis (6–22 weeks), but not on a long-term basis (34 weeks) [41].

Intense Pulsed Light: 590–1,200 nm
A variety of different IPL systems are available using different cutoff filters to match the patient's skin type and hair color. IPL devices are best reserved for patients with skin types I–IV due to a potential risk of side effects in darker individuals (table 1). The hair removal efficacy after treatment with IPL has been evaluated on a short-term basis up to 6 months postoperatively, whereas no long-term data are available so far. Three treatments with IPL (94% of patients had a mean hair reduction of 49%) were superior to three treatments with the ruby laser (55% of patients had a mean hair reduction of 21%) evaluated at 6 months postoperatively by hair counts and patients' subjective evaluations [14]. In 2 studies, IPL was found to be similar to the alexandrite laser at 6 months after treatment (50–68.8% hair reduction after 2–7 treatments) [23, 36], whereas one RCT reported a better outcome from the alexandrite laser than IPL (43 vs. 27% after 6 months) [40]. IPL and diode laser were found equally effective at 6 months after final treatment (50–72% hair reduction after 2–7 treatments) [23, 36]. Similarly, one treatment with long-pulse Nd:YAG laser and IPL resulted in equal hair removal efficacies at 6 weeks postoperatively (73 and 64%), whereas the IPL treatment more often resulted in post-inflammatory pigmentation as compared with the long-pulse Nd:YAG laser [21].

Hair Removal in Hirsute Patients
Despite the fact that many women suffer from hirsutism and hair removal efficacies are considered inferior due to hormonal imbalances, so far there have only been a few controlled studies looking specifically at hair removal efficacy in this population [45–48]. These studies have reported hair reductions in a range from 27% (IPL) to 46–49% (alexandrite laser) at 6 months after final treatment. A recent split-face RCT compared IPL and long-pulse diode laser in hirsute women with verified normal testosterone levels (both intrinsically normal and medically normalized), and found that efficacies declined during the postoperative period with 77–40% hair reduction for IPL and 68–34% for long-pulse diode laser during the 1- to 6-month follow-up period [48]. These data indicate that the 6-month facial treatment outcome in hirsute women may be inferior compared to treatment outcomes in non-hormonal-sensitive zones of non-hirsute patients. However, it currently remains unknown whether hairs may be regrowing more rapidly in hirsute patients than when treatments are given outside the facial area to non-hormonal-sensitive zones. Recent data have documented that increasing insulin sensitivity by metformin enhances IPL-assisted hair removal efficacy in patients with polycystic ovary syndrome [47].

Adverse Events, Complications, and Safety

Safety issues have to be addressed and patients should be informed of the risk of adverse effects and complications before committing to laser- and IPL hair removal procedures (table 2). Hair removal with lasers and IPL sources are generally regarded as safe treatment procedures when performed by properly educated and trained operators. However, treatments have expanded outside dermatology clinics, which means that treatments are offered without supervision from trained physicians. Unfortunately, adverse events and injuries do occur, which may be explained by insufficient training and education of laser operators. It is mandatory that the practical aspects of laser procedures in terms of patient selection, pre-treatment care, treatment procedures, and post-treatment care are carried out with high standards in order to obtain successful treatment outcomes without complications and adverse effects (table 2). Risks and adverse effects include burns with blistering, ulceration, and crusting (fig. 1d). Moreover, postoperative infections, post-inflammatory hyperpigmentation, hypopigmentation, scarring, eye injury, paradoxical hair stimulation, and livedo reticularis may occur [49–51]. Higher incidences of laser-induced pigment changes are associated with darker skin types and the use of shorter wavelength lasers and IPL devices [51]. Pigment changes are usually transient and resolve spontaneously without sequelae, although long-term or even permanent dyschromia may occur [49, 51]. The risk of eye injury is severe when any hair removal laser or IPL device is used within the bony orbit, and should therefore be avoided. If treating near the eyes, the cornea must be covered with metal eye shields. Paradoxical hair stimulation has been noted after laser and IPL treatments in approximately 5% of women of Mediterranean and Middle Eastern descent [52]. The mechanism for this observation is unknown. However, the reactive hair growth is usually long-lasting or permanent and may appear in previously treated areas and in adjacent non-treated skin, mainly on the face and neck. Patients with skin types III and IV and those with ill-defined frontal hairlines seem to be at risk of developing paradoxical hair growth, and it is recommended to avoid treating these patients. Patients have to be instructed to avoid sun exposure before, during, and after treatment sessions for hair removal in order to reduce the risk of postoperative complications and adverse events.

New and Alternative Treatment Strategies

Hair removal with optical devices has become a popular mainstream treatment and new treatment procedures are evolving. Consumer-based treatments with portable home-based devices are rapidly evolving, and presently include low-level diode lasers and IPL devices [53, 54]. Data support a transient short-term hair reduction from repetitive and continuous use with these low-level devices and the most prominent safety issue is regarding ocular damage [55]. Potentially, photodynamic therapy may be a useful future approach for hair removal since photosensitizers tend to accumulate in the follicular epithelium [2, 56].

Topical application of eflornithine (VaniqaT, a medical enzyme that irreversibly inhibits the enzyme ornithine decarboxylase involved in controlling hair growth and proliferation) may be a useful adjuvant for laser and IPL hair removal, especially in patients with facial hairiness resistant to optical procedures [57]. Controlled studies have shown that combined treatment with eflornithine and laser procedures may increase the efficacy of laser hair removal [29, 41]. However, clinical controlled trials and long-term data are needed to substantiate the efficacy and safety of these new modalities.

Key Messages

- Laser and IPL treatment for hair removal depends on the presence of melanin in the hair shaft.
- Treatment parameters must be adjusted to the patient skin type and chromophore.
- Longer wavelengths are safer for patients with darker skin types.
- Literature search reveals substantial evidence for laser and photoepilation.
- Short-tem hair removal efficacy up to 6 months after treatment is well documented in controlled trials with visible and infrared laser systems and with IPL sources.
- Evidence exists for long-term hair removal efficacy beyond 6 months after repetitive treatments with alexandrite, diode, and long-pulse Nd:YAG lasers.
- Consumer-based home devices are evolving.

References

1 Dierickx C: Laser-assisted hair removal: state of the art. Dermatol Ther 2000;13:80–89.
2 Dierickx CC, Grossman MC: Laser hair removal; in Goldberg DJ, Dover JS: Laser and Lights. Amsterdam, Elsevier Saunders, 2005, vol 2.
3 Goldberg DJ: Laser and light-based hair removal: an update. Expert Rev Med Devices 2007;4:253–260.
4 Goldberg DJ: Laser hair removal; in Goldberg DJ: Laser Dermatology: Pearls and Problems. Hoboken, Blackwell Publishing, 2008.
5 Olsen EA: Methods of hair removal. J Am Acad Dermatol 1999;40:143–155.
6 Neithardt AB, Barnes RB: The diagnosis and management of hirsutism. Semin Reproductive Med 2003;21:285–293.
7 Wendelin DS, Pope DN, Mallory SB: Hypertrichosis. J Am Acad Dermatol 2003;48:161–179.
8 Bridgeman-Shah S: The medical and surgical therapy of pseudofolliculitis barbae. Dermatol Ther 2004;17:158–163.
9 Conroy FJ, Mahaffey PJ: Intraoral flap depilation using the long-pulsed alexandrite laser. J Plast Reconstruct Aesthet Surg 2009;63:e421–e423.
10 Anderson RR: Selective photothermolysis: precise microsurgery by selective absorption of pulsed radiation. Science 1983;220:524–527.
11 Tierney EP, Goldberg DJ: Laser hair removal pearls. J Cosmet Laser Ther 2008;10:17–23.
12 Nelson JS, Majaron B, Kelly KM: Active skin cooling in conjunction with laser dermatologic surgery. Semin Cutan Med Surg 2000;19:253–266.
13 Drosner M, Adatto M: Photoepilation: Guidelines for care from the European Society for Laser Dermatology (ESLD). J Cosmet Laser Ther 2005;7:33–38.
14 Bjerring P, Cramers M, Egekvist H, et al: Hair reduction using a new intense pulsed light irradiator and a normal mode ruby laser. J Cutan Laser Ther 2000;2:63–71.
15 Görgü M, Aslan G, Aköz T, et al: Comparison of alexandrite laser and electrolysis for hair removal. Dermatol Surg 2000;26:37–41.
16 Lou WW, Quintana AT, Geronemus RG, et al: Prospective study of hair reduction by diode laser (800 nm) with long-term follow-up. Dermatol Surg 2000;26:428–432.
17 US Food and Drug Administration: Hair removal (updated 5/17/2002). www.fda.gov/cdrh/consumer/laserfacts.html.
18 Haedersdal M, Wulf HC: Evidence-based review of hair removal using lasers and light sources. J Eur Acad Dermatol Venereol 2006;20:9–20.
19 Haederdal M, Gøtzsche PC: Laser and photoepilation for unwanted hair growth. Cochrane Database Syst Rev 2006(4):CD004684.
20 Allison KP, Kiernan MN, Waters RA, et al: Evaluation of the ruby 694 Chromos for hair removal in various skin sites. Lasers Med Sci 2003;18:165–170.
21 Goh CL: Comparative study on a single treatment response to long pulse Nd:YAG lasers and intense pulse light therapy for hair removal on skin type IV to VI – Is longer wavelengths lasers preferred over shorter wavelengths lights for assisted hair removal. J Dermatol Treatment 2003;14:243–247.
22 Lorenz S, Brunnberg S, Landthaler M, et al: Hair removal with the long pulsed Nd:YAG laser: a prospective study with one year follow-up. Lasers Surg Med 2002;30:127–134.
23 Toosi P, Sadighha A, Sharifian A, et al: A comparison study of the efficacy and side effects of different light sources in hair removal. Lasers Med Sci 2006;21:1–4.
24 Baugh WP, Trafeli JP, Barnette DJ, et al: Hair reduction using a scanning 800 nm diode laser. Dermatol Surg 2001;27:358–364.
25 Fournier N, Aghajan-Nouri N, Barneon G, et al: Hair removal with an Athos Nd:YAG 3.5 ms pulse laser: a 3-month clinical study. J Cutan Laser Ther 2000;2:125–130.
26 Grossman MC, Dierickx C, Farinelli W, Flotte T, Anderson RR: Damage to hair follicles by normal-mode ruby laser pulses. J Am Acad Dermatol 1996;35:889–894.
27 Nanni CA, Alster TS: Long-pulsed alexandrite laser-assisted hair removal at 5, 10, and 20 millisecond pulse durations. Lasers Surg Med 1999;24:332–337.

28 Polderman MCA, Pavel S, Le Cessie S: Efficacy, tolerability, and safety of a long-pulsed ruby laser system in the removal of unwanted hair. Dermatol Surg 2000;26:240–243.
29 Hamzavi I, Tan E, Shapiro J, et al: A randomized bilateral vehicle-controlled study of eflornithine cream combined with laser treatment versus laser treatment alone for facial hirsutism in women. J Am Acad Dermatol 2007;57:54–59.
30 Davoudi SM, Behnia F, Gorouhi F, et al: Comparison of long-pulsed alexandrite and Nd:YAG lasers, individually and in combination, for leg hair reduction: an assessor-blinded, randomized trial with 18 months follow-up. Arch Dermatol 2008;144:1323–1327.
31 Eremia S, Li C, Newman N: Laser hair removal with alexandrite versus diode laser using four treatment sessions: 1-year results. Dermatol Surg 2001;27:925–930.
32 Hee Lee J, Huh CH, Yoon HJ, et al: Photo-epilation results of axillary hair in dark-skinned patients by intense pulsed light: comparison between different wavelengths and pulse width. Dermatol Surg 2006;32:234–240.
33 Walther T, Bäumler W, Wenig M, et al: Selective photothermolysis of hair follicles by normal-mode ruby laser treatment. Acta Derm Venereol 1998;78:443–444.
34 Dierickx C, Grossman MC, Farinelli WA, et al: Permanent hair removal by normal-mode ruby laser. Arch Dermatol 1998;134:837–842.
35 Hædersdal M, Egekvist H, Efsen J, et al: Skin pigmentation and texture changes after hair removal with the normal-mode ruby laser. Acta Derm Venereol 1999;79:465–468.
36 Amin SP, Goldberg DJ: Clinical comparison of four hair removal lasers and light sources. J Cosmet Laser Ther 2006;8:65–68.
37 Handrick C, Alster TS: Comparison of long-pulsed diode and long-pulsed alexandrite lasers for hair removal: a long-term clinical and histologic study. Dermatol Surg 2001;27:622–626.
38 Rao J, Goldman MP: Prospective, comparative evaluation of three laser systems used individually and in combination for axillary hair removal. Dermatol Surg 2005;31:1671–1676.
39 Khoury JG, Saluja R, Goldman MP: Comparative evaluation of long-pulse alexandrite and long-pulse Nd:YAG laser systems used individually and in combination for axillary hair removal. Dermatol Surg 2008;34:665–670.
40 McGill DJ, Hutchison C, McKenzie E, et al. A randomised, split-face comparison of facial hair removal with the alexandrite laser and intense pulsed light system. Lasers Surg Med 2007;39:767–772.
41 Smith SR, Piacquadio DJ, Beger B, et al: Eflornithine cream combined with laser therapy in the management of unwanted facial hair growth in women: a randomized trial. Dermatol Surg 2006;32:1237–1243.
42 Chan HH, Ying S-Y, Ho W-S, et al: An in vivo study comparing the efficacy and complications of diode laser and long-pulsed Nd:YAG laser in hair removal in Chinese patients. Dermatol Surg 2001;21:950–954.
43 Sand M, Bechara FG, Sand D, et al: A randomized, controlled, double-blind study evaluating melanin-encapsulated liposomes as a chromophore for laser hair removal of blond, white, and gray hair. Ann Plast Surg 2007;58:551–554.
44 Bodenhof MO, Grunewald S, Preuss M, et al: Comparison between axillary hair removal with a continuously scanned diode laser and a spot-to-spot scanned alexandrite laser (Epicon study) (abstract). Lasers Surg Med 2009;41:112.
45 Clayton WJ, Lipton M, Elford J, et al: A randomized controlled trial of laser treatment among hirsute women with polycystic ovary syndrome. Br J Dermatol 2005;152:986–992.
46 McGill DJ, Hutchisin C. McKenzie E, et al: A randomized, split-face comparison of facial hair removal with the alexandrite laser and intense pulsed light system. Laser Surg Med 2007;39:767–772.
47 Rezvanian H, Adibi N, Siavash M, et al: Increased insulin sensitivity by metformin enhances intense-pulsed-light-assisted hair removal in patients with polycystic ovary syndrome. Dermatology 2009;218:231–236.
48 Haak CS, Nymann P, Pedersen AT, et al: Hair removal in hirsute women with normal testosterone levels: a randomized controlled trial of long-pulsed diode laser versus intense pulsed light. Br J Dermatol 2010;163:1007–1013.
49 Goldberg DJ: Complications in laser treatment of unwanted hair; in Goldberg DJ: Complications in Cutaneous Laser Surgery. London, Taylor & Francis, 2004.
50 Goldberg DJ: Laser complications: hair removal. J Cosmet Laser Ther 2006;8:197–202.
51 Lim SPR, Lanigan SW: A review of the adverse effects of laser hair removal. Lasers Med Sci 2006;21:121–125.
52 Willey A, Torrontequi J, Azpiazu J, et al: Hair stimulation following laser and intense pulsed light photo-epilation: review of 543 cases and ways to manage it. Lasers Surg Med 2007;39:297–301.
53 Wheeland RG: Simulated consumer use of a battery-powered, hand-held, portable diode laser (810 nm) for hair removal: a safety, efficacy and ease-of-use study. Lasers Surg Med 2007;39:476–493.
54 Elm CM, Wallander ID, Walgrave SE, Zelickson BD: Clinical study to determine the safety and efficacy of a low-energy, pulsed light device for home use hair removal. Laser Surg Med 2010;42:287–291.
55 Town G, Ash C: Are home-use intense pulsed light (IPL) devices safe? Lasers Med Sci 2010;25:773–780.
56 Hædersdal M, Sakamoto FS, Farinelli WA, et al: Fractional CO_2 laser assisted drug delivery. Laser Surg Med 2010;42:113–122.
57 Lapidoth M, Dierickx C, Lanigan S, et al: Best practice options for hair removal in patients with unwanted facial hair using combination therapy with laser: guidelines drawn up by an expert working group. Dermatology 2010;221:34–42.

Merete Haedersdal, MD
Department of Dermatology, Bispebjerg Hospital, University of Copenhagen
Bispebjerg Bakke 23
DK–2400 Copenhagen NV (Denmark)
Tel. +45 35 31 60 02, E-Mail mhaedersdal@dadlnet.dk

Bogdan Allemann I, Goldberg DJ (eds): Basics in Dermatological Laser Applications.
Curr Probl Dermatol. Basel, Karger, 2011, vol 42, pp 122–130

Photoaging

Lisa Danielle Grunebaum[a] · Ryan N. Heffelfinger[b]

[a]Miller School of Medicine, University of Miami, Miami, Fla., and [b]Thomas Jefferson University, Philadelphia, Pa., USA

Abstract

Treatments for photoaging are commonly requested by cosmetic patients. Laser resurfacing treatment addresses the myriad aspects of photoaging, including fine rhytides, dyspigmentation, and abnormal texture. Recent developments in laser medicine – such as fractional resurfacing in both ablative and non-ablative wavelengths – have improved the safety and side effect profile of laser resurfacing. Both ablative and non-ablative wavelengths are effective. The use of lasers should be limited to experienced practitioners to minimize the risks of scarring, pigment alterations, and other unwanted effects.

Causes of aging skin are multifactorial, and are typically the sequelae of environmental, behavioral, and general medical assaults on the skin. Photoaging of the skin, a common cosmetic complaint, is characterized by dyschromia, elastosis, fine rhytides, erythema, telangiectasias, textural changes, and keratoses [1]. Photoaging is also referred to as dermatoheliosis. Signs of sun-induced skin damage may be visible as early as age 15 years in cases of heavy childhood sun exposure. Photoaging is not analogous to normal aging [2], yet it exacerbates the undesirable effects of the normal aging process of dermal thinning and atrophic rhytide formation. Photoaging also creates specific molecular and histopathological changes in the skin architecture, including epidermal thickening that induces a rough dry leathery texture. Dermatoheliosis is most characterized histologically by 'elastosis', a disorganized hypertrophy of elastic fibers that leads to wrinkles.

Fully Ablative Skin Resurfacing

Laser resurfacing was initially popularized in the 1980s as an effective treatment for photoaging/facial rejuvenation. Introduced in 1968, the fully ablative CO_2 laser was developed and became the original gold standard for improvements in photoaging. The CO_2 laser emits monochromatic light with a wavelength of 10,600 nm, which is highly absorbed by water and results in ablation of the epidermis and part of the dermis. Upon re-epithelialization and deeper induction of neocollagenesis, obvious improvement in photoaging is achieved. The erbium:YAG (Er:YAG) 2,940 nm was later introduced as another fully ablative option. The Er:YAG laser is similar to CO_2 lasers in that it is ablative with minimal heat dissipation [3]. However, because the Er:YAG emits light at a wavelength of 2,940 nm, which is absorbed by water ten times more than the CO_2 lasers, ablation is limited to a more superficial layer [4]. The thermal damage induced by CO_2 has been the explanation for greater skin tightening versus the Er:YAG or other resurfacing lasers. The more recently introduced Er:YSGG 2,790 nm ablative laser is considered a hybrid between CO_2 and Er:YAG lasers, with a tissue absorption coefficient approximately 5 times that of CO_2 and 1/3 of Er:YAG. This wavelength has also shown promise in the treatment of photoaging [5]. The improvement achieved with traditional resurfacing required significant trade-off in long social downtime and the risk of potential complications, such as infection, dyspigmentation, prolonged erythema, and scarring [6].

Non-Ablative Skin Resurfacing

Non-ablative skin rejuvenation (NASR) was introduced in an attempt to mitigate the undesirable downtime and side effect profile of ablative skin resurfacing. This approach non-specifically heated the dermis, and all approaches required some form of epidermal cooling. The first reported attempt at NASR was with the Q-switched Nd:YAG laser in 1997 [7]. Studies that followed generally described histological changes in the dermis similar to ablative therapy, but limited clinical improvements. Also, when using the fluences needed to yield clinical improvement, the risk for significant epidermal injury increases [8, 9]. Studies with the 1,320 nm Nd:YAG laser followed, with slightly more reported success. The theorized reason being that the 1,320 nm wavelength is more optimal to penetrate to the desired dermal depths of 100–400 um [10]. Nonetheless, the objective measures of collagen deposition and reorganization in the dermis did not correspond to significant improvements in the epidermis [11]. An increased number of treatments was reported to improve clinical outcomes, yet it remains to be determined how many treatments are needed to obtain desired outcomes [12].

The 1,540 nm Er:glass laser was introduced in an attempt to obtain appropriate dermal penetration while minimizing melanin absorption. Initial studies reported a good dermal response, but were overshadowed by significant epidermal heating with resultant erythema, edema, and scarring [13]. Surrounding thermal injury was lessened by using multiple short pulses over a single long pulse, though clinical results remained modest [14].

The pulsed dye laser and intense pulsed light (IPL) have also been purported to hold promise in NASR. However, as with the previously mentioned laser modalities (1,064, 1,320, and 1,540 nm), none have been reported to approach the efficacy of ablative laser resurfacing for skin rejuvenation. Although most have been reported to effect dermal changes, clinical outcomes remain suspect, and they are not the treatment of choice for photodamaged skin.

Fractional Photothermolysis

In 2004, fractional photothermolysis using a non-ablative wavelength of 1,550 nm was introduced as an alternative to traditional fully ablative laser resurfacing [15]. Fractional photothermolysis treats only a definable and usually programmable portion of skin, leaving behind intact bridges of healthy skin that serve as cell donors for re-epithelialization. Fractional photothermolysis in multiple wavelengths, both non-ablative and ablative, has been shown to be efficacious and potentially lower in risk than fully ablative treatments. Fractional resurfacing has become the new gold standard in the treatment of photoaging. Due to the success of fractional resurfacing, the number of laser resurfacing procedures performed increased 154% from 2000 to 2009 [16]. This chapter focuses on the literature supporting the use and efficacy of different available fractional treatments.

Patient Selection and Evaluation for Laser Treatment

As with any cosmetic treatment, successful outcome completely relies on proper patient selection. A full history and physical examination should be taken during the in-office consult. Although some evidence supports treatment of darker skin types [22], the ideal resurfacing patient has a Fitzpatrick skin type of I–III. Specifically, any history of pigment abnormality – such as previous incidence of post-inflammatory hyperpigmentation or melasma – is important to note prior to treatment. Patients with pigment abnormalities should be warned about the increased risk of post-inflammatory pigmentation. Although controversial, the authors warn against almost all laser approaches for the treatment of melasma [23, 24], given that an unpredictable subset of these patients, especially those with an Asian skin type, may have disease exacerbation. Any history of perioral or facial herpes simplex needs to be noted, and many physicians provide these patients with prophylaxis for any treatment – ablative or non-ablative. Valacyclovir 500 mg twice per day starting the night before treatment is recommended. A history of poor healing or propensity for keloids are relative contraindications to treatment. Autoimmune diseases, such as scleroderma, should be considered a relative contraindication as well, due to the propensity for decreased neocollagenesis and poor healing [4]. Any patient who has been treated with oral retinoids within the past 6 months should not be treated with ablative lasers. Importantly, realistic expectations for improvement must be reviewed and documented.

A challenging aspect of the photoaging consult is recommending the most effective laser type treatment for any individual patient with photoaging. Because photoaging encompasses a multitude of common cosmetic problems, each treatment plan should be tailored to address as many of the issues as possible with one laser type. Frequently, combination treatment with more than one wavelength may be necessary for optimal results. Separating and addressing each specific photoaging-related patient complaint may be useful in choosing the optimal treatment plan.

Non-Ablative Fractional Resurfacing

The underlying skin histological changes which follow non-ablative fractional laser treatment leading to rejuvenation have been described from forearm biopsy samples after treatment with 1,550 nm at baseline and at defined times following treatment [17]. Different energy levels were examined. An acute inflammatory response similar to wound healing is indicated by an increase in proinflammatory cytokines and several metalloproteinases. HSP 70 (heat-shock protein response) is also transiently expressed. Ultimately, the expected neocollagenesis and dermal remodeling is indicated by increased procollagen I throughout the dermis. Interestingly, in this study, no difference in molecular changes were noted between low- and high-energy treatments.

Ablative Fractional Resurfacing

In 2007, Hantash et al. [18] described the first use of an ablative CO_2 fractional resurfacing device. They used an array of microscopic treatment zones (MTZ) of customizable density and depth, which extended through the stratum corneum, epidermis, and dermis. Using forearm skin in vivo, they also demonstrated with immunohistochemistry that persistent collagen remodeling occurred for at least 3 months after treatment. With the greater degree of injury associated with fractionated CO_2, there would potentially be a greater and prolonged effect on new and remodeled dermal collagen.

The histology associated with ablative fractional photothermolysis (AFP) has also been further elucidated [18]. Immediately after treatment, there is a zone of epidermal and dermal necrosis, which corresponds to the MTZ. There is a documented loss of epidermal and dermal cell viability within the MTZ. Within 24 h, these 'wounds' are repaired by rapid keratinocyte movement and extrusion moderated by intact keratinocytes at the margins of the MTZ. Melanin and necrotic epidermal debris are also found within the extruded material of each MTZ [18]. In addition, histological and ultrastructural evidence points to a wound repair mechanism and the deposition of new collagen [19]. Other changes include an increase in mucin in the papillary and superficial reticular dermis and an increased undulating rete ridge pattern (loss of rete ridges is seen in aged skin). Both are desirable markers of skin rejuvenation [18]. Reilly et al. [20] suggested similar molecular mechanisms of action for both fractional and fully ablative CO_2 laser resurfacing, finding statistically significant changes in the gene expression of several matrix metalloproteinases with CO_2 ablative fractional resurfacing which had previously been reported for fully ablative CO_2. Other studies have shown that AFP using the CO_2 laser resulted in increased skin thickness and decreased subepidermal low echogenic band thickness (i.e. the portion of the papillary dermis filled with solar elastosis) [21].

Other fractional ablative wavelengths have subsequently become available and are discussed later in the chapter.

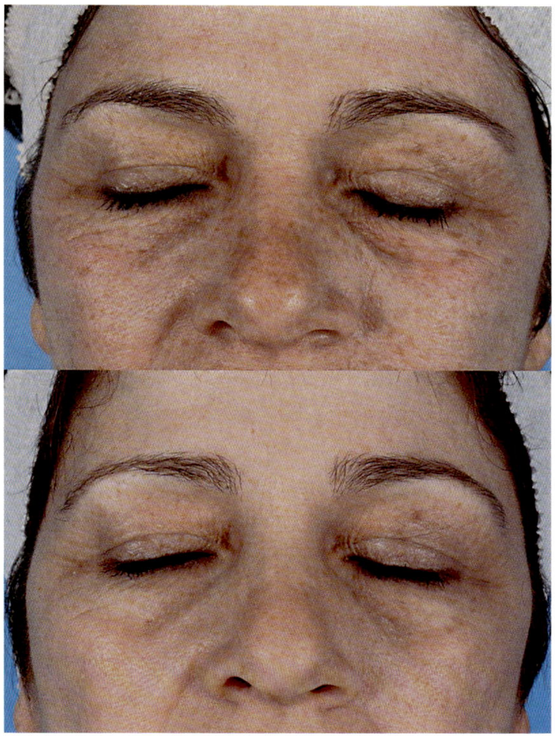

Fig. 1. Long-term outcome in dyschromia treated with a fractional CO_2 laser.

Laser Treatment of Specific Aspects of Photoaging

Telangiectasias

Telangiectasias are a common aspect of photoaging. A vascular laser or light treatment that can address and/or resolve redness should be used to treat this issue. IPL (discussed elsewhere in this volume) and PDL (585 nm) are likely the most effective treatments for telangiectasia in Fitzpatrick skin types I–III. In addition, KTP (532 nm) has shown good results for telangiectasias as well, although there is an increased risk of edema and erythema with this modality [25].

Dyschromia

Dyschromia, such as lentigines, is essentially a problem of abnormal pigmentation. In the case of dyschromia, pigment-specific lasers and a variety of IPL that treat excess melanin are useful for dyschromia. Obviously, the treating practitioner must be well versed in diagnosing causes of dyschromia other than photoaging, so that lesions such as malignant disorders are not mistreated as cosmetic concerns. Although initial reports have indicated success with non-ablative fractional treatment – particularly 1,550 nm for melasma and dyschromia [1, 7] via a 'melanin-shuttle' process – in the authors' experience, laser treatment

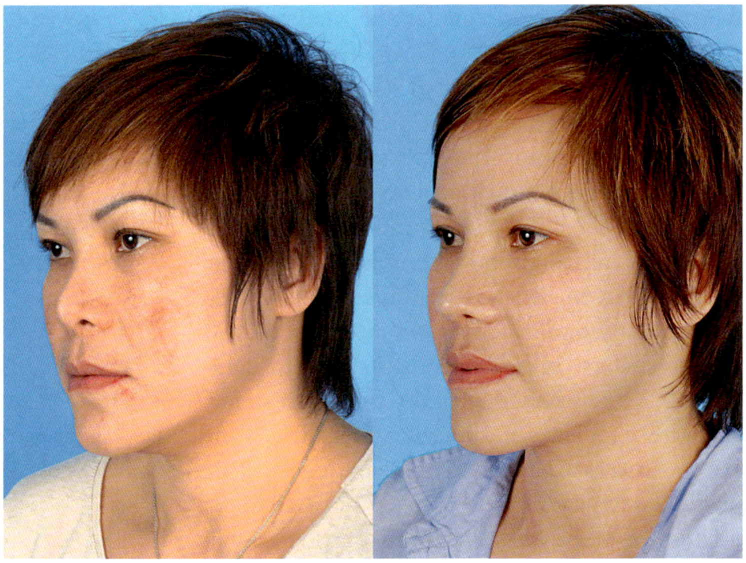

Fig. 2. Successful long-term result with CO_2 fractional resurfacing for treatment of melasma.

of melasma is occasionally unsuccessful and may lead to disease exacerbation. One commercially available 1,550 nm device is now sold as a combination device with a 1,927 nm attachment. The 1,927 nm portion of the treatment is superficial and has shown promise for greater improvement in dyspigmentation [26]. Fully ablative treatments have success with dyschromia, due to epidermal removal (fig. 1, 2). Tierney et al. [21] reported a pilot study using CO_2 AFP that showed pigmentary and textural improvements in photoaged skin. Specifically, they saw mean score improvements in dyschromia, skin laxity, and rhytides of 61, 54, and 52%, respectively. The mean improvement in overall cosmetic outcome was 60%. Results were tallied using pre- and post-laser photographs analyzed by a blinded physician.

Rhytides

Rhytides have successfully been treated with both non-ablative and ablative fractional resurfacing. In general, the clear advantage of non-ablative fractional resurfacing versus ablative fractional resurfacing is reduced social downtime. However, non-ablative resurfacing requires more treatments for an efficacious outcome. In addition, the results of non-ablative treatments specifically for moderate-to-severe wrinkling are generally not deemed as effective as outcomes after ablative treatments [27–29]. The original fractional prototype device described by Manstein et al. [15] produced significant improvements in periorbital rhytides and skin texture. The 1,550 nm non-ablative wavelength has proven useful in improvements for upper lip rhytides [28], hand skin wrinkling [30], and general improvement in rhytides [31]. Although large scale studies are lacking, the authors note that the microsecond Q-switched 1,064 nm non-ablative treatment may be useful for mild rhytides if multiple treatments are used [32].

Others have confirmed that CO_2 ablative fractional resurfacing treatments are safe and very effective for periorbital and perioral rhytidosis. Kotlus [33] analyzed 6-month post-laser photographs and reported a 53% improvement in eyelid skin rhytidosis and a 42% improvement in eyelid skin redundancy. Patients received a dual-depth treatment (1 superficial pass and 1 deep pass) .

The 2,790 nm YSGG laser has also shown promise in the treatment of rhytides and photoaging in general; 8 patients receiving 2 treatments each showed general safety and efficacy in wrinkle reduction, pigmentation, and tone and texture [5]. Lee et al. [34] reported a safe clinical improvement in periorbital wrinkles in ten Korean patients using a 2,790 nm YSGG laser.

Outcomes and histological findings from human and animal skin with both 2,940 and 2,790 nm lasers [35] were investigated establishing both safety and general efficacy with wrinkle reduction with these wavelengths. Furthermore, this same study found greater thermal damage with the 2,790 nm handpiece which supporting its possible use for increased skin tightening.

Although early split-face Er:YAG versus CO_2 fully ablative treatment comparisons are available, few comparison studies have been performed with fractional resurfacing [36]. Weiss et al. [37] reported a split-face trial with half the face treated by AFP (with CO_2) and the contralateral half by non-ablative fractional photothermolysis (erbium). Results included significantly better improvement in periocular rhytides in the AFP group. One study (abstract) presented by Abbasi et al. [38] was performed using 3 different CO_2 devices and one 2,790 nm device. Each study patient had one facial quadrant treated with one of each of the lasers. No differences were ultimately found in 6-month follow-up data in photoaging improvement.

Other investigators have publicized results using 1,440 nm for rhytide improvement. Lloyd [39] showed 'overall improvement' in patients with rhytides, with 42% (5/12) of subjects having 'good' results using the 1,440 nm laser with CAP (combined apex pulse) technology at moderate fluences. Neither increasing nor decreasing laser fluence improved efficacy. A later study supported these findings for both a 1,440 nm and a 1,320/1,440 nm multiplex laser. Both lasers safely and effectively improved photodamage, and histopathology supported these findings [40].

Although touted for their high safety profile, recent reports involving ablative fractionated lasers have proven that relatively serious complications can occur [41–43]. Luckily, the majority of problems associated with laser treatments can be avoided by careful patient and laser parameter selection.

Complications

Campbell and Goldman [44] reviewed 373 patients treated by fractionated CO_2 laser for adverse events. Overall, 14% of patients experienced an adverse event. The most frequent were allergic or contact dermatitis (4.6%), acneiform breakout (3.5%), prolonged erythema (1.1%), and herpes simplex outbreaks (1.1%). This study also indicated that the complication rate increases with the number of sites and amount of skin surface treated.

Metelitsa and Alster [45] published the most comprehensive review of fractional laser treatment complications. They divide complications into mild, moderate, and severe. Mild complications include prolonged erythema, acne and milia, delayed purpura, superficial erosions, contact dermatitis, and the recall phenomenon. Moderate complications include infection, pigmentary alterations, anesthesia toxicity, and eruptive keratoacanthomas. Severe complications include hypertrophic scarring, ectropion formation, and disseminated infection. Most reported complications involving scarring occurred on the neck. The thin skin of the neck, with its relative paucity of pilosebaceous units, makes this area much more sensitive to laser resurfacing [36]. Ectropion has similarly been reported in the thin skin of the periorbital area [35]. In addition, previous facial or neck surgery may be a significant risk factor for laser complications given that surgical undermining may create vascular compromise and reduced healing potential [pers. commun., Dr. Avram]. The laser surgeon should understand the differences in wound healing in these areas and more cautious treatment parameters should be used.

Pre- and Post-Treatment Procedures

Regardless of laser choice, our pre- and post-treatment procedures are standard. Prior to any treatment, patients should be warned against sun exposure. Tanned patients should not be treated, as this increases the risk of a temporary or permanent pigment abnormality. The authors place all patients on a regimen of SPF 55 or greater to the treatment area for 2 weeks prior to treatment. Patients at increased risk of hyperpigmentation, such as those with darker skin types or with a history of hyperpigmentation, may benefit from daily pretreatment with 4% hydroquinone. Although counterintuitive, the use of a retinoid may actually encourage better and more rapid healing when used up to 5–7 days prior to treatment [46].

For non-ablative treatments topical anesthetic 5% lidocaine EMLA (for 30–45 min) or a compounded topical is used for superficial anesthesia. Treatment is delivered with simultaneous forced cooling using the Zimmer system (Zimmer MedizinSystems, Irvine, Calif., USA). We have found the forced cooling to be the single biggest factor in improving patient tolerance and comfort. For ablative treatments, nerve blocks may be useful in addition to topical anesthetic. Some practitioners use IV or more commonly oral sedation. Similar to non-ablative protocol, forced cooling is also applied. We use laser shields to protect the eyes of all patients receiving ablative procedures. If we plan on treating the eyelids, a topical anesthetic is given and metal corneal shields are placed.

Following non-ablative treatment, ice packs may provide immediate pain relief. High SPF sunscreen is applied and used for at least 2 weeks; however, long-term sun avoidance is recommended. After ablative treatments, Aquaphor or a marketed occlusive topical is used for 48–72 h. Vinegar washes/soaks are also recommended. Patients are re-evaluated at 48–72 h. A moisturizing cream with SPF is usually tolerated at 72 h. Again, long-term sun avoidance is recommended. Retinoids and glycolic creams are generally avoided for 4 weeks after treatment. Patients at risk for post-inflammatory hyperpigmentation are considered for hydroquinone prophylactic treatment once reepithelialization is confirmed.

Conclusion

Since the recent advent of fractional laser resurfacing, treatment of photoaging with this modality has become less risky and remains highly effective. Careful patient selection and proper choice of wavelength choice increases the likelihood of good outcome. Knowledge of possible complications and techniques for avoidance is essential for the laser surgeon.

References

1 Tierney E, Hanke W: Review of the literature: treatment of dyspigmentation with fractionated resurfacing. Dermatol Surg 2010;36:1499–1508.

2 Weller R, Hunter J, Savin J, Dahl M: Clinical Dermatology, ed 5. London, Blackwell, 2010, chapter 19 (accessed online at www.mdconsult.com on 12/13/10).

3 Goldberg DJ: Facial Rejuvenation. Berlin, Springer, 2007.

4 Brightman LA, Brauer JA, Anolik R, Weiss E, Karen J, Chapas A, Hale E, Bernstein L, Geronemus RG: Ablative and fractional ablative lasers. Dermatol Clin 2009;27:478–489.

5 Ross EV, Swann M, Soon S, Izadpanah A, Barnette D, Davenport S: Full-face treatments with the 2790-nm Erbium:YSGG laser system. J Drugs Dermatol 2009;8:248–252.

6 Alster T, Lupton J: Prevention and treatment of side effects and complications of cutaneous laser resurfacing. Plast Reconstr Surg 2002;109:308–316.

7 Goldberg DJ, Whitworth J: Laser skin resurfacing with the Q-switched Nd:YAG laser. Dermatol Surg 1997;23:121–127.

8 Goldberg DJ, Silapunt S: Histologic evaluation of a q-switched Nd:YAG laser in the nonablative treatment of wrinkles. Dermatol Surg 2001;27:744–46.

9 Newman J, Lord J, Mcdaniel DH: Nonablative laser therapy in skin types I-IV: clinical evaluation of facial treatment using QS 1064nm Nj:YAG laser combined with carbon suspension lotion. Lasers Surg Med 2000;12:17.
10 Menaker GM, Wrone DA, Williams RM, Moy RL: Treatment of facial rhytids with a nonablative laser: a clinical and histologic study. Dermatol Surg 1999;25:440–444.
11 Trelles MA, Allones I, Luna R: Facial rejuvenation with a nonablative 1320 nm Nd:YAG laser: a preliminary clinical and histologic evaluation. Dermatol Surg 2001;27:111–116.
12 Goldberg DJ: Full-face nonablative dermal remodeling with a 1320 nm Nd:YAG laser. Dermatol Surg 2000;26:915–918.
13 Ross EV, Miller C, Meehan K, McKinlay J, Sajben P, Trafeli JP, et al: One-pass CO_2 versus multiple-pass Er:YAG laser resurfacing in the treatment of rhytids: a comparison side-by-side study of pulsed CO_2 and Er:YAG lasers. Dermatol Surg 2001;27:709–715.
14 Fournier N, Dahan S, Barneon G, Diridollou S, Lagarde JM, Gall Y, Mordon S: Nonablative remodeling: clinical, histologic, ultrasound imaging, and profilometric evaluation of a 1540 nm Er:glass laser. Dermatol Surg 2001;27:799–806.
15 Manstein D, Herron GS, Sink RK, Tanner H, Anderson R: Fractional photothermolysis: a new concept for cutaneous remodeling using microscopic patterns of thermal injury. Lasers Surg Med 2004;34:426–438.
16 American Society of Plastic Surgeons: 2010 report of the 2009 statistics: national clearinghouse of plastic surgery statistics. www.plasticsurgery.org/News-and-Resources/2009-Statistics.html.
17 Orringer JS, Rittie L, Baker D, Voorhees JJ, Fisher G: Molecular mechanisms of nonablative fractionated laser resurfacing. Br J Dermatol 2010;163:757–768.
18 Hantash BM, Bedi VP, Kapadia B, Rahman Z, Jiang K, Tanner H, Chan KF, Zachary CB: In vivo histological evaluation of a novel ablative fractional resurfacing device. Lasers Surg Med 2007;39:96–107.
19 Berlin AL, Hussain M, Phelps R, Goldberg DJ: A prospective study of fractional scanned nonsequential carbon dioxide laser resurfacing: a clinical and histopathologic evaluation. Dermatol Surg 2009;35:222–228.
20 Reilly M, Cohen M, Hokugo A, Keller G: Molecular effects of fractional carbon dioxide laser resurfacing on photodamaged human skin. Arch Facial Plast Surg 2010;12:321–325.
21 Tierney EP, Hanke CW, Petersen J, Bartley T, Eckert JR, McCutchen C: Clinical and echographic analysis of ablative fractionated carbon dioxide laser in the treatment of photodamaged facial skin. Dermatol Surg 2010;36:2009–2021.
22 Shah S, Alster TS: Laser treatment of dark skin: an updated review. Am J Clin Dermatol 2010;11:389–397.
23 Rokshar CK, Fitzpatrick RE: The treatment of melasma with fractional photothermolysis: a pilot study. Dermatol Surg 2005;31:1645–1650.
24 Wind BS, Kroon MW, Meesters AA, Beek JF, van der Veen JP, Nieuweboer-Krobotová L, Bos JD, Wolkerstorfer A: Non-ablative 1550 nm fractional laser therapy versus triple topical therapy for the treatment of melasma: a randomized controlled split-face study. Lasers Surg Med 2010;42:607–612.
25 Weiss R, Weiss M, Beasley K, Munavalli G: Our approach to non-ablative treatment of photoaging. Lasers Surg Med 2005;37:2–8.
26 Weiss E: Nonablative 1927 nm fractional resurfacing effective for facial actinic keratoses. Annu Meet Am Soc Dermatol Surg, Chicago, 2010.
27 Narurkar VA: Skin rejuvenation with microthermal fractional photothermolysis. Dermatol Ther 2007;20(suppl 1):S10–S13.
28 Geronemus R: Fractinal photothermolysis: current and future applications. Lasers Surg Med 2006;38:169–176.
29 Alster TS, Tanzi EL: Laser skin resurfacing: ablative and nonablative; in Robinson J, Sengelman R, Siegel DM, Hanke CM (eds): Surgery of the Skin. Philadelphia, Elseiver, 2005, pp 611–624.
30 Jih MH, Goldberg LH, Kimyai-Asadi A: Fractional photothermolysis for photoaging of hands. Dermatol Surg 2008;34:73–78.
31 Rahman Z, Alam M, Dover JS: Fractional laser treatment for pigmentation and texture improvement. Skin Ther Lett 2006;11:7–11.
32 Schmults CD, Phelps R, Goldberg D: Erythema reduction and histologic evidence of new collagen formation using a 300-microsecond 1064-nm Nd:YAG laser. Arch Dermatol 2004;140:1373–1376.
33 Kotlus BS: Dual-depth fractional carbon dioxide laser resurfacing for periocular rhytidosis. Dermatol Surg 2009;35:1158–1159.
34 Lee JW, Beom JK, Kim MN, Lee CK: Treatment of periorbital wrinkles using a 2790-nm yttrium scandium gallium garnet laser. Dermatol Surg 2010;36:1382–1389.
35 Dierickx CC, Khatri KA, Tannous ZS, Childs JJ, Cohen RH, Erofeev A, Tabatadze D, Yaroslavsky IV, Altshuler GB: Micro-fractional ablative skin resurfacing with two novel erbium laser systems. Lasers Surg Med 2008;40:113–123.
36 Khatri K, Ross V, Grevelink J, Magro C, Anderson R: Comparison of erbium:YAG and carbon dioxide lasers in resurfacing of facial rhytides. Arch Dermatol 1999;135:391–397
37 Weiss R, Weiss M, Beasley K: Prospective split-face trial of a fixed spacing array computed scanned fractional CO_2 laser versus hand scanned 1,550 nm fractional for rhytids. Abstract 28th Annu Conf Am Soc Laser Med Surg, Kissimee, 2008.
38 Abbasi N, LaTowsky B, Arndt K, Dover J, Kaminer M, Rohrer T, Wesley N, MacGrego J: Comparison of four ablative fractional devices in the treatment of photoaging. Abstract 30th Annu Conf Am Soc Laser Med Surg, Phoenix, 2010.
39 Lloyd JR: Effect of fluence on efficacy using the 1440 nm laser with CAP technology for the treatment of rhytids. Lasers Surg Med 2008;40:387–389.
40 Geraghty LN, Beisman B: Clinical evaluation of a single-wavelength fractional laser and a novel multi-wavelength fractional laser in the treatment of photoaged skin. Lasers Surg Med 2009;41:408–416.
41 Ross RB, Spencer J: Scarring and persistent erythema after fractionated ablative CO_2 laser resurfacing. J Drugs Dermatol 2008;77:1072–1073.

42 Fife DJ, Firzpatrick RE, Zachary CB: Complications of fractional CO_2 laser resurfacing: four cases. Laser Surg Med 2009;41:179–184.
43 Avram MM, Tope WD, Yu T, Szachowicz E, Nelson JS: Hypertrophic scarring of the neck following ablative fractional carbon dioxide laser resurfacing. Laser Surg Med 2009;41:185–188.
44 Campbell T, Goldman M: Adverse events of fractionated carbon dioxide laser: review of 373 treatments. Dermatol Surg 2010;36:1645–1650.
45 Metelitsa A, Alster T: Fractionated laser skin resurfacing treatment complications: a review. Dermatol Surg 2010;36:299–306.
46 Alt T: Technical aids for dermabrasion. J Dermatol Surg Oncol 1987;13:638–648.

Lisa Danielle Grunebaum, MD
University of Miami Miller School of Medicine
1450 NW 10th St., Suite 2044
Miami, FL 33136 (USA)
E-Mail lgrunebaum@med.miami.edu

Bogdan Allemann I, Goldberg DJ (eds): Basics in Dermatological Laser Applications.
Curr Probl Dermatol. Basel, Karger, 2011, vol 42, pp 131–139

Scars

Mohamed L. Elsaie[a,b] · Sonal Choudhary[a] · Michael McLeod[a] · Keyvan Nouri[a]

[a]Department of Dermatology and Cutaneous Surgery, University of Miami, Miller School of Medicine, Miami, Fla., USA;
[b]NRC Department of Dermatology and Venereology, Cairo, Egypt

Abstract

Scars are usually classified into atrophic scars, hypertrophic scars, or keloids. A number of lasers, both with known and unknown mechanisms of action, have been investigated for improving the appearance of such scars, with variable outcomes. Currently, multiple lasers are being utilized for the treatment of scars. The 585 nm pulsed dye laser is the gold standard of laser therapy for keloid and hypertrophic scars. The ablative lasers such as the CO_2 and Er:YAG lasers are best suited for atrophic scars. Non-ablative approaches have also become popular and are being widely used to achieve reduced downtimes yet still provide excellent outcomes. This chapter discusses these lasers and their role in scar improvement.

Classification of Scars

Keloids

Keloids are clinically described as purple nodules that extend beyond the original boundaries of injury. They can be associated with abnormal sensations such as pain, pruritus, or burning, and are frequently disfiguring and may result in functional impairment. Keloids appear within weeks or even years of the original cutaneous insult. The most common locations for keloids are the earlobes, anterior chest, shoulders, and upper back. Besides the usual cutaneous injuries that cause scarring, keloids may also result from ear piercing, abrasions, tattooing, vaccinations, and acne scars, amongst other causes. Complete resolution of a keloid scar is rare. They are not known to spontaneously regress and have high recurrence rates with most therapies. There is a familial and ethnic predisposition to the formation of keloids, with darker-skinned individuals having a higher incidence. The fibroblasts taken from keloids are known to produce excessive amounts of collagen type I.

Hypertrophic Scars

In contrast to keloid scars, hypertrophic scars do not extend beyond the original cutaneous injury's border. Additionally, they tend to be raised above the surrounding cutaneous topography and firm in consistency. Their etiology is likely derived from too much collagen production by fibroblasts during the wound healing process, coupled with not enough collagenolysis during wound remodeling [1].

Multiple disturbances in the wound healing process have been implicated in the development of hypertrophic scars and keloids; however, the precise mechanism has not been clearly established [2]. TGF-β may be present in increased amounts or perhaps the immunological cells responsible for collagen and fibronectin deposition may be more sensitive to TGF-β in patients who are predisposed towards forming hypertrophic scars [3, 4].

Atrophic Scars

Atrophic scars occur as a result of a lack of collagen, usually caused by a destructive process, such as the localized inflammatory response that occurs with cystic acne, varicella lesions, or traumatic conditions such as surgery. Typically, atrophic scars resemble depressions in the skin's cutaneous surface [4].

There are generally 4 types of scars that can occur following the active inflammatory response of cystic acne: ice pick, boxcar, saucerized, or atrophic [4]. Acne scarring originates from a deep inflammatory reaction. It involves the destruction or loss of connective tissue, with associated dermal atrophy and fibrosis. During the maturation phase, the scar contracts and pulls the surface layers of skin causing indentation. Even though atrophic scars are the most common type of post-acne scarring, some patients may also present with hypertrophic scars and keloids. Lasers are known to be effective for atrophic scars. Dermal fillers and/or punch excision are usually more effective approaches in the treatment of ice pick and boxcar scars. Previous studies using fractional photothermolysis have demonstrated its effectiveness in the treatment of atrophic and hypopigmented scars, although only preliminary results on its use in hypertrophic scars have been reported [5–7].

Striae Distensae

Stria distensae are commonly referred to as 'stretch marks' in everyday language. These lesions result when there is excessive tension imposed upon the integument, such as during a rapid change in body morphology (pregnancy), or in certain endocrinopathies (Cushing's disease) [8]. The exact pathogenesis leading to stria distensae is largely unknown; however, the formation of stria distensae is marked by two phases. The earlier phase of the lesion is termed stria rubra and manifests clinically as red-to-pink linear streaks. It is believed that these lesions progress to stria alba, which are characterized by linear streaks of hypopigmentation, likely due a fibrotic response stimulated from the inflammatory reaction during the stria rubra phase. Stria distensae primarily constitute a cosmetic issue, and generally do not cause further medical problems.

Burn Scars

Each burn scar has its own unique characteristics. There is marked heterogeneity in the number of functional pilosebaceous units remaining following the insult to the cutaneous surface as well as the level of burn from one burn scar to another [9–11].

Anatomical Location and Prophylactic/Early Laser Treatment

Skin in certain anatomical locations, such as the presternal area, is more likely to heal with a poorer outcome than other anatomical locations. Laser treatment during the wounding process has been shown to reduce the amount of scar tissue formed [8, 11].

Therapeutic Considerations Prior to Laser Surgery for Scars

Patient-specific factors, such as skin type, as well as lesion-specific factors, including age of the scar, color, width, depth, location, orientation, and number of lesions need to be considered to ensure the optimal outcome [12].

Skin color, as measured by the Fitzpatrick classification, influences the outcome of treatment to a great extent. Fair-skinned individuals have an overall better response with fewer side effects (most notably pigmentation changes). When used in dark-skinned individuals (Fitzpatrick skin types IV–VI), the epidermal melanin may compete for laser light absorption, which may result in reduced targeting of the scar and unwanted skin pigmentation changes. The majority of patients suffering from hypertrophic and keloid scars have darker skin types (skin types IV–VI). Although the scars of darker-skinned patients may benefit from laser treatment, there is a high risk of pigmentation alteration as a side effect, which the patients should be warned about. In patients with darker skin, a test spot is useful to anticipate or predict the likelihood of any side effects as well as to determine the optimal laser parameters to be used for that individual scar. Many authors agree that when treating patients with skin types IV–VI, lower energies should be used, which may consequently result in more treatment sessions to achieve the desired result.

The feasibility of scar revision also depends on the patient's expectations. When evaluating a patient with a functionally or cosmetically undesirable facial scar, the laser specialist must be prepared to deal with the technical, physical, and psychological issues in order to determine whether the patient is a candidate for revision.

The age of the scar is an important first consideration. A scar may take 1–2 years to mature. If the injury is recent, the patient should be encouraged to allow adequate time for scar maturation before considering revisions. Erythema and raised borders are hallmark indicators of an immature scar. Newly formed scars that are only red may not need further treatment due to the high likelihood of clearance with time. However, newly formed scars that are red and start becoming hypertrophic may greatly benefit from early laser treatment intervention. Ideally, the color of a scar should closely match that of the surrounding tissue. For optimal aesthetic outcome, the scar should be narrow, flat, and positioned between skin folds or wrinkles.

Patients may present with a history of previously failed attempts to improve the appearance of their scars. Previous treatments, such as cryotherapy, may cause increased fibrosis of the scarring tissue; therefore, adjustment of the laser parameters and the number of sessions may be appropriate when treating these patients.

Treatment of Hypertrophic Scars and Keloids

Just as the precise mechanisms leading to keloid and hypertrophic scars are not known, the exact mechanisms by which lasers improve the appearance of hypertrophic and keloid scare are also not specifically known. The plausible mechanisms of action upon which laser energy acts on hypertrophic scars and keloids to achieve an improvement in clinical appearance include laser-induced tissue hypoxia leading to neocollagenesis from decreased microvascular perfusion, collagen fiber heating with dissociation of disulfide bonds and subsequent collagen realignment, selective photothermolysis of vasculature, and mast cell factors (including histamine, interleukins, and various immunofactors) that could affect collagen and extracellular matrix metabolism [2].

From a historical perspective, laser treatment of hypertrophic and keloid scars began with the argon, CO_2, and Nd:YAG lasers. The results achieved with these lasers were transient, and associated with a high level of recurrence, especially within the first postoperative year [13]. Apfelberg et al. [14, 15] first reported favorable preliminary findings using the argon and CO_2 lasers on keloids; however, follow-up studies demonstrated high postoperative recurrence rates. Hence, the use of these lasers soon fell out of favor.

The principle upon which the Nd:YAG laser was used to treat keloid scars is based on in vivo and in vitro studies demonstrating that collagen deposition could be inhibited [16]. These studies lacked clear outcome definitions as well as long-term follow-up [17].

Subsequent studies with the vascular-specific 585 nm flashlamp-pumped PDL over the past decade have reported that the 585 nm pulsed dye laser (PDL) is effective for improving the vascularity, color, height, texture, and pliability of existing hypertrophic and keloids scars [18–22].

The PDL is noninvasive, painless, requires no anesthesia, and has few risks, including a low risk of scarring. It is known to primarily operate according to Anderson and Parrish's principle of selective photothermolysis of blood vessels. Although the exact mechanism by which the PDL affects scarring is unknown (as previously mentioned), the destruction of the scar's microvasculature may lead to ischemia. It is theorized that this ischemia may modulate collagen or collagenase release or it may also deprive a scar of nutrients, thereby preventing scar hypertrophy [23]. Multiple treatment sessions are suggested to achieve a greater response [24].

Alster and Williams [25] reported a significant improvement in pruritus, erythema, scar height, and skin surface texture in the treated half compared with the untreated control half of the scars on 16 patients with hypertrophic or keloidal median sternotomy scars treated with the 585 nm flashlamp-pulsed PDL [19]. The clinical improvement in response to the PDL treatments persisted for at least 6 months. Goldman and Fitzpatrick evaluated hypertrophic scars treated with 585 nm PDL and found that scars present for less than 1 year responded better and required fewer treatments than older scars.

Manuskiatti et al. [26] also demonstrated laser-derived clinical improvement of keloid and hypertrophic sternotomy scars, including 2 patients with darker skin. They found that lower fluences tended to have a more rapid onset of benefits; multiple treatments allowed for greater clinical improvement [27].

It is important to point out that not all studies have shown a significant response to the 585 nm PDL in the treatment of keloids. Paquet et al. [27] found only minimal effects, if any, on the erythema of keloids with 3 treatments by 585 nm PDL. Clinical improvement of erythema could not be confirmed by objective measurements with remittance spectroscopy.

Fitzpatrick recommended a multifactorial approach to the treatment of hypertrophic and keloid scars [28]. He suggested that treatment of scars with concomitant use of the PDL and simultaneous injection therapy is most effective in the elimination of erythema and improving scar texture [4].

To further support his claim, Manuskiatti and Fitzpatrick [29] carried out a study to compare intralesional triamcinolone, intralesional triamcinolone + 5-fluorouracil, 5-fluorouracil, and PDL on hypertrophic and keloids scars. They demonstrated that the combination intralesional treatments were associated with a more rapid improvement in scar appearance than just the PDL treatment alone [29]. Alster [30] evaluated whether a combination of 585 nm PDL and intralesional corticosteroids produced better hypertrophic scar improvement than 585 nm PDL alone in 22 females with bilateral hypertrophic inframammary scars. Increased pliability alongside decreased pruritus and sclerosis were observed in all scars after PDL treatment with or without the concomitant use of corticosteroids. Adjunctive corticosteroid use only significantly enhanced clinical outcome by decreasing pruritus experienced by patients in the most symptomatic scars.

Nouri et al. [31] found that the 585 nm effectively improved postoperative scars with treatments starting on the day of suture removal. When comparing the effects of 585 and 595 nm on postoperative linear scars beginning on the day of suture removal, both showed improvement in the visual appearance of the scars more than controls with a trend towards greater improvement when using the 585 nm wavelength, as it substantially normalized the height in addition to the vascularity and pliability in a significant number of scars [32].

The comparison of different pulse durations (450 μs vs. 1.5 ms) using the 585 nm PDL on the effect of the treatments of postsurgical linear scars immediately after suture removal demonstrated that both short-pulse (450 μs) and long-pulse (1.5 μs) PDL are safe and effective in improving the quality and cosmetic appearance of surgical scars in skin types I–IV with no significant difference between the 2 pulse durations [33].

Treatment Protocol

The PDL treatment for hypertrophic scars and keloids is an outpatient procedure. The following is a step-by-step explanation of the entire process:

(1) The laser procedure is well-tolerated, and usually does not require anesthesia.

(2) Makeup must be removed from the area to be treated, so it does not interfere with laser light absorption.

(3) Assessment of the scar to be treated must be carried out, noting the size, color, height, and pliability. It is recommended that the assessment always be undertaken by the same person and with the same objective parameters used as a type of scale. For this purpose, you may use the Vancouver Scar Scale or the Patient and Observer Scar Assessment Scale. The presence and degree of any associated symptoms must also be recorded.

(4) Pre-laser pictures of the lesion should be taken prior to the procedure for every session. We also recommend that the pictures always be taken with the same camera, lighting, distance, and angle in order to maintain a standardized image.

(5) The patient and all personnel present should be provided with the appropriate eyewear protection.

(6) Calibrate the laser and set parameters to be used for treating the scar. For treating scars in general, we most commonly use the following parameters at our clinic: wavelength 585 nm, pulse duration 450 μs, spot size 10 mm, and fluences ranging from 3–4 J/mm^2. When using smaller spot sizes, the energy must be increased. Initial treatment visits should be started with low fluences, which can be adjusted according to the response in subsequent sessions. Energy densities are also adjusted according to the patient's as well as the scar's characteristics (i.e. reduced fluence in darker-skinned individuals).

(7) The patient should be informed that the procedure is about to commence. He or she should also be reminded that they will experience a feeling similar to a 'rubber band snapping' during the treatment.

(8) Place the handpiece over one end of the scar and start applying laser pulses over the entire scar surface in a continuous pattern until reaching the opposite end. Overlapping of 10% is generally accepted. The clinical endpoint is redness of the scar area.

(9) Postoperative care instructions should include strict sun avoidance to avoid pigmentation alterations. The treated area may be cleansed normally with soap and water. Trauma to the site should be avoided.

(10) The next treatment can be performed within 4–6 weeks.

Fractional Photothermolysis

Fractional photothermolysis is a relatively novel concept using arrays of microscopic thermal damage to stimulate a therapeutic response. It has the potential to avoid the extended periods of recovery and prolonged erythema associated with ablative lasers, yet achieves more noticeable clinical improvement than non-ablative lasers. The array laser configuration creates very small columns of thermal damage commonly known as microthermal zones in both the epidermis as well as the dermis. These small columns of thermal injury are created amongst undamaged surrounding cutaneous tissue [27, 34, 35].

Some preliminary data on the safety and efficacy of 1,550 nm Er-doped fiber laser treatment (35–50 mJ and 8–10 passes) of hypertrophic scars showed significant improvement of pigmentation and pliability and a less significant improvement in vascularity and height. Clinical response was observed even after the first treatment session, particularly in pigmentation, although further improvement occurred with additional sessions [36].

Treatment of Striae Distensae

Stria distensae can be treated with lasers using 1–2 sessions of the PDL and lower fluences than when treating hypertrophic and keloid scars. Each lesion is treated with no overlap, making certain to irradiate the entire stria distensae with the laser beam. Since lower fluences are used, the side effects of crusting and vesicle formation are not usually observed [8]. The patient should avoid direct exposure to sunlight after the treatment to prevent alterations in pigmentation. Mild soap and water can be used to clean the treated site, as in normal daily cleansing. Fractional photothermolysis has recently been investigated in an attempt to treat striae rubra and reported in 2 cases who underwent successful treatments by Katz et al. [37].

The 585 nm PDL improves the appearance of mature striae [38]; however, Jiménez et al. [39] recommend its use only for striae rubra in patients with Fitzpatrick skin phototypes II–IV. A study comparing the 585 nm PDL and the short-pulse CO_2 laser suggests that both lasers should be avoided or used with extreme caution [40] when treating patients with skin phototypes IV–VI. According to Tay et al. [41], the non-ablative 1,450 nm diode laser is also not to be recommended for the treatment of striae in patients with the same phototypes. Studies have shown an induced repigmentation of striae distensae with the use of the ultraviolet B laser and a light source, and that the use of the 308 nm excimer laser is effective in the repigmentation of mature hypopigmented striae [42, 43].

Treatment of Atrophic Scars

Ablative Scar Remodeling

A number of treatments for atrophic scars involve injecting 'filling' material such as fat, fibrin, or collagen to fill in the atrophic or depressed portion of the scar [8]. Unfortunately, most of these materials only temporarily achieve the desired effect due to reabsorption and need to be reinjected over periods of time. Injections of more permanent materials, which can include silicone, are associated with the formation of granulomas and additionally can migrate from the injected site. Dermabrasion can cause fibrosis and cannot treat more delicate anatomical locations. Lasers have their own side effects and complications; however, they are able to offer unique advantages over the above therapeutic approaches [8].

The CO_2 and Er:YAG lasers have successfully been used for ablative atrophic scar remodeling [8]. Lasers injure the tissue in a very selective manner. For example, by selectively injuring the collagen through denaturation, the CO_2 laser has been shown to increase myofibroblast production of matrix proteins [44]. Remarkably, the increase in matrix production has been shown to continue for at least 18 months following treatment with the laser. In addition to the production of new matrix, the laser also causes shrinkage of existing collagen, thereby 'tightening' the integument [8]. The Er:YAG laser offers an additional advantage over the CO_2 laser in that more of the laser energy is selectively absorbed by water leading to a higher amount of tissue vaporization and less residual thermal injury [45]. From a clinical perspective, the CO_2 laser is known to be more effective, but also has more side effects and a longer recovery time in comparison to the Er:YAG laser. Therefore, the Er:YAG laser should be used for the less severe atrophic scar, while the CO_2 laser should be reserved for widespread areas of severe atrophic scaring [46, 47].

Non-ablative Laser Scar Remodeling

Since ablative lasers are associated with some risk and side effects, a number of non-ablative lasers have been investigated for treating mild atrophic scars located on the face. The primary non-ablative lasers that have been used for this role include the 1,320 nm Nd:YAG and the 1,450 nm diode lasers. These lasers do not injure the epidermis, instead targeting the water-containing tissues in the dermis. Generally, these lasers are used once per month for 3 months to treat mild atrophic scars with approximately 40–50% improvement noted [35, 45]. It should be noted that the degree of clinical improvement is markedly less than with ablative laser treatments [8].

Fractional Treatment

A number of reports on the use of fractional technology in the treatment of atrophic scars have been published [48–50].

Non-Ablative Fractional Laser Treatments

Similar to non-ablative non-fractional lasers, non-ablative fractional lasers do not injure the epidermis, but instead target the dermis, leading to neocollagenesis [51]. Acne scarring has become one of the most widely treated conditions with non-ablative fractional resurfacing (NFR) lasers and the data shows impressive results [52–58].

NFR has been also investigated for the treatment of hypopigmented scars, and was found to be potentially effective for the treatment of hypopigmented scarring on the face, with no adverse effects being observed [6]. On the basis of the clinical trials looking at NFR for treating scars and resurfacing, it has been demonstrated to be a safe and effective modality. On the other hand, the drawback of NFR is the multiple treatment sessions required to achieve desirable effect.

Ablative Fractional Laser Treatments

The conventional ablative laser systems using the CO_2 and Er:YAG lasers have been modified to generate fractional ablative laser systems, in an attempt to achieve the clinical results observed with traditional ablative lasers [58–61].

These lasers resurface the skin by causing true ablation of the epidermis in addition to variable depths of ablative damage to the dermis. This results in an enhanced wound healing response of the skin with dermal fibrosis, and is likely the reason why a more noticeable clinical outcome is observed with these lasers. Additionally, the microscopic zones of ablated tissue leave behind normal skin in adjacent areas. Healing time as well as scarring risk is minimized compared with conventional ablative lasers.

Combination Treatment

Complete or near-complete resolution of many types of scars has been demonstrated using a combination of ablative lasers (pulsed CO_2 and Er:YAG) followed by the fractional (1,550 nm) and non-ablative (1,450 nm) lasers alternatively every 2–3 weeks [4, 61].

Although the use of the ablative fractional laser with high energy resulted in an improvement in patients' acne scars, the combination of ablative fractional laser resurfacing and non-ablative laser resurfacing yielded the best results with fewer complications observed [62].

Conclusions

The availability of several lasers in the present market offers a wide range of options for the dermatologist to choose from when managing different types of scars. The pulsed dye laser has proved to be the laser of choice for surgical scars, while ablative lasers, CO_2, and Er:YAG lasers have remained the gold standard for improving atrophic and acne scars. The new generation of fractional lasers – both ablative and non-ablative – have revolutionized laser treatments for scars by considerably reducing the recovery time, while maintaining the efficacy. By using a combination of laser approaches, the benefits of each type of laser therapy can be exploited in an attempt to achieve the best possible outcome: complete or near-complete resolution of the scar [62, 63].

References

1 Laserase: Scar treatment. 2011. www.laserasescotland.com (accessed on May 13, 2011).

2 Gauglitz GG, Korting HC, Pavicic T, Ruzicka T, Jeschke MG: Hypertrophic scarring and keloids: pathomechanisms and current and emerging treatment strategies. Mol Med 2011;17:113–125.

3 Tredget EE, Nedelec B, Scott PG, Ghahary A: Hypertrophic scars, keloids, and contractures: the cellular and molecular basis for therapy. Surg Clin North Am 1997;77:701–730.

4 Voraphol VV, Nouri K, Patel AR: Treatment of keloids and hypertrophic scars in ethnic skin; in Alam M, Bhatia A, Yoo S, Kundu R (eds): Cosmetic Dermatology for Skin of Color. New York, McGraw-Hill, 2008, pp 136-155.

5 Alster TS, Tanzi EL, Lazarus M: The use of fractional laser photothermolysis for the treatment of atrophic scars. Dermatol Surg 2007;33:295–299.

6 Glaich AS, Rahman Z, Goldberg LH, Friedman PM: Fractional resurfacing for the treatment of hypopigmented scars: a pilot study. Dermatol Surg 2007;33:289–294.

7 Behroozan DS, Goldberg LH, Dai T, et al: Fractional photothermolysis for the treatment of surgical scars: a case report. J Cosmet Laser Ther 2006;8:35–38.

8 Alster TS, Greenberg HL: Laser treatment of scars and striae; in Kauvar A, Hruza G (eds): Principles and Practices in Cutaneous Laser Surgery. London, Informa Healthcare, 2005, pp 619–636.

9 Ito M, Cotsarelis G: Is the hair follicle necessary for normal wound healing? J Invest Dermatol 2008;128:1059–1061.

10 Langton AK, Herrick SE, Headon DJ: An extended epidermal response heals cutaneous wounds in the absence of a hair follicle stem cell contribution. J Invest Dermatol 2008;128:1311–1318.

11 Lee Y: Combination treatment of surgical, post-traumatic and post-herpetic scars with ablative lasers followed by fractional laser and non-ablative laser in Asians. Lasers Surg Med 2009;41:131–140.

12 Alster T, Zaulyanov L: Laser scar revision: a review. Dermatol Surg 2007;33:131–140.

13 Alster TS: Laser treatment of hypertrophic scars, keloids, and striae. Dermatol Clin 1997;15:419–429.

14 Apfelberg DB, Maser MR, Lash H, et al: Preliminary results of argon and carbon dioxide laser excision of keloids. Lasers Surg Med 1984;4:283.

15 Apfelberg DB, Maser MR, White DN, et al: Failure of carbon dioxide laser excision of keloids. Lasers Surg Med 1989;9:382.

16 Castro DJ, Abergel RP, Meeker CA, Dwyer RM, Lesavoy MA, Uitto J: Effects of the Nd:YAG laser on DNA synthesis and collagen production in human skin fibroblast cultures. Ann Plast Surg 1983;11:214–222.

17 Shaffer JJ, Taylor SC, Cook-Bolden F: Keloidal scars: a review with a critical look at therapeutic options. J Am Acad Dermatol 2002;46:2.

18 Bowes LE, Nouri K, Berman B, et al: Treatment of pigmented hypertrophic scars with the 585 nm pulsed dye laser and the 532 nm frequency-doubled Nd:YAG laser in the Q-switched and variable pulse modes: a comparative study. Dermatol Surg 2002;28:714–719.

19 Alster TS, Williams CM: Improvement of keloid sternotomy scars with 585 nm flashlamp-pumped pulsed-dye laser. Lancet 1995;345:1198–1200.

20 Alster TS: Improvement of erythematous and hypertrophic Scars by 585-nm flashlamp-pumped pulse dye laser. Ann Plast Surg 1994;32:186–190.

21 Dierickx C, Goldman MP, Fitzpatrick RE: Laser treatment of erythematous/hypertrophic and pigmented scars in 26 patients. Plast Reconstr Surg 1995;95:84–90.

22 Wittenberg GP, Fabian BG, Bogomilsky JL, Schultz LR, Rudner EJ, Chaffins ML, et al: Prospective single-blind, randomized, controlled study to assess the efficacy of the 585-nm flashlamp-pumped pulsed-dye laser and silicone gel sheeting in hypertrophic scar treatment. Arch Dermatol 1999;135:1049–1055.

23 Zelickson BD, Kist DA: Pulsed dye laser and photoderm treatment stimulates production of type-I collagen and collagenase transcripts in papillary dermis fibroblasts. Lasers Surg Med 2001;27:33S.

24 Alster TS, Lewis AB, Rosenbach A: Laser scar revision: comparison of CO_2 laser vaporization with and without simultaneous pulsed dye laser treatment. Dermatol Surg 1998;24:1299–1302.

25 Reiken SR, Wolfort SF, Berthiaume F, Compton C, Tomkins RG, Yarmush ML: Control of hypertrophic scar growth using selective photothermolysis. Lasers Surg Med 1997;21:7–12.

26 Manuskiatti W, Fitzpatrick RE, Goldman MP: Energy density and numbers of treatment affect response of keloidal and hypertrophic sternotomy scars to the 585-nm flashlamp-pumped pulsed-dye laser. J Am Acad Dermatol 2001;45:557–565.

27 Paquet P, Hermanns JF, Pierard GE: Effect of the 585 nm flashlamp-pumped pulsed dye laser for the treatment of keloids. Dermatol Surg 2001;27:171–174.

28 Fitzpatrick RE: Treatment of inflamed hypertrophic scars using intralesional 5- FU. Dermatol Surg 1999;25:224–231.

29 Manuskiatti W, Fitzpatrick RE: Treatment response of keloidal and hypertrophic sternotomy scars: comparison among intralesional corticosteroid, 5-fluorouracil, and 585-nm flashlamp-pumped pulsed-dye laser treatments. Arch Dermatol 2002;138:1149–1155.

30 Alster TS: Scar revision: comparison study of 585-nm pulsed laser with and without intralesional corticosteroids. Dermatol Surg 2003;29:25–29.

31 Nouri K, Jimenez GP, Harrison-Balestra C, Elgart GW: 585-nm pulsed dye laser in the treatment of surgical scars starting on the suture removal day. Dermatol Surg 2003;29:65–73, discussion 73.

32 Nouri K, Rivas MP, Stevens M, Ballard CJ, Singer L, Ma F, Vejjabhinanta V, Elsaie ML, Elgart GW: Comparison of the effectiveness of the pulsed dye laser 585 nm versus 595 nm in the treatment of new surgical scars. Lasers Med Sci 2009;24:801–810.

33 Nouri K, Elsaie ML, Vejjabhinanta V, Stevens M, Patel SS, Caperton C, Elgart G: Comparison of the effects of short- and long-pulse durations when using a 585-nm pulsed dye laser in the treatment of new surgical scars. Lasers Med Sci 2010;25:121–126.

34 Manstein D, Herron GS, Sink RK, et al: Fractional photothermolysis: a new concept for cutaneous remodeling using microscopic patterns of thermal injury. Laser Surg Med 2004;34:426–438.

35 Vasily DB, Cerino ME, Ziselman EM, Zeina ST: Non-ablative fractional resurfacing of surgical and post-traumatic scars. J Drugs Dermatol 2009;8,:998–1005.
36 Niwa AB, Mello AP, Torezan LA, Osório N: Fractional photothermolysis for the treatment of hypertrophic scars: clinical experience of eight cases. Dermatol Surg 2009;35:773–778.
37 Katz TM, Goldberg LH, Friedman PM: Nonablative fractional photothermolysis for the treatment of striae rubra. Dermatol Surg 2009;35:1430–1433.
38 Nehal KS, Lichtenstein DA, Kamino H, et al: Treatment of mature striae with the pulsed dye laser. J Cutan Laser Ther 1999;1:41–44.
39 Jiménez GP, Flores F, Berman B, Gunja-Smith Z: Treatment of striae rubra and striae alba with the 585-nm pulsed-dye laser. Dermatol Surg 2003;29:362–365.
40 Nouri K, Romagosa R, Chartier T, Bowes L, Spencer JM: Comparison of the 585 nm pulse dye laser and the short pulsed CO_2 laser in the treatment of striae distensae in skin types IV and VI. Dermatol Surg 1999;25:368–370.
41 Tay YK, Kwok C, Tan E: Non-ablative 1,450 nm diode laser treatment of striae distensae. Lasers Surg Med 2006;38:196–199.
42 Goldberg DJ, Marmur ES, Schmults C, et al: Histologic and ultrastructural analysis of ultraviolet B laser and light source treatment of leukoderma in striae distensae. Dermatol Surg 2005;31:385–387.
43 Goldberg DJ, Sarradet D, Hussain M: 308 nm excimer laser treatment of mature hypopigmented striae. Dermatol Surg 2003;29:596–599.
44 Alster TS: Increased smooth muscle actin, factor XIIIa, and vimentin-positive cells in the papillary dermis of carbon dioxide laser-debrided porcine skin (commentary). Dermatol Surg 1998;24:155.
45 Walia S, Alster TS: Prolonged clinical and histological effects from CO_2 laser resurfacing of atrophic scars. Dermatol Surg 1999;25:926–930.
46 Alster TS: Cutaneous resurfacing with Er:YAG lasers. Dermatol Surg 2000;26:73–75.
47 Walia S, Alster TS: Prolonged clinical and histological effects from CO_2 laser resurfacing of atrophic scars. Dermatol Surg 1999;25:926–930.
48 Behroozan DS, Goldberg LH, Dai T, Geronemus RG, Friedman PM: Fractional photothermolysis for the treatment of surgical scars: a case report. J Cosmet Laser Ther 2006;8:35–38.
49 Alster TS, Tanzi EL, Lazarus M: The use of fractional laser photothermolysis for the treatment of atrophic scars. Dermatol Surg 2007;33:295–299.
50 Hu S, Chen MC, Lee MC, Yang LC, Keoprasom N: Fractional resurfacing for the treatment of atrophic facial acne scars in Asian skin. Dermatol Surg 2009;35:826–832.
51 Narurkar VA: Nonablative fractional laser resurfacing. Dermatol Clin 2009;27:473–478.
52 Weiss R, Weiss M, Beasley K: Long-term experience with fixed array 1540 fractional erbium laser for acne scars. Abstract Am Soc Laser Med Surg Conf, Kissimmee, April 2008.
53 Alster TS, Tanzi EL, Lazarus M: The use of fractional laser photothermolysis for the treatment of atrophic scars. Dermatol Surg 2007;33:295–299.
54 Lee HS, Lee JH, Ahn GY, et al: Fractional photothermolysis for the treatment of acne scars: a report of 27 Korean patients. J Dermatol Treat 2008;19:45–49.
55 Glaich AS, Goldberg LH, Friedman RH, Friedman PM: Fractional photothermolysis for the treatment of postinflammatory erythema resulting from acne vulgaris. Dermatol Surg 2007;33:842–846.
56 Chapas AM, Brightman L, Sukal S, et al: Successful treatment of acneiform scarring with CO_2 ablative fractional resurfacing. Lasers Surg Med 2008;40:381–386.
57 Lloyd J, Tanghetti E: Comparison of Affirm 1320/1440nm versus 1320nm for the treatment of acne scars: – a clinical and histologic study. Abstract Am Soc Laser Med Surg Conf, Kissimmee, April 2008.
58 Kim S: Clinical trial of a pinpoint irradiation technique with the CO(2) laser for the treatment of atrophic acne scars. J Cosmet Laser Ther 2008;29:1–4.
59 Ortiz A, Elkeeb L, Truitt A, et al: Evaluation of a novel fractional resurfacing device for the treatment of acne scarring. Abstract Am Soc Laser Med Surg Conf, Kissimmee, April 2008.
60 Chapas AM, Brightman L, Sukal S, et al: Successful treatment of acneiform scarring with CO_2 ablative fractional resurfacing. Lasers Surg Med 2008;40:381–386.
61 Lloyd J, Tanghetti E: Comparison of Affirm 1320/1440nm versus 1320nm for the treatment of acne scars: a clinical and histologic study. Abstract Am Soc Laser Med Surg Conf, Kissimmee, April 2008.
62 Lee Y: Combination treatment of surgical, post-traumatic and post-herpetic scars with ablative lasers followed by fractional laser and non-ablative laser in Asians. Lasers Surg Med 2009;41:131–140.
63 Kim S, Cho KS: Clinical trial of dual treatment with an ablative fractional laser and a nonablative laser for the treatment of acne scars in Asian patients. Dermatol Surg 2009;35:1089–1098.

Sonal Choudhary, MD
Department of Dermatology and Cutaneous Surgery, University of Miami, Miller School of Medicine
1475 NW 12th Ave, Suite 2175
Miami, FL 33136 (USA)
Tel. +1 305 243 3996, E-Mail drsonalchoudhary@gmail.com

Bogdan Allemann I, Goldberg DJ (eds): Basics in Dermatological Laser Applications.
Curr Probl Dermatol. Basel, Karger, 2011, vol 42, pp 140–146

Acne Vulgaris

Severin Läuchli

Department of Dermatology, University Hospital, Zurich, Switzerland

Abstract

Acne vulgaris is one of the most common dermatological diseases, and caused by a combination of pathogenetic factors including follicular hyperkeratosis, seborrhea, colonization with *Propionibacterium acnes* and inflammatory phenomena. Various medical treatments address these causative factors, but they are not always well tolerated. In the last decades, there has been increased interest in laser and light-based treatments. These optical devices mainly target *Propionibacterium acnes* by activating porphyrins produced by the bacterium, whereas some treatments directly lead to the destruction of sebaceous glands. Photodynamic therapy can enhance these effects through the prior application of a photosensitizing agent. Many laser and light-based treatments have shown excellent clinical results in the reduction of inflammatory acne lesions and improved acne severity. Although relapses are frequent, results can be improved by combination therapy with conventional medical treatments. Data of randomized controlled studies or studies with a split-face design are contradictory and difficult to compare. Furthermore, the cost of the treatments is often a limiting factor. In summary, some optical devices are an important additional tool in the treatment of patients with mainly inflammatory acne where conventional treatments are not well tolerated.

Acne vulgaris is one of the most common skin diseases, and responsible for up to 30% of dermatological consultations [1]. Typically, teenagers are most commonly affected by the disease – up to 85% of youths between 12 and 19 years show signs of acne [2]. However, it is often not a self-limiting disease; some studies indicate that up to 50% of patients show signs of acne far beyond the age of 20 years. We must assume that a substantial proportion of adults are affected by the disease, one study indicates that 12% of women at the age of 44 years still suffer from acne [3]. Acne is now regarded as a chronic disease showing many parallels to other chronic diseases, such as atopic dermatitis [4], and involves a substantial psychosocial impact, including social phobia, lowered self-esteem and a higher incidence of depression [5]. Furthermore, moderate and severe acne can lead to permanent scarring, which underlines the necessity for early and effective treatment.

Pathogenesis

Pathogenesis of acne is based on four main factors which can be targeted by different treatments: follicular hyperkeratosis, increased sebum production, follicular colonization with *Propionibacterium acnes* (*P. acnes*) and subsequent inflammatory phenomena [6, 7]. Follicular hyperkeratosis is primarily genetically determined, the desquamation of epithelial cells leads to plugging of follicles and to the formation of microcomedones, the primary acne lesion. Mainly during puberty, imbalances in sex hormones with an increase in androgens lead to greater sebum production. This in turn favors the colonization with *P. acnes*, a Gram-positive bacterium which is part of the normal skin flora thriving in a sebaceous milieu. This increased presence of *P. acnes*, along with cleavage products of fatty acids induced by it, results in various inflammatory phenomena including the induction of interleukin 1a release and the attraction of leucocytes.

Treatment

Many of these pathogenetic factors can be effectively addressed with the available medical therapies for acne [8]. Topical retinoids, the mainstay of acne treatment, reduce follicular plugging and prevent the formation of microcomedones. They can be combined with topical or systemic antibacterial substances, chiefly benzoyl peroxide and antibiotics, to reduce bacterial colonization and inflammatory phenomena. Systemic retinoids permanently reduce hyperseborrhea, counteract follicular plugging and have anti-inflammatory properties. However, the usage of many of these treatments is limited by their side effects and frequent relapses. Antibiotics have the disadvantage of possible induction of resistance of the targeted pathogens, and can therefore only be used for a limited duration. Furthermore, their effects usually do not last beyond discontinuation. Topical retinoids often cause skin irritation. Systemic retinoids are teratogenic and the list of their potential side effects is quite extensive; therefore, their use is limited to severe forms of acne and those that lead to scarring. For these reasons, physical treatment modalities are welcome alternatives for many patients, especially if they do not respond well to conventional medical therapies or if their use is limited by their side effects.

Laser and Light-Based Treatments

There are different physical treatment modalities for acne based on the application of light or lasers that address the various pathogenetic factors of acne [1, 9–11]. Laser and light-based devices target a chromophore which selectively absorbs a certain wavelength emitted by the device. The main targets of these treatments are *P. acnes* and the sebaceous glands. *P. acnes* produces several porphyrins, amongst them protoporphyrin ix and coproporphyrin III [12]. When irradiated with visible light of various wavelengths, the porphyrin compound gets excited, causing the formation of singlet oxygen and reactive free radicals. These damage lipids in the cell wall of *P. acnes*, which ultimately leads to the destruction of the bacterium, consequently reducing the bacterial load [1]. A variety of low-level-light-based devices and higher energy lasers and intense pulsed light (IPL) devices can elicit this inhibitory effect on *P. acnes*. When adding an external photosensitizer – for example, methyl aminolevulinate (MAL) or 5-aminolaevulinic acid (ALA) – corresponding to the basic principle of photodynamic therapy (PDT), the destruction of *P. acnes* can be increased. In addition, when using an external photosensitizer, which is absorbed into the pilosebaceous unit, the free oxygen radicals generated may also damage the sebaceous gland itself resulting in decreased sebum production [13].

Sebaceous glands can also be directly targeted with laser devices emitting longer wavelengths, leading to thermal damage of the glands. Lasers with wavelengths of 1,320–1,540 nm as well as radiofrequency-based devices can lead to heating of the sebaceous glands, resulting in decreased glandular size and sebum production, whilst preserving the epidermis. It is speculated that the destruction of sebaceous glands may be achieved by heating water in the surrounding tissue rather than direct heating of the sebaceous gland [14]. Treatments targeting the sebaceous glands may lead to a longer duration of the effects on acne.

Table 1. Overview of optical devices used for the treatment of acne and their targets

Device	Wavelength nm	Target chromophore and mechanism of action
Blue light	405–420	protoporphyrin IX (antibacterial?)
Red light	660	protoporphyrin IX (anti-inflammatory?)
KTP laser	532	protoporphyrin IX (vasculature of inflammatory lesions and sebaceous glands)
Pulsed dye laser	585/595	protoporphyrin IX (vasculature of inflammatory lesions and sebaceous glands)
Diode laser	1,450	sebaceous glands (thermal damage)
Er:glass laser	1,540	sebaceous glands (thermal damage)
IPL	various cutoffs (530–1,200)	protoporphyrin IX (vasculature of inflammatory lesions and sebaceous glands)
ALA/MAL-PDT	various (red light, PDL, IPL)	protoporphyrin IX; MAL more lipophilic (sebaceous glands)
ICG-PDT	810	sebaceous glands

PDL = Pulsed dye laser; ICG = indocyanine green.

One aspect that most of these physical treatment modalities have in common is that there are good hypotheses for their mechanism of action and their clinical effectiveness has been demonstrated in many case series. However, there is very limited evidence of their effectiveness in randomized controlled studies [9–11]. Most studies regarding physical treatment modalities of acne are underpowered, showing results of small patient samples. Many are not controlled studies or have inadequate control groups and there are only a few split-face trials which could be considered the gold standard in this area. In order to get approval by health authorities, medical treatments must undergo extensive scrutiny of safety and effectiveness and present results in large multicenter studies. Physical devices on the other hand only have to show their safety and a proof of principle in order to be approved by the FDA and most other national health authorities.

Treatment Devices

An overview of the optical devices used for the treatment of acne and their targets are presented in table 1.

Blue and Red Light

Both blue (405–420 nm) and red (633–660 nm) light are effective in activating the porphyrins produced by *P. acnes*, and thus reducing the bacterial load and sebum production [13]. Peak absorption of porphyrins occurs around 415 nm; therefore, blue light activates porphyrins most effectively. However, red light penetrates the skin more deeply and may have direct anti-inflammatory effects by influencing mediators. The treatment is generally delivered over 10–20 min twice a week for 4–12 weeks. Side effects of this treatment are minimal. There are several small studies supporting the effectiveness of blue and red light, reporting reductions in lesion counts between 60 and 80%. However, only one study used an active standard treatment as a comparison and only a few trials used randomization or a split-face design [15–23]. There is some indication that the effect of mixed blue-red light is greater than the effect of blue light alone (both treatments applied daily over 3 months) or a standard treatment (benzoyl peroxide) [24, 25]. It was shown that the effect is greatest on inflammatory lesions. Altogether, narrow-band light therapy seems to have an effect on inflammatory lesions of mild-to-moderate acne that is at least comparable to standard anti-inflammatory treatments, but the evidence for this statement is so far insufficient and there is no effect on hyperkeratinization and microcomedo formation, thus leading to high relapse rates after discontinuation of treatment.

Infrared Lasers: Diode Laser (1,450 nm), Er:Glass Laser (1,540 nm)
The diode laser targets water in the upper dermis, thus leading to collagen remodeling, and is mainly used for facial rejuvenation and treatment of rhytides. It also causes thermal damage to sebaceous glands and has been shown to improve inflammatory acne lesions as well as acne scars, which is why it was approved for treatment of acne. Some studies show a marked reduction in inflammatory and non-inflammatory acne lesions after 3 treatments, lasting several months [26–30]. Side effects were minimal and include pain during the treatments, erythema and hyperpigmentation. Similar results could be achieved with a 1,540 nm Er:glass laser [31].

Pulsed Dye Laser (585 nm)
The pulsed dye laser (PDL) targets oxyhemoglobin and is traditionally used for treating cutaneous vascular lesions and dermatologic conditions where the vasculature plays an important role, such as warts and scars. There are reports about successful treatments of inflammatory acne lesions [32], but these results could not be confirmed in a large study with a split-face design [33]. Furthermore, PDL can result in significant side effects, mainly discomfort during treatments and hyperpigmentation, especially in dark-skinned individuals.

KTP Laser (532 nm)
The KTP laser also mainly targets vascular lesions along with stimulating collagen production. Its effectiveness in treating inflammatory acne lesions has been shown in a large study involving 175 subjects, where a 30–95% reduction in lesion counts could be achieved after 6 treatments, especially in combination with topical therapies [34, 35].

Intense Pulsed Light
IPL devices use flashlamps with bandpass filters to emit certain wavelengths of incoherent pulsed light targeting a variety of structures. They can have positive effects on acne both through the photodynamic effects on the porphyrins produced by *P. acnes* as well as through their effects on the vasculature of inflammatory lesions and the possible destruction of sebocytes. Several non-controlled trials have shown significant reductions in inflammatory and non-inflammatory acne lesions, but study protocols and results vary greatly. Split-face studies have shown improvements in skin tone and irregular pigmentation, but not acne lesions [36], when compared to topical benzoyl peroxide, and less persistent results compared to PDL [37].

Photodynamic Therapy
Adding a photosensitizer to light therapy can greatly enhance its efficacy [38]. There are three substances which have been used successfully in the treatment of acne: ALA, its methylated ester

MAL and indocyanine green. ALA is a prodrug of protoporphyrin IX and can be activated by blue (415 nm), red (635 nm) or green (800 nm) light [13, 39, 40]. MAL is more lipophilic and can therefore better penetrate sebaceous glands and may be more selective for acne lesions [41, 42]. For both substances, accumulation into the glands occurs within 3–4 h. Indocyanine green effectively accumulates in sebaceous glands if applied in an emulsion for up to 24 h, and can be activated with light sources around 800 nm [43, 44]. There are protocols with shorter application times [45–48]. Activation of these substances leads to production of singlet oxygen radicals which can disrupt cellular membranes, resulting in reducing the load of *P. acnes* and destruction or shrinkage of sebaceous glands, with little damage to the surrounding tissue. Light sources that can be used for PDT in acne include continuous-wave light sources, IPL and some lasers. Continuous-wave light sources include lamps emitting blue (415 nm), red (635 nm) or green (800 nm) light. Some protocols use pulsed light sources with bandwidths such as 420–950, 550–700 and 580–1,000 nm or lasers such as the pulsed dye laser (595 nm). Side effects of all modalities of PDT are common and include urticarial erythema and significant pain for some patients. Study results with the different modalities are hard to compare as they have used widely varying protocols. Several studies report a significant decrease in acne severity compared to placebo-treated sites and a lasting effect for more than 20 weeks after multiple treatments [13, 41, 44, 46–53].

Combination with Medical Therapy

Whereas light and laser-based therapies have the potential to effectively reduce bacterial load and sebum production, they do not address follicular hyperkeratinization, which is one of the main pathogenetic factors in acne. This may partially explain the high relapse rate seen after many physical treatments. If tolerance allows, it would therefore be ideal to combine laser and light-based therapies with a medical therapy regulating keratinocyte differentiation, such as a topical retinoid. This was adequately demonstrated in a large study where adding concomitant standard topical medical therapy (such as topical retinoids or benzoyl peroxide) to KTP laser treatments increased the improvement to 80–95%, a result which could be maintained for over 4 months in >50% of the subjects [35].

Conclusions

Many laser- and light-based therapies have shown promising results in the treatment of acne. They have the potential to address several pathogenetic factors involved in the development of acne, namely a reduction in the bacterial load of *P. acnes* and of the number and size of sebaceous glands. Their effect is comparable to oral antibiotics and patient satisfaction is usually high [1]. However, as none of these treatments address all pathogenetic factors, they often only improve inflammatory lesions and, in many cases, the positive results cannot be maintained over an extended time period.

In accordance with the Global Alliance to improve outcomes in Acne Group, treatments with narrow-band light sources (blue and red light, 415 and 635 nm) can be recommended for some patients, as they are easy to administer and have few side effects, but they have to be applied over extended time periods. However, the most convincing and long-lasting results have been shown for treatments with infrared lasers such as the diode laser (1,450 nm) that lead to a reduction in sebaceous glands, photodynamic treatments that combine laser or light therapy with the application of a topical photosensitizer, and treatments that are combined with a topical medical treatment. Hence, these treatments can be most recommended [4]. Cost is often a limiting factor for many of these treatments, especially as they are not covered by most health insurance schemes. They have an important place in the treatment of acne especially in cases where inflammatory lesions are predominant and medical treatments cannot be tolerated due to their side effects or where the patient specifically requests such treatment.

References

1 Elman M, Lebzelter J: Light therapy in the treatment of acne vulgaris. Dermatol Surg 2004;30(part 1):139–146.
2 Swinyer LJ, Swinyer TA, Britt MR: Topical agents alone in acne: a blind assessment study. JAMA 1980;243:1640–1643.
3 Goulden V, Stables GI, Cunliffe WJ: Prevalence of facial acne in adults. J Am Acad Dermatol 1999;41:577–580.
4 Thiboutot D, Gollnick H, Bettoli V, et al: New insights into the management of acne: an update from the Global Alliance to Improve Outcomes in Acne group. J Am Acad Dermatol 2009;60:S1–S50.
5 Koo JY, Smith LL: Psychologic aspects of acne. Pediatr Dermatol 1991;8:185–188.
6 Gollnick H, Cunliffe W, Berson D, Dreno B, Finlay A, Leyden JJ, Shalita AR, Thiboutot D, Global Alliance to Improve Outcomes in Acne: Management of acne: a report from a Global Alliance to Improve Outcomes in Acne. J Am Acad Dermatol 2003;49(suppl):S1–S37.
7 Leyden JJ: Therapy for acne vulgaris. N Engl J Med 1997;336:1156–1162.
8 Hamilton F, Car J, Layton A: Acne vulgaris. BMJ 2009;338:a2738.
9 Hamilton FL, Car J, Lyons C, Car M, Layton A, Majeed A: Laser and other light therapies for the treatment of acne vulgaris: systematic review. Br J Dermatol 2009;160:1273–1285.
10 Nouri K, Ballard CJ: Laser therapy for acne. Clin Dermatol 2006;24:26–32.
11 Taub AF: Procedural treatments for acne vulgaris. Dermatol Surg 2007;33:1005–1126.
12 Ashkenazi H, Malik Z, Harth Y, Nitzan Y: Eradication of *Propionibacterium acnes* by its endogenic porphyrins after illumination with high intensity blue light. FEMS Immunol Med Microbiol 2003;35:17–24.
13 Hongcharu W, Taylor CR, Chang Y, Aghassi D, Suthamjariya K, Anderson RR: Topical ALA-photodynamic therapy for the treatment of acne vulgaris. J Invest Dermatol 2000;115:183–192.
14 Ross EV: Acne, lasers, and light. Adv Dermatol 2005;21:1–32.
15 Elman M, Lask G: The role of pulsed light and heat energy (LHE) in acne clearance. J Cosmet Laser Ther 2004;6:91–95.
16 Elman M, Slatkine M, Harth Y: The effective treatment of acne vulgaris by a high-intensity, narrow band 405–420 nm light source. J Cosmet Laser Ther 2003;5:111–117.
17 Gold MH, Rao J, Goldman MP, Bridges TM, Bradshaw VL, Boring MM, Guider AN: A multicenter clinical evaluation of the treatment of mild to moderate inflammatory acne vulgaris of the face with visible blue light in comparison to topical 1% clindamycin antibiotic solution. J Drugs Dermatol 2005;4:64–70.
18 Kawada A, Aragane Y, Kameyama H, Sangen Y, Tezuka T: Acne phototherapy with a high-intensity, enhanced, narrow-band, blue light source: an open study and in vitro investigation. J Dermatol Sci 2002;30:129–135.
19 Morton CA, Scholefield RD, Whitehurst C, Birch J: An open study to determine the efficacy of blue light in the treatment of mild to moderate acne. J Dermatolog Treat 2005;16:219–223.
20 Omi T, Bjerring P, Sato S, Kawana S, Hankins RW, Honda M: 420 nm intense continuous light therapy for acne. J Cosmet Laser Ther 2004;6:156–162.
21 Sigurdsson V, Knulst AC, van Weelden H: Phototherapy of acne vulgaris with visible light. Dermatology 1997;194:256–260.
22 Tremblay JF, Sire DJ, Lowe NJ, Moy RL: Light-emitting diode 415 nm in the treatment of inflammatory acne: an open-label, multicentric, pilot investigation. J Cosmet Laser Ther 2006;8:31–33.
23 Tzung TY, Wu KH, Huang ML: Blue light phototherapy in the treatment of acne. Photodermatol Photoimmunol Photomed 2004;20:266–269.
24 Goldberg DJ, Russell BA: Combination blue (415 nm) and red (633 nm) LED phototherapy in the treatment of mild to severe acne vulgaris. J Cosmet Laser Ther 2006;8:71–75.
25 Papageorgiou P, Katsambas A, Chu A: Phototherapy with blue (415 nm) and red (660 nm) light in the treatment of acne vulgaris. Br J Dermatol 2000;142:973–978.
26 Alam M, Peterson SR, Silapunt S, Chopra K, Friedman PM, Goldberg LH: Comparison of 1450-nm diode laser and 595-nm pulsed-dye laser for treatment of facial acne: a left-right randomized trial of efficacy and adverse effects. Lasers Surg Med 2003;(suppl 15):97.
27 Friedman PM, Jih MH, Kimyai-Asadi A, Goldberg LH: Treatment of inflammatory facial acne vulgaris with the 1450-nm diode laser: a pilot study. Dermatol Surg 2004;30(part 1):147–151.
28 Jih MH, Friedman PM, Goldberg LH, Robles M, Glaich AS, Kimyai-Asadi A: The 1450-nm diode laser for facial inflammatory acne vulgaris: dose-response and 12-month follow-up study. J Am Acad Dermatol 2006;55:80–87.
29 Paithankar DY, Ross EV, Saleh BA, Blair MA, Graham BS: Acne treatment with a 1,450 nm wavelength laser and cryogen spray cooling. Lasers Surg Med 2002;31:106–114.
30 Wang SQ, Counters JT, Flor ME, Zelickson BD: Treatment of inflammatory facial acne with the 1,450 nm diode laser alone versus microdermabrasion plus the 1,450 nm laser: a randomized, split-face trial. Dermatol Surg 2006;32249–255, discussion 255.
31 Kassir M: ErGlass (1.54um) laser for the treatment of facial acne vulgaris. Lasers Surg Med 2004;34:S65.
32 Seaton ED, Charakida A, Mouser PE, Grace I, Clement RM, Chu AC: Pulsed-dye laser treatment for inflammatory acne vulgaris: randomised controlled trial. Lancet 2003;362:1347–1352.

33 Orringer JS, Kang S, Hamilton T, Schumacher W, Cho S, Hammerberg C, Fisher GJ, Karimipour DJ, Johnson TM, Voorhees JJ: Treatment of acne vulgaris with a pulsed dye laser: a randomized controlled trial. JAMA 2004;291:2834–2839.
34 Bowes LE, Rox Andersen R: Effects of 532 nm KTP laser exposure on acne and sebaceous glands. Lasers Med Sci 2003;18(suppl 1):S6–S7.
35 Nataloni R: Laser treatment comparable to oral antibiotics: 532nm laser addresses multiple acne pathogens. Dermatol Times 2003;S32–S36.
36 Chang SE, Ahn SJ, Rhee DY, Choi JH, Moon KC, Suh HS, Soyun-Cho: Treatment of facial acne papules and pustules in Korean patients using an intense pulsed light device equipped with a 530- to 750-nm filter. Dermatol Surg 2007;33:676–679.
37 Choi YS, Suh HS, Yoon MY, Min SU, Lee DH, Suh DH: Intense pulsed light vs. pulsed-dye laser in the treatment of facial acne: a randomized split-face trial. J Eur Acad Dermatol Venereol 2010;24:773–780.
38 Taylor MN, Gonzalez ML: The practicalities of photodynamic therapy in acne vulgaris. Br J Dermatol 2009;160:1140–1148.
39 Horfelt C, Stenquist B, Larkö O, Faergemann J, Wennberg AM: Photodynamic therapy for acne vulgaris: a pilot study of the dose-response and mechanism of action. Acta Derm Venereol 2007;87:325–329.
40 Peng Q, Moan J, Warloe T, Nesland JM, Rimington C: Distribution and photosensitizing efficiency of porphyrins induced by application of exogenous 5-aminolevulinic acid in mice bearing mammary carcinoma. Int J Cancer 1992;52:433–443.
41 Hörfelt C, Funk J, Frohm-Nilsson M, Wiegleb Edström D, Wennberg AM: Topical methyl aminolaevulinate photodynamic therapy for treatment of facial acne vulgaris: results of a randomized, controlled study. Br J Dermatol 2006;155:608–613.
42 Yung A, Stables GI, Fernandez C, Williams J, Bojar RA, Goulden V: Microbiological effect of photodynamic therapy (PDT) in healthy volunteers: a comparative study using methyl aminolaevulinate and hexyl aminolaevulinate cream. Clin Exp Dermatol 2007;32:716–721.
43 Lloyd JR, Mirkov M: Selective photothermolysis of the sebaceous glands for acne treatment. Lasers Surg Med 2002;31:115–20.
44 Tuchin VV, Genina EA, Bashkatov AN, Simonenko GV, Odoevskaya OD, Altshuler GB: A pilot study of ICG laser therapy of acne vulgaris: photodynamic and photothermolysis treatment. Lasers Surg Med 2003;33:296–310.
45 Goldman MP, Boyce SM: A single-center study of aminolevulinic acid and 417 NM photodynamic therapy in the treatment of moderate to severe acne vulgaris. J Drugs Dermatol 2003;2:393–396.
46 Rojanamatin J, Choawawanich P: Treatment of inflammatory facial acne vulgaris with intense pulsed light and short contact of topical 5-aminolevulinic acid: a pilot study. Dermatol Surg 2006;32:991–996, discussion 996–997.
47 Wiegell SR, Wulf HC: Photodynamic therapy of acne vulgaris using 5-aminolevulinic acid versus methyl aminolevulinate. J Am Acad Dermatol 2006;54:647–651.
48 Yeung CK, Shek SY, Bjerring P, Yu CS, Kono T, Chan HH: A comparative study of intense pulsed light alone and its combination with photodynamic therapy for the treatment of facial acne in Asian skin. Lasers Surg Med 2007;39:1–6.
49 Genina EA, Bashkatov AN, Simonenko GV, Odoevskaya OD, Tuchin VV, Altshuler GB: Low-intensity indocyanine-green laser phototherapy of acne vulgaris: pilot study. J Biomed Opt 2004;9:828–834.
50 Haedersdal M, Togsverd-Bo K, Wiegell SR, Wulf HC: Long-pulsed dye laser versus long-pulsed dye laser-assisted photodynamic therapy for acne vulgaris: a randomized controlled trial. J Am Acad Dermatol 2008;58:387–394.
51 Pollock B, Turner D, Stringer MR, Bojar RA, Goulden V, Stables GI, Cunliffe WJ: Topical aminolaevulinic acid-photodynamic therapy for the treatment of acne vulgaris: a study of clinical efficacy and mechanism of action. Br J Dermatol 2004;151:616–622.
52 Taub AF: A comparison of intense pulsed light, combination radiofrequency and intense pulsed light, and blue light in photodynamic therapy for acne vulgaris. J Drugs Dermatol 2007;6:1010–1016.
53 Wiegell SR, Wulf HC: Photodynamic therapy of acne vulgaris using methyl aminolaevulinate: a blinded, randomized, controlled trial. Br J Dermatol 2006;154:969–976.

Severin Läuchli, MD
Department of Dermatology, University Hospital Zurich
Gloriastrasse 31
CH–8091 Zurich (Switzerland)
Tel. +41 44 255 2811, E-Mail severin.laeuchli@usz.ch

Bogdan Allemann I, Goldberg DJ (eds): Basics in Dermatological Laser Applications.
Curr Probl Dermatol. Basel, Karger, 2011, vol 42, pp 147–152

Skin Tightening

Heather Woolery-Lloyd[a] · Jenna N. Kammer[b,c]

[a]Department of Dermatology and Cutaneous Surgery and [b]Cosmetic Medicine and Research Institute, University of Miami, Miami Beach, Fla., and [c]Drexel University College of Medicine, Philadelphia, Pa., USA

Abstract

Skin tightening describes the treatment of skin laxity via radiofrequency (RF), ultrasound, or light-based devices. Skin laxity on the face is manifested by progressive loss of skin elasticity, loosening of the connective tissue framework, and deepening of skin folds. This results in prominence of submandibular and submental tissues. Genetic factors (chronological aging) and extrinsic factors (ultraviolet radiation) both contribute to skin laxity. There are many RF, ultrasound, and light-based devices directed at treating skin laxity. All of these devices target and heat the dermis to induce collagen contraction. Heating of the dermis causes collagen denaturation and immediate collagen contraction in addition to long-term collagen remodeling. Via RF, light, or ultrasound, these skin tightening devices deliver heat to the dermis to create new collagen and induce skin tightening. This chapter will provide an overview of the various skin tightening devices.

Skin laxity on the face is manifested by progressive loss of skin elasticity, loosening of the connective tissue framework, and deepening of skin folds. This results in prominence of submandibular and submental tissues [1]. Skin laxity on the body occurs with normal aging on the upper arms, thighs, and legs. Skin laxity on the body is also commonly observed after significant rapid weight loss. In addition, abdominal skin laxity can occur after pregnancy. Inherent genetic factors (chronological aging) and extrinsic factors (ultraviolet radiation) contribute to skin laxity. Unlike photoaging, skin laxity is a cosmetic complaint observed in all skin types.

There are many radiofrequency (RF)-, light-, and ultrasound-based devices directed at treating skin laxity. All of these devices are based on the notion that collagen contracts when heat is applied. Heating of the dermis causes collagen denaturation and immediate collagen contraction, in addition to long-term collagen remodeling [2, 3]. Via RF, light, or ultrasound, these skin tightening devices deliver heat to the dermis to create new collagen and induce skin tightening. Some of the available skin tightening devices on the market are listed in table 1.

Patient Selection

Patient selection is very important in treating skin laxity. Older patients with excessive skin laxity are less likely to achieve the results that a younger patient with mild-to-moderate skin laxity will achieve. Also, patients with significant fat underlying the loose skin are less likely to get

Table 1. Skin tightening devices

Technology	Device
Infrared light (1,100–1,800)	Cutera (Titan)
Infrared light (850–1,350 nm)	Palomar (Lux IR and Lux Deep IR)
Infrared light (800–1,400)	Sciton (SkinTyte)
RF: monopolar	Thermage (Theramcool)
RF: bipolar with vacuum	Lumenis (Aluma)
RF: bipolar with optical energy	Syneron (ReFirme)
RF: unipolar and bipolar	Alma (Accent)
Ultrasound	Ulthera P (Ulthera)

optimal results. Predictability of patient response to treatment has been a concern with skin tightening procedures. It is clear that the level of patient response can vary widely and be difficult to predict. Therefore, appropriate selection of patients and managing realistic patient expectations are fundamental considerations to ensure satisfaction with results. It is important to note that it can take up to 6 months to achieve the best results and this aspect of skin tightening treatments should also be discussed with patients [4]. As with all cosmetic procedures, management of patient expectations is essential. Patients must realize that results can be noticeable, but often are subtle. Since the changes are gradual and the patient sees themselves every day, before and after pictures can be helpful.

Technologies

All skin tightening devices work via a similar mechanism. They all strive to deliver enough heat through ultrasound, RF, or light-based energy deep into the dermis, causing an initial contraction of the collagen fibers followed by dermal remodeling and in the end neocollagen synthesis. Several light-based, ultrasound, and RF devices have been introduced.

Laser Devices: CO_2 and Er:YAG Lasers

Many studies have shown that treatments with ablative CO_2 and Er-YAG lasers selectively heat dermal collagen, resulting in a dermal neocollagen formation and tightening of the facial skin [5–7]. However, the major disadvantage of ablative treatment methods is epidermal damage, resulting in a recuperation period of many weeks. Additionally, there are potential risks such as hyper- or hypopigmentation, infections, and scarring [5, 8, 9]. For this reason, non-ablative skin rejuvenation has become popular over the recent years.

Infrared Tightening

Infrared skin tightening utilizes infrared light to heat the dermis. One of the first infrared tightening devices introduced was the Titan (Cutera, Inc., Brisbane, Calif., USA). It utilizes a long pulse of infrared light with a spectrum of 1,100–1,800 nm, with the greatest intensity in the range of 1,400–1,500 nm. The penetration depth is 1–2 mm, which is best for targeting the reticular dermis [10]. Light is emitted in multisecond cycles in order to sufficiently heat the dermis while providing appropriate cooling of the epidermis. Like other skin tightening devices, it is designed to thermally induce immediate collagen contraction, followed by the induction of collagen remodeling and the synthesis of new collagen. Water is the target chromophore which allows for uniform heating of the targeted area. The procedure causes mild-to-moderate discomfort and mild temporary erythema. The epidermis is protected through contact cooling with a sapphire tip. The procedure is associated with minimal downtime and is safe to use on all skin types. Two initial passes are performed over the entire designated treatment area,

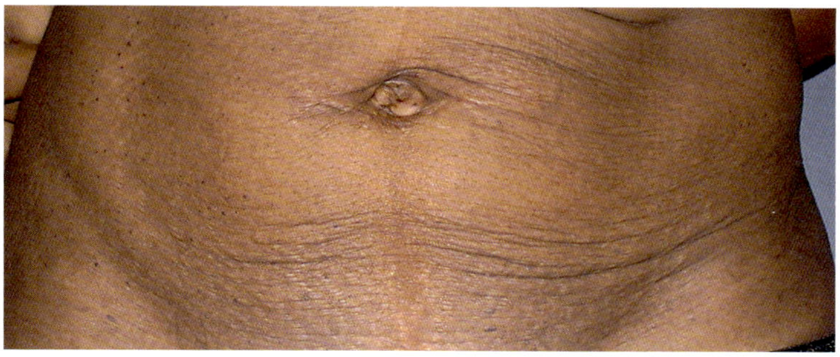

Fig. 1. Baseline.

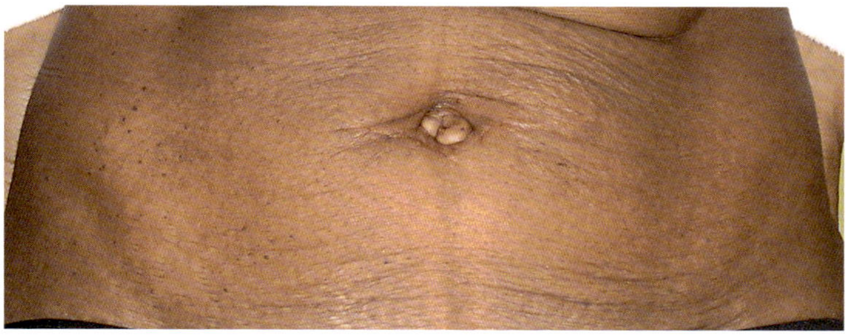

Fig. 2. After 3 treatments with an infrared tightening device.

with additional passes over areas of great concern. A series of 3–4 treatments, 4–6 weeks apart, are recommended for the best results. Although an immediate tightening effect is achieved, the full effect is observed 6 months after the last treatment [11] (fig. 1, 2).

In an open-label trial in 21 Asian patients, skin tightening was observed in 86% of patients as measured by 3 independent physician observers. This improvement ranged from mild to excellent (mild 29%, moderate 38%, excellent 19%) [12]. Variable response rates with skin tightening technologies may be due to the patient selection, energies used, and overall skin quality [13].

Another infrared device, the SkinTyte (Sciton, Inc., Palo Alto, Calif., USA) delivers infrared light in the range of 800–1,400 nm. It utilizes uniform targeted deep dermal heating to achieve skin tightening. Continuous cooling of the epidermis is achieved with thermoelectric sapphire plates.

The LuxIR and the LuxDeep IR (Palomar Medical Technologies, Inc., Burlington, Ma., USA) deliver infrared light in the range of 850–1,350. The LuxDeepIR Fractional Infrared Handpiece (Palomar Medical Technologies) utilizes fractional infrared light and was designed to deliver infrared deep light extending up to 4 mm into the dermis and fat layer without damaging the epidermis. The LuxDeep IR Handpiece safely delivers more effective heat to greater depths, and offers longer pulse durations. The energy is delivered through a sapphire crystal with contact cooling to prevent epidermal injury. Several treatments are required. Various clinical trials have demonstrated improvement in skin tightening [14–16].

Radiofrequency Devices

Radiofrequency

RF devices heat, lift, and tighten dermal tissue to treat skin laxity [17]. RF energy is an alternating current that runs from the tip of an electrode to tissue that it contacts [13]. In contrast to direct current, with alternating current, the flow of electrons cycles back and forth at a certain frequency. The frequency of the alternating current in RF devices is typically 0.3–10 MHz [18]. These devices generate dermal heating, denature collagen, and induce collagen remodeling. Wound healing mechanisms trigger wound contraction, which enhances the appearance of mild to moderate skin laxity. RF devices are either monopolar (unipolar), bipolar, or both.

Monopolar Radiofrequency

In monopolar RF systems, current flows from the electrode in contact with the treatment area to a grounding electrode on the skin at a distant site. The current flows from the electrode contact area through the body to the grounding pad. Most of the heat is generated just beneath the treating electrode. One of the first monopolar RF devices ThermaCool (Thermage, Inc., Hayward, Calif., USA) was developed to specifically tighten deeper dermal structures without damaging the epidermis. The benefit of monopolar RF devices is that the current penetrates to the deeper layers of skin [18, 19]. RF treatments can be mild-to-moderately painful and patients may require some pain management [19].

The ThermaCool (Thermage) is a non-ablative monopolar RF device that heats the deep tissues utilizing a capacitive coupling membrane. The capacitive coupling causes volumetric tissue heating, allowing large amounts of energy to be distributed over a three-dimensional volume of dermal tissue. This device uses monopolar RF with a grounding pad placed on the patient to deliver heat deep in the dermis (volumetric deep dermal heating). In addition to the dermis, monopolar RF energy delivers heat to the fibrous septae within the fat. A cryogen system simultaneously cools the epidermis. Disposable single-use tips are available in different sizes to treat different body parts. Originally, high energies were delivered in 1–2 passes to achieve clinical results. Later treatments evolved to utilize lower energies with multiple passes to improve tolerability. Fat necrosis, which had been originally described as a possible adverse effect, is now rare with the newer parameters [20]. Clinical studies show efficacy in treating skin laxity with monopolar RF in all skin types [3, 21, 22].

Bipolar Radiofrequency

Bipolar RF devices use two electrodes positioned at fixed distances from each other, and both electrodes are in contact with the area being treated [11]. Alternating current enters the skin and passes only through the tissue between the two electrodes. In these devices, the penetration depth of the electrical current is approximately half the distance between the electrodes [18]. With bipolar systems the distribution of current can be controlled, but the depth of the treatment is less than that with monopolar RF. For this reason, many bipolar RF devices have been coupled with other modalities to address this issue.

Bipolar Radiofrequency: Vacuum-Assisted

The Aluma (Lumenis, Santa Clara, Calif., USA) uses its bipolar RF energy with an accompanying vacuum apparatus, which allows energy to selectively target the deep dermis. With this system there is little or no pain associated with the procedure, and it utilizes less energy because the vacuum brings the electrodes closer to the dermal tissue. The disposable treatment tip can be used for multiple treatment sessions. Clinical trials have demonstrated safety and efficacy for skin tightening [19, 23].

Combined Unipolar and Bipolar Radiofrequency
The Accent (Alma Lasers, Inc., Ft. Lauderdale, Fla., USA) RF system includes both a unipolar handpiece for volumetric deep heating and a bipolar handpiece for superficial heating of the dermis. One of the first studies demonstrated its effectiveness in reducing the appearance of cellulite and also confirmed its effect on tissue tightening [24]. Since then, a number of reports have verified its safety and efficacy in treating wrinkles and skin laxity [25, 26].

Combined Bipolar Radiofrequency and Optical Energy
The ReFirme (Syneron Medical Ltd., Yokneam, Israel) is a combined bipolar RF and optical energy device (700–2,000 nm infrared light). The theory is that with these combined energy sources, less RF energy is necessary. A grounding plate is not required. The RF energy is more superficial and the cooling apparatus is in the handpiece. Clinical studies have demonstrated improvement in skin laxity with this device [27, 28]. In one study, blinded evaluation of treated subjects revealed statistically significant improvements in the cheeks, jowls, and nasolabial folds [28].

Ultrasound Tightening
Intense ultrasound is an energy that can propagate through tissue up to several millimeters. Ultrasound waves induce a vibration in the molecules of a target tissue during propagation, and the thermoviscous losses in the medium lead to tissue heating [29]. When the beam is directed to focus on the tissue at a certain depth, it produces thermal coagulative necrosis, leaving the superficial layers unaffected.

The Ulthera P (Ulthera, Mesa, Ariz., USA) ultrasound device utilizes intense focused ultrasound energy to achieve skin tightening. Epidermal injury is minimized, and thermal energy is directed into the reticular dermis and subcutis, where immediate tissue contraction and delayed remodeling are believed to collectively cause tightening. It has been developed specifically for treating facial soft tissues and targeting the superficial musculoaponeurotic system [30].

It is a micro-ablation device that non-invasively treats multiple layers of tissue. The device uses standard ultrasound to visualize the target layers of the sub-epidermis immediately before the therapeutic level of focused ultrasonic energy is delivered. This gives the clinician the ability to visualize and target the desired treatment level of the skin. The outer epidermal layer of skin is spared, while deeper treatment immediately triggers the body's natural healing effect, resulting in a gradual lifting and tightening of the face and neck. This procedure can be painful and typically requires pain management. Clinical studies have demonstrated improvement in skin laxity in addition to brow elevation [31].

Conclusion

Non-invasive skin tightening is a highly sought after procedure and new devices in this arena continue to be developed. There are many commercially available devices that have offered patients effective results without surgery. However, ultrasound, laser, and light-based devices do not deliver surgical results and patient selection is key. Furthermore, assessing of the action of tightening is difficult, as there is no standardized measure available. Photographic evaluations are sometimes helpful but can be challenging, as a minor change in position will alter the results. Additionally, it is difficult to measure if slight tightening occurs, especially if it occurs over a long period of time. There is a definite need for more blinded controlled trials to assess these technologies more thoroughly. In addition it is still uncertain how long the results will last before maintenance therapies are necessary. Further long-term clinical trials are warranted to look into the skin tightening longevity with the current tightening devices available.

References

1 Alexiades-Armenakas M: Assessment of the mobile delivery of infrared light (1,100–1,800 nm) for the treatment of facial and neck skin laxity. J Drugs Dermatol 2009;8:221–226.
2 Zelickson BD, Kist D, Bernstein E, et al: Histological and ultrastructural evaluation of the effects of a radiofrequency based nonablative dermal remodeling device. Arch Dermatol 2004;140:204–209.
3 Fitzpatrick R, Geronemus R, Goldberg D, et al: Multicenter study of noninvasive radiofrequency for periorbital tissue tightening. Lasers Surg Med 2003;33:232–242.
4 Gold MH: Tissue tightening: a hot topic utilizing deep dermal heating. J Drugs Dermatol 2007;6:1238–1242.
5 Fitzpatrick RE, Goldman MP, Sotur NM, et al: Pulsed carbon dioxide laser resurfacing of photoaged skin. Arch Dermatol 1996;132:395–402.
6 Kauvar AN, Geronemus RG: Histology of laser resurfacing. Dermatol Clin 1997;15:459–467.
7 Majaron B, Kelly KM, Park HB, et al: Er:YAG laser skin resurfacing using repetitive long-pulse exposure and cryogen spray cooling. I. Histological study. Lasers Surg Med 2001;28:121–130.
8 Greve B, Raulin C: Professional errors caused by laser and IPL technology in dermatology and aesthetic medicine: preventive strategies and case studies. Dermatol Surg 2001;28:156–161.
9 Nanni CA, Alster TS: Complications of carbon dioxide laser resurfacing: an evaluation of 500 patients. Dermatol Surg 1998;24:315–320.
10 Dierickx CC: The role of deep heating for noninvasive skin rejuvenation. Lasers Surg Med 2006;38:799–807.
11 Sadick NS: Tissue tightening technology: fact or fiction. Aesthet Surg J 2008;28:180–188.
12 Chua SH, Ang P, Khoo LS, Goh CL: Nonablative infrared skin tightening in type IV to V Asian skin: a prospective clinical study. Dermatol Surg 2007;33:146–151.
13 Bunin LS, Carniol BJ: Cervical facial skin tightening with an infrared device. Facial Plast Surg Clin North Am 2007;15:179–184.
14 Dierickx C, Altshuler G, Erofeev A, et al: Skin tightening with Palomar's StarLux IR deep dermal, fractional heating. Lasers Surg Med 2008;40:1–132.
15 Weiss R: Fractional infrared light for skin tightening: six month prospective study. Lasers Surg Med 2007;39:15–34.
16 Dierickx C, Altshuler G, Childs J, et al: Dermal/hypodermal junction as a new target for deep fractional thermal treatment. Lasers Surg Med 2007;39:1–14.
17 Alster TS, Lupton JR: Nonablative cutaneous remodeling using radiofrequency devices. Clin Dermatol 2007;25:487–491.
18 Sadick NS, Makino Y: Selective electrothermolysis in aesthetic medicine: a review. Lasers Surg Med 2004;34:91–97.
19 Montesi G, Calvieri S, Balzani A, et al: Bipolar radiofrequency in the treatment of dermatologic imperfections: clinicopathological and immunohistochemical aspects. J Drugs Dermatol 2007;6:890–896.
20 Dover JS, Zelickson B: Results of a survey of 5,700 patient monopolar radiofrequency facial skin tightening treatments: assessment of a low-energy, multiple-pass technique leading to a clinical point algorithm. Dermatol Surg 2007;33:900–907.
21 Alster TS, Tanzi E: Improvement of neck and cheek laxity with a nonablative radiofrequency device: a lifting experience. Dermatol Surg 2004;30:503–507.
22 Kushikata N, Negeshi K, Takeuchi K, Wakamatsu S: Nonablative skin tightening with radiofrequency in Asian skin. Lasers Surg Med 2005;36:92–97.
23 Gold MH, Goldman MP, Rao J, et al: Treatment of wrinkles and elastosis using vacuum-assisted bipolar radiofrequency heating of the dermis. Dermatol Surg 2007; 33:300–309.
24 Emelia del Pino M, Rosado RH, Azuela A, et al: Effect of controlled volumetric tissue heating with radiofrequency on cellulite and the subcutaneous tissue of the buttocks and thighs. J Drugs Dermatol 2006;5:714–722.
25 Freidman DJ, Gilead LT: The use of hybrid radiofrequency device for the treatment of rhytides and lax skin. Dermatol Surg 2007;33:543–551.
26 Mayoral FA: Skin tightening with a combined unipolar and bipolar radiofrequency device. J Drugs Dermatol 2007;6:212–215.
27 Sadick NS: Electro-optical synergy in aesthetic medicine: novel technology, multiple applications. Cosmet Dermatol 2005;18:201–206.
28 Yu, CS, Yeung CK, Shek SY, Tse RK, Kono T, Chan HH: Combined infrared light and bipolar radiofrequency for skin tightening in Asians. Lasers Surg Med 2007;39:471–475.
29 Kennedy JE, terHaar GR, Cranston D: High intensity focused ultrasound: surgery of the future? Br J Radiol 2003;76:590–599.
30 White WM, Makin IR, Barthe PG, Slayton MH, Gliklich RE: Selective creation of thermal injury zones in the superficial musculoaponeurotic system using intense ultrasound therapy: a new target for noninvasive facial rejuvenation. Arch Facial Plast Surg 2007;9:22–29.
31 Alam M, White LE, Martin N, Witherspoon J, Yoo S, West DP: Ultrasound tightening of facial and neck skin: a rater-blinded prospective cohort study. J Am Acad Dermatol 2010;62:262–269.

Heather Woolery-Lloyd, MD
University of Miami Department of Dermatology and Cutaneous Surgery
1295 Northwest 14th Street
South Building, Suite K
Miami, FL 33125 (USA)
Tel. +1 305 243 6704, E-Mail woolerylloyd@yahoo.com

Bogdan Allemann I, Goldberg DJ (eds): Basics in Dermatological Laser Applications.
Curr Probl Dermatol. Basel, Karger, 2011, vol 42, pp 153–159

Special Considerations for Darker-Skinned Patients

Henry H.L. Chan

Division of Dermatology, Department of Medicine, University of Hong Kong, Hong Kong, SAR, China;
Wellman Center of Photomedicine, Massachusetts General Hospital, Harvard Medical School, Boston, Mass., USA

Abstract

Today, lasers are being used to treat an increasing number of medical and aesthetic challenges with great success. However, the effects of laser treatments upon darker-skinned patients differ from those on Caucasians in several aspects; this makes their use not without risk. First of all, skin of color has a greater epidermal melanin content than Caucasian skin, and this increase in epidermal pigmentation acts as a competing chromophore. Furthermore, with epidermal melanin acting as a competing chromophore, the light dosage that reaches the actual target is reduced, and higher fluence is necessary to produce a desirable effect. Although effective cooling and longer wavelengths with decreased melanin absorption have been used, the treatment of patients with Fitzpatrick skin types >IV remains associated with a significant risk of side effects. Thus, careful selection of laser and light source parameters in conjunction with skin cooling are important in obtaining desirable clinical outcomes in ethnic skin.

An ever-increasing number of medical and aesthetic challenges are nowadays treated with laser or light sources with great success, and normally great safety. The majority of treatment parameters have however been tested and established in light-skinned patients, and hence mainly apply to skin phototypes I–III. Of the limited number of available studies performed on darker skin, most have been conducted on Asian skin types.

Treatment for darker-skinned patients with laser or light devices differs greatly from treatment for Caucasians, and is not without risk.

The difference in the skin color between various ethnic groups and populations is not due to an increased number of melanocytes, but to the increased production and amount of melanin within the melanocytes [1]. Melanin production increases upon exposure to ultraviolet radiation, which is the basis upon which Fitzpatrick categorized the skin types into phototypes I–VI [2] (table 1).

Hence, as the epidermal melanin content is increased in darker-skinned patients, and melanin is one of the main chromophores in laser surgery, the understanding of the absorption curve of melanin and its relationship to wavelength is crucial.

Table 1. Fitzpatrick skin phototypes

Fitzpatrick phototype	Response to sunlight exposure	Constitutive color/hair color/eye color
I	always burns; never tans	lightest skin/light hair/light eyes
II	always burns; tans minimally with difficulty	light skin/fair to dark hair/light eyes
III	burns minimally; tans gradually and uniformly	medium tone skin/brown hair/dark hazel-brown eyes
IV	burns minimally; tans easily	tan skin/dark hair/brown eyes
V	rarely burns; tans profusely	brown skin/dark hair/brown eyes
VI	never burns; tans profusely	dark brown/dark hair/dark eyes

The absorption spectrum of melanin lies within a wide spectrum ranging from 200 to 1,200 nm, with the strongest absorption within the UV and visible light range. The absorption curve constantly decreases; hence, shorter wavelengths are strongly absorbed by melanin, while longer wavelengths get absorbed less.

As the epidermal melanin acts as a competing chromophore, treatment may result in excessive non-specific thermal damage of surrounding tissue, leading to post-inflammatory dyspigmentation, blistering or even scarring.

In addition, because the epidermal melanin acts as a competing chromophore, light energy that reaches the targeted blood vessels or follicular melanocytes is reduced, and higher fluences are necessary to produce the desired clinical effect.

Based on the above-mentioned facts, in order to prevent adverse events when treating skin of color:

- Longer laser/light source wavelengths should be used to reduce absorption by competing epidermal melanin. In addition, longer wavelengths penetrate deeper into the tissue.
- Skin cooling and longer laser pulse widths to protect the epidermis are crucial.
- Besides strict sun avoidance and sun protection before and after laser/intense pulsed light (IPL) treatment, topical bleaching agents are important as an adjunctive therapy.
- Test spots should be made before treating the entire lesion.

All of these measures can further enhance clinical outcomes as well as reduce the adverse events.

Pigmented Lesions

There are several acquired pigmentary conditions that are particularly common in darker-skinned patients. Lentigines and seborrheic keratoses are frequent signs of photoaging among Asians. Melasma and Hori's macules are particularly common among female dark-skinned patients. Furthermore, post-inflammatory hyperpigmentation (PIH) is a frequent concern in patients with darker skin types.

Given the low melanoma risk in colored skin, lasers are frequently used to treat congenital melanocytic lesions, such as nevus of Ota and even congenital melanocytic nevi. All Q-switched (QS) lasers have been shown to be effective in the treatment of nevus of Ota, including the QS ruby, QS alexandrite and QS 1,064 nm Nd:YAG [3]. Data indicate that early treatment at a younger age group can be associated with less treatment sessions as well as a lower risk of complications [4]. While different lasers have been used for the removal of congenital melanocytic nevi, the use of a long-pulse pigmented laser to first remove the epidermis, immediately followed by multiple passes of the QS laser, was found to be particularly effective [5]. For café au lait patches and Becker's nevi, a long-pulse alexandrite laser (10-mm spot size, 20–30 J/cm^2) can be used, but relapse remains an issue with 40% of the lesions recurring after 4 treatment sessions [6].

While QS lasers can be most effective in the treatment of freckles and lentigo in skin of color, the risk of PIH is an issue and can affect about 10% of treated patients. Long-pulse pigmented laser treatment can also be used, and by matching the pulse duration to the thermal relaxation time of the epidermis, the complication risk is much lower (with a PIH rate of about 1% [7]). Multiple treatment sessions at monthly intervals are necessary to obtain the desirable clinical outcome. Another important factor to take into consideration is the contrast between lesional and non-lesional skin. If the contrast is low, then large-spot-size lasers or IPL are inappropriate, as with such a narrow therapeutic window either under-treatment would occur or complications would develop as a result of above-threshold injury to the non-lesional skin.

Melasma is a common acquired pigmentary condition that can be most difficult to treat, and laser/light sources should be reserved for resistant cases. In recent years, fractional resurfacing has largely replaced laser resurfacing in skin rejuvenation as well as treatment of melasma. Initial studies indicated the effectiveness of fractional resurfacing for the treatment of melasma to be excellent in 60%, mild-to-moderate in 30% and worsening in 10% of patients [8]. However, more recent data revealed the efficacy to be much less impressive and most cases tend to recur 6 months after treatment [9]. It is the opinion of the author that fractional resurfacing fails to achieve complete removal of the abnormal melanocytes in diseased skin, and its role is therefore mainly adjunctive.

Previous studies have examined the use of IPL (570 nm and 590 to 615 nm filters in 4-week intervals for a total of 4 treatments) for the treatment of melasma in Asian patients, and showed improvement in 39.8% of those treated compared to only 11.6% of the control group [10]. Two patients in the IPL group experienced transient PIH, and partial repigmentation was noted 24 weeks after the last treatment session. In a more recent study, 89 Chinese patients with melasma were treated with an IPL source device with a uniform pulse profile. They were able to achieve excellent results (77.5% of treated subjects had over 50% improvement 3 months after treatment) with a low risk of increased pigmentation (3%) [11]. This was despite the lack of topical bleaching agents during the study period. Their findings differ significantly from previous data. Although the uniform profile of this IPL device may be one of the reasons for such observations, as suggested by the investigators, other factors are more likely to contribute to the differences in clinical outcome, including the geographical location of the tested site and recruitment criteria.

In recent years, use of the large-spot-size QS 1,064 nm Nd:YAG laser (6–8 mm spot size, 1.6–3.1 J/cm^2) has gained a great deal of popularity in the treatment of melasma. Mild erythema has been used as the clinical endpoint, and treatment is often performed in 1- to 2-weekly intervals with patients undergoing many treatment sessions (10 or more). While melasma can certainly be lightened with such an approach, repeated treatments can lead to mottled depigmentation that can be resistant to therapy [12] (fig. 1). Furthermore, relapse of melasma (and in some cases, rebound) can occur. The use of such a laser should therefore be utilized as an adjunctive therapy in the treatment of melasma.

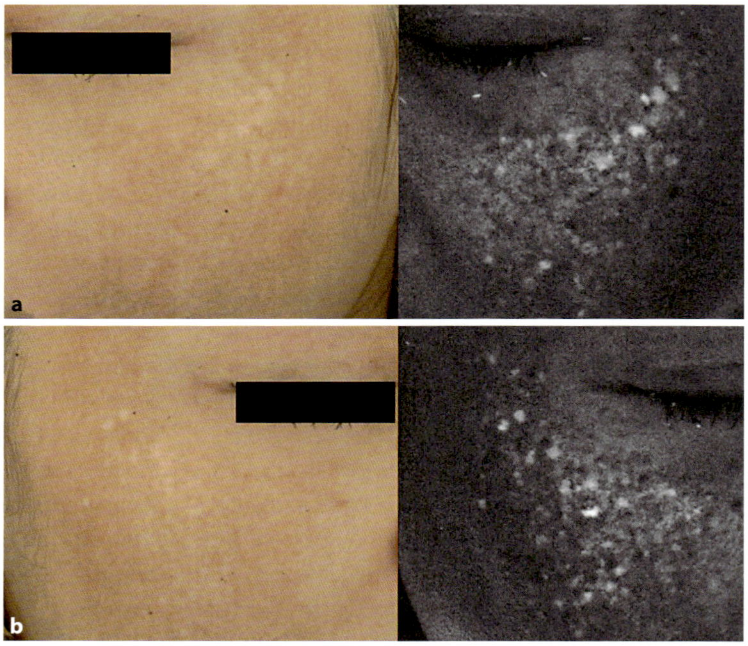

Fig. 1. Cross-polarized (left) and UV (right) images of depigmented macules over left (**a**) and right (**b**) cheeks after 9 sessions of QS 1,064 nm Nd:YAG laser for melasma.

Hair Removal

In darker-skinned patients, especially African-Americans, laser- or light-assisted hair removal is often requested to treat a medical instead of a cosmetic concern, e.g. pseudofolliculitis barbae.

The target chromophore in hair removal is the melanin found in hairs. Hence, melanin in the epidermis will act as a competing chromophore, increasing the risk of epidermal injury, with post-inflammatory dyspigmentation or even blistering and scarring.

Using devices with long wavelengths, where melanin absorption is reduced, as well as long pulse durations alongside effective epidermal cooling will reduce the risk of epidermal injury [13, 14]. Among the laser and light devices used for hair removal, either an 810 nm diode laser or a 1,064 nm Nd:YAG laser should be used for darker-skinned patients. These are the only devices approved by the Food and Drug Administration for photoepilation in darker skin types [15].

Photoaging

The clinical presentation of photoaging differs between skin of color and Caucasians in that pigmentary changes tend to occur with a greater incidence than skin wrinkling in darker-skinned patients [16, 17]. Chung [18] found both pigmentary changes and wrinkling to be major features of photoaging in Asians. However, moderate-to-severe wrinkling becomes apparent only at about 50 years, which is a decade or two later than age-matched Caucasians [19].

Besides pigment reduction, improvements in skin texture, pore size and facial erythema are common requests among patients with moderate degrees of photoaging. The use of non-ablative

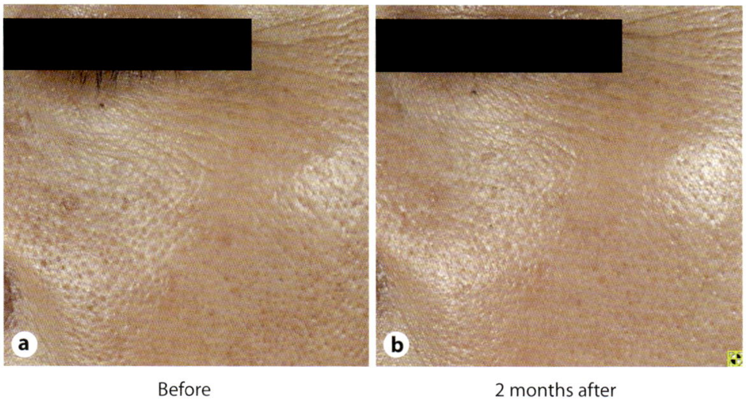

Fig. 2. Before (**a**) and 2 months after (**b**) a single treatment with non-ablative dual-wavelength (1,550 and 1,927 nm) fractional resurfacing showing improvement in pore size.

skin rejuvenation is particularly applicable to darker-skinned patients given the lesser degree of wrinkling among this population. Other advantages include lack of downtime and lower risk of adverse effects. A combination approach using multiple devices, with the intention of optimizing the clinical outcome, has gained a great deal of popularity in recent years [20]. Such treatment includes the use of vascular lasers, infrared lasers and pigment lasers or IPL – all in the same session. The aims of vascular laser/IPL are to reduce facial erythema, induce microvascular injury and, in doing so, promote new collagen formation. Infrared lasers heat up the dermis and promote collagen damage; thus, leading to new collagen formation [21]. To protect the epidermis, skin cooling is extremely important in darker-skinned patients.

The delay in the onset of wrinkling in skin of color and the greater incidence of adverse effects also imply that more aggressive approaches for skin rejuvenation, such as non-ablative or ablative fractional resurfacing, are less applicable (fig. 2). The newer, especially non-ablative, fractional devices might be an option for skin resurfacing in darker-skinned patients.

Acne Scars

Treatment parameters for acne scarring in skin of color depend on several factors. While a more aggressive approach can lead to a greater degree of improvement, the downtime and the risk of PIH can also increase. For non-ablative fractional resurfacing, while both the density and energy can have an impact on the risk of PIH, density is of particular importance [22]. A recent study in Asian patients indicated that by halving the density but doubling the number of treatment sessions, the risk of PIH can be lowered threefold. Ablative fractional resurfacing has the benefit of a single treatment session, and has been shown to be effective for acne scarring and skin rejuvenation for skin of color. Patients should be aware of the high PIH rate which tends to be acceptable among male patients [23].

Skin Cooling for the Treatment of Vascular Lesions and Laser Hair Removal

With a greater epidermal melanin content, skin cooling is extremely important in the laser treatment of vascular lesions and hair removal among skin of color. Previous studies have indicated contact cooling to be less effective in the prevention of complications as compared with dynamic cooling in the treatment of port wine stains in skin of color [24, 25]. For laser treatment in skin types III and IV, the typical dynamic cooling setting is 40-ms spray time with 20-ms delay; a 50-ms spray time can be used for darker skin types. Air cooling can also be used, particularly in conjunction with nonablative fractional resurfacing, as a means of reducing bulk tissue heating. Its role in the treatment of vascular lesions and hair removal in skin of color is less documented than dynamic cooling.

Conclusions

Lasers and light sources can be used effectively and safely in skin of color if the appropriate considerations are taken into account. The use of long-pulse lasers – preferentially with long wavelengths, skin cooling and the selection of less aggressive parameters – can ensure patient satisfaction as well as a desirable clinical outcome.

References

1 Alaluf S, Atkins D, Barrett K, Blount M, Carter N, Heath A: Ethnic variation in melanin content and composition in photoexposed and photoprotected human skin. Pigment Cell Res 2002;15:112–118.
2 Fitzpatrick TB: The validity and practicality of sun-reactive skin types I through VI. Arch Dermatol 1988;124:869–871.
3 Chan HH, Kono T: Nevus of Ota: clinical aspects and management. Skinmed 2003;2:89–98.
4 Kono T, Chan HH, Erçöçen AR, Kikuchi Y, Uezono S, Iwasaka S, Isago T, Nozaki M: Use of Q-switched ruby laser in the treatment of nevus of Ota in different age groups. Laser Surg Med 2003;32:391–395.
5 Kono T, Nozaki M, Chan HH, Sasaki K, Kwon SG: Combined use of normal mode and Q-switched ruby lasers in the treatment of congenital melanocytic naevi. Br J Plast Surg 2001;54:640–643.
6 Choi JE, Kim JW, Seo SH, Son SW, Ahn HH, Kye YC: Treatment of Becker's nevi with a long-pulse alexandrite laser. Dermatol Surg 2009;35:1105–1108.
7 Chan HH, Fung WK, Ying SY, Kono T: An in vivo trial comparing the use of different types of 532 nm Nd:YAG lasers in the treatment of facial lentigines in Oriental patients. Dermatol Surg 2000;26:743–749.
8 Rokhsar C, Fitzpatrick RE: The treatment of melasma with fractional photothermolysis: a pilot study. Dermatol Surg 2005;31:1645–1650.
9 Lee HS, Won CH, Lee DH, An JS, Chang HW, Lee JH, Kim KH, Cho S, Chung JH: Treatment of melasma in Asian skin using a fractional 1,550-nm laser: an open clinical study. Dermatol Surg 2009;35:1499–1504.
10 Wang CC, Hui CY, Sue YM, Wong WR, Hong HS: Intense pulsed light for the treatment of refractory melasma in Asian patients. Dermatol Surg 2004;30:1196–1200.
11 Li YH, chen JZ, Wei HC, Liu M, Xu YY, Dong GH, Chen HD: Efficacy and safety of intense pulsed light in treatment of melasma in Chinese patients. Dermatol Surg 2008;34:693–700.
12 Wattanakrai P, Mornchan R, Eimpunth S: Low-fluence Q-switched neodymium-doped yttrium aluminum garnet (1,064 nm) laser for the treatment of facial melasma in Asians. Dermatol Surg 2010;36:76–87.
13 Battle EF Jr, Hobbs LM: Laser-assisted hair removal for darker skin types. Dermatol Ther 2004;17:177–183.
14 Alster TS, Bryan H, Williams CM: Long-pulsed Nd:YAG laser-assisted hair removal in pigmented skin: a clinical and histological evaluation. Arch Dermatol 2001;137:885–889.
15 Battle E, Suthamjariya K, Alora M, et al: Very long pulses (20–200 ms) diode laser for hair removal on all skin types (abstr). Lasers Surg Med 2000;12:21–24 (suppl).
16 Griffiths CEM, Wang TS, Hamilton TA, Voorhees JJ, Ellis CN: A photonumeric scale for the assessment of cutaneous photodamage. Arch Dermatol 1992;128:347–351.
17 Larnier C, Ortonne JP, Venot A, Faivre B, Béani JC, Thomas P, Brown TC, Sendagorta E: Evaluation of cutaneous photodamage using a photographic scale. Br J Dermatol 1994;130:167–173.
18 Chung JH: Photoaging in Asians. Photodermatol Photoimmunol Photomed 2003;19:109–121.
19 Chan HH, Jackson B: Laser treatment on Ethnic Skin; in Lim HW, Honigsmann H, Hawk JLM (eds): Photodermatology. New York, Informa Healthcare, 2007, pp 417–432.
20 Chan HH: Effective and safe use of lasers, light sources, and radiofrequency devices in the clinical management of Asian patients with selected dermatoses. Laser Surg Med 2005;37:179–185.
21 Chan HH, Lam LK, Wong DS, Kono T, Trend-Smith N: Use of 1,320 nm Nd:YAG laser for wrinkle reduction and the treatment of atrophic acne scarring. Lasers Surg Med 2004;34:98–103.

22 Chan NP, Ho SG, Yeung CK, Shek SY, Chan HH: The use of non-ablative fractional resurfacing in Asian acne scar patients. Lasers Surg Med 2010;42:710–715.
23 Chan NP, Ho SG, Yeung CK, Shek SY, Chan HH: Fractional ablative carbon dioxide laser resurfacing for skin rejuvenation and acne scars in Asians. Lasers Surg Med 2010;42:615–623.
24 Chan HH, Chan E, Kono T, Ying SY, Ho WS: The use of variable pulse width frequency doubled Nd:YAG 532 nm laser in the treatment of port-wine stain in Chinese patients. Dermatol Surg 2000;26:657–661.
25 Chiu CH, Chan HH, Ho WS, Yeung CK, Nelson JS: Prospective study of pulsed dye laser in conjunction with cryogen spray cooling for treatment of port wine stains in Chinese patients. Dermatol Surg 2003;29:909–915.

Henry H.L. Chan, MBBS, MD, PhD, FRCP
13th Floor, Club Lusitano
16 Ice House Street, Central
Hong Kong, SAR (China)
Tel. +852 2290 1717, E-Mail hhlchan@hkucc.hku.hk

Bogdan Allemann I, Goldberg DJ (eds): Basics in Dermatological Laser Applications.
Curr Probl Dermatol. Basel, Karger, 2011, vol 42, pp 160–165

At-Home Laser and Light-Based Devices

Alia S. Brown

Skin Laser & Surgery Specialists of New York & New Jersey, Hackensack, N.J., USA

Abstract

With the advancement in laser technology, there has been a recent trend in laser and light devices towards 'do it yourself' at-home use. There has been an exponential growth in devices aimed at the treatment of acne, hair removal, and psoriasis. With the exception of the at-home UVB box and excimer lasers, these systems do not require a prescription and patients may purchase the systems as over-the-counter products. Medical treatment still remains the standard of care for most conditions, but these at-home devices play a significant role in the treatment of various conditions. Media and marketing have driven the popularity of these devices and have encouraged patients to try them out. Patients are attracted to at-home devices because of the one-time cost, ease of use, convenience, and privacy of use. Naturally, despite these devices being for at-home use, there should be a level of caution adhered to when using any laser or light device.

With the advent of do-it-yourself procedures, at-home lasers and light-based technologies have become increasingly more popular. Patients are attracted to the affordability, convenience, ease of use, and promising results. There is a wide range of dermatological disorders that may benefit from at-home laser/light devices, including acne treatment, hair removal, hair loss, anti-aging, psoriasis, and wound healing. Within recent years, the Food and Drug Administration (FDA) has approved many laser/light devices for at-home use, allowing easier patient access [1].

Acne Treatment with At-Home Devices

Acne is problematic for millions of patients worldwide. With the overwhelming emphasis on one's appearance by society, cosmesis has become a concern for most patients with acne, and they have been seeking methods to improve their appearance. At-home laser and light devices for acne have gained popularity by promising patients tangible results as well as saving them an office visit. Though not as effective as office-based laser procedures, patients may be attracted by the onetime cost of at-home devices and the ease of use (table 1).

Acne has been treated in the office with lasers that emit longer wavelengths or a continuous-wave blue light output source. Blue light devices emit 25–400 mW/cm^2 and are painless. The

Table 1. At-home acne treatment devices

Device	Mechanism	Cost USD	Proven efficacy	FDA approved
Tända	blue light: 414 nm; red light: 600 nm; use twice daily; 4-min cycles	200–300	reduces acne lesions within 24 h	yes
Baby Quasar and Baby Blue	blue light: 415 nm; 2–3 days/week; 15 min twice daily; no set timer	349–400	reduces lesion count within 24 h	yes
Tria	400 mw/cm^2 of 415 nm blue light for 5 min	395	reduction in frequency of breakouts within 1 week	yes
Claro	IPL (400–700 nm) and heat delivered in 6-second pulses	275–300	reduces lesion count in 24 h	yes
Zeno Hot Spot and Zeno Pro	direct heat for 2.5 min per cycle	30 (Hot Spot) and 150–200 (Pro)	reduces acne lesions within 24 h; multiple treatments to a single lesion may be required	yes
Thermaclear	95°C on the low setting and 115°C on the high setting; short auto pulses	150	reduction in lesion count 4 times faster than lesions treated with placebo device	yes

mechanism of action involves *P. acnes*, which produces endogenous porphyrins when metabolized; therefore, when excited by the blue light source, singlet oxygen is formed [1–3]. At-home laser/light devices aim to achieve the same benefits by targeting *P. acnes*, sebaceous glands, and the infundibulum of the sebaceous glands. In a study by Gold et al. [2], they examined the efficacy of a self-applied 414 nm blue light-emitting diode (LED) for the improvement of mild-to-moderate facial acne in 21 patients over 8 weeks. Pertinent endpoints included: (1) time to improvement and/or resolution of the number of blemishes and lesions on the face; (2) quality of skin condition; (3) occurrence and count of the number of new blemishes and lesions; (4) ease of product use; patient comfort, wellbeing, and satisfaction during the treatment period; and safety of treatment. They found at days 7 and 28 that the total number of open/closed comedones and papules were significantly reduced; no statistically significant change in the number of pustules and nodules was observed. Patients were found to be comfortable and confident when using the at-home blue light modality and regarded this treatment as superior to other treatments previously rendered. At-home 414 nm blue LED (Tända Clear) treatment has been shown to kill the *P. acnes* bacteria that causes mild-to-moderate acne. This device helps improve existing blemishes as well as prevent future outbreaks. It is indicated in the treatment of mild-to-moderate inflammatory acne. There are various at-home devices (Baby Quasar, Baby Blue, and Tria) utilizing the blue LED as their primary mechanism of action to destroy *P. acnes*.

Other examples of at-home laser/light devices for acne include a clearing device (Zeno) available without a prescription. It is a handheld portable electronic device approved for the treatment of patients with mild-to-moderate acne. The treatment tip is made from a biocompatible material and delivers a specific low-level heat to each individual acne lesion. This device, as well as another one called Thermaclear, uses timed direct heating to reduce inflammation within an acne lesion. Claro is the only FDA-approved device that utilizes intense pulsed light (IPL; 400–700 nm) including the blue, red, and infrared spectrums.

Hair Removal

Numerous lasers have been proven to be efficacious for the removal of unwanted hair. Wavelengths between 650 and 1,100 nm are absorbed by melanin, the main chromophore used in hair removal. Initially, physician based lasers including the 695 nm ruby lasers were used, but with the growth of technology, the 755 nm alexandrite, 800 to 810 nm diode, and 1,064 nm Nd:YAG lasers have also been found to be efficacious and have better side effect profiles. Now IPL platforms with epidermal cooling and filtering mechanisms are widely being utilized to remove unwanted hair. There are several factors to take into account when considering an appropriate laser for hair removal. Fitzpatrick skin type, location of hair, cooling mechanism, and choosing an adequate pulse duration all play a significant role in choosing an ideal laser and setting its parameters for treatment. Home devices are becoming popular because they are able to utilize higher fluences with longer pulse widths. There are currently only two FDA-approved at-home devices for the treatment of unwanted hair. These include an 810 nm diode laser (Tria, SpectraGenics, Inc.) and an IPL with 475–1,200 nm (Silk'n, Home Skinovations, Ltd; fig. 1). The efficacy of hair removal with home-use IPL was evaluated by Alster and Tanzi [4]. They looked at 20 female patients with skin types I–IV who treated dark terminal hair in non-facial sites (axilla, forearms, inguinal region, legs) by 3 self-administered treatments (2 weeks apart) using a handheld IPL device. Matched untreated skin sites were also studied. Hair counts and clinical photographs were obtained before treatment and at 1, 3, and 6 months after the third treatment. All patients showed a positive clinical response to treatment, with a reduction in unwanted hair. No reduction in hair was noted in untreated matched areas. Six months after the 3 treatments, hair counts were reduced by 37.8–53.6%. They concluded home IPL use demonstrated a significant continued reduction in hair growth.

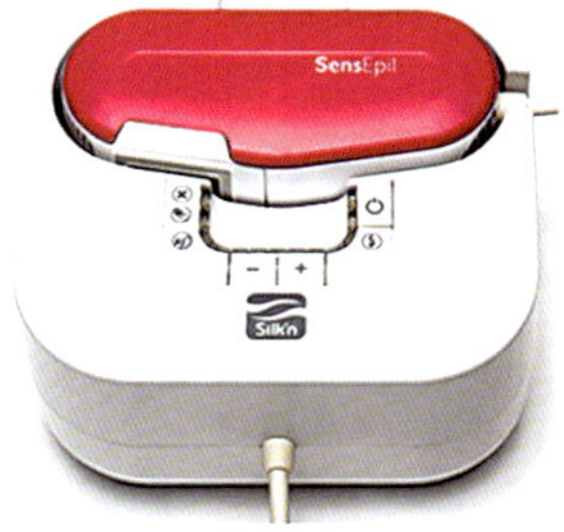

Fig. 1. Silk'n SensEpil at-home hair removal system.

At-home laser devices are effective, but drawbacks include lower fluences, a much smaller spot size than traditional in-office lasers, no epidermal cooling, not being FDA-approved for facial hair (except Silk'n), and taking much longer to achieve similar results to in-office laser procedures. Of note, there are multiple non-FDA-approved devices available for purchase on the internet, advertising low costs, small side effect profiles, and convenience. With any at-home laser procedure, patients must be aware that despite the ease of use there is still an acknowledgeable risk of burning, erythema, and post-inflammatory hyperpigmentation. Silk'n and Tria laser/light devices both aim to target skin types I–IV. Silk'n SensEpil has a built-in smart skin sensor which detects user skin tone, and will not fire if the candidate's skin tone contains too much pigment. The Tria device has a pigment indicator that must register the presence of proper skin types each time it is unlocked to initiate treatment. These safety measures should greatly reduce the risk of unwanted side effects which occur when the devices are used in improper settings [1–3, 5]. Eadie et al. [5] assessed the optical radiation hazard from at-home IPL sources. They aimed to study the spectral and temporal optical radiation output from at-home IPL. They used the iPulse Personal which is an at-home IPL system for hair reduction. Optical radiation was measured using a calibrated diode array spectrometer. Pulse duration was measured using a fast photodiode. Retinal thermal hazard, blue light hazard, and infrared radiation hazard were all determined. They found that neither the blue light hazard nor the infrared radiation hazard exceeded the exposure limit values set by the International Committee on Non-Ionizing Radiation Protection. They concluded the home-use IPL device (at 10 J/cm^2 and a pulse duration of 26 ms) does not present an optical hazard.

Anti-Aging At-Home Devices

Numerous at-home laser modalities have come on the market in the past 10 years (table 2). Consumers are searching for the latest greatest device that can preserve their youth or turn back the hands on the clock. LED devices have been reported and proven to be safe and effective at regenerating collagen, reducing fine lines and wrinkles, suppressing acne/rosacea, and improving overall skin integrity. These devices have been used in the dermatologist office setting for years. At-home lasers use lower energy, smaller spot size, and longer pulse durations making them safer to use in both at-home and out-of-the-office settings [6].

The PaloVia™ Skin Renewing Laser™ (Palomar) is a non-ablative fractional laser that has been FDA-cleared for the at-home treatment of periorbital fine lines and wrinkles. Two trials were performed, both included an active phase and a maintenance phase. Patients (n = 124) were treated and followed for up to 6 months after completion of the active phase. Fitzpatrick wrinkle scores were assigned by 3 blinded physician evaluators: 90% of study participants had improved by the end of the active phase and 79% by the end of the maintenance phase [6].

At-home devices such as the Omnilux New-U LED light and the Red and Blue LED Light Therapy (from Light Therapy) are two among the

Table 2. At-home anti-aging devices

Device	Mechanism
Tända Regenerate red light kit	660 nm red or 414 nm blue LED
Omnilux U LED light	red 660 nm LED
DPL Systems	combined red and blue LED
Red And Blue Light Therapy	combined red and blue LED
Baby Quasar	red/infrared LED (632–950 nm)
Evis Beauty MD Platinum LED	red light 625–635 nm

many that have gained increasing popularity as at-home photorejuvenation devices [6–9]. Many companies are introducing at-home treatments as this market continues to grow. Silk'n FaceFX is pending FDA clearance.

Lasers/Light Devices for Hair Loss Treatment

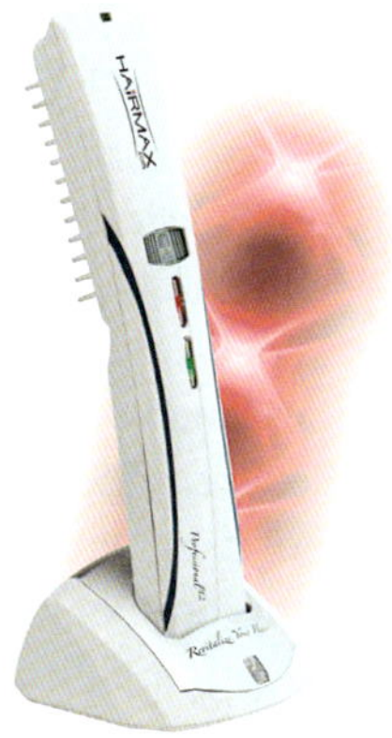

Fig. 2. HairMax Laser Comb.

Hair loss has become an increasingly common problem among men and women, and some are being affected at young ages. There are many products to treat hair loss, ranging from topical medicinal products (such as minoxidil) to oral medications (including finasteride and dutasteride) to topical covering agents such as Toppik.

The FDA has cleared the HairMax LaserComb (fig. 2) for patients with androgenetic alopecia who have a Norwood-Hamilton classification of I–IV. It was found to be clinically effective in over 90% of patients in their clinical trials. The HairMax LaserComb is thought to show results in as little as 8 weeks after treatment. Its mechanism of action is by low-level light at 655 nm stimulating the anagen phase of hair growth, and thereby accelerating hair growth. The comb has 9 beams emitting at that wavelength. When the hair is parted, there is better penetration to the scalp by the laser comb. In a 26-week randomized double-blind controlled trial, 110 males with androgenetic alopecia were treated. Patients in the HairMax group showed a statistically significant and visible increase in hair regrowth compared to baseline. It was also found to be well tolerated, and there were no serious or adverse events reported [10, 11].

At-home laser/light devices have found their niche amongst cosmetic consumers. They are relatively safe when used properly; most available in the United States are FDA-approved, and may provide patients with some benefit and improvement. Nevertheless, these devices do not replace an in-office visit to a dermatologist, nor are they as effective as those devices found in office settings.

References

1 Lopez-Esterbaranz JL, Cuerda E: Medical-cosmetic devices for home use: present and future considerations. Acta Dermatol 2010;1011:223–229.

2 Gold M, Andriessen A, Biron J, Andriessen H: Clinical efficacy of self-applied blue light therapy for mild-to-moderate facial acne. J Clin Aesthet Dermatol 2009;2:44–50.

3 Hodson DS: Current and future trends in home laser devices. Semin Cutan Medi Surg 2008;27:292–300.

4 Alster T, Tanzi E: Effect of novel low-energy pulsed light device for home use hair removal. Dermatol Surg 2009;35:483–489.

5 Eadie E, Miller P, Goodman T, Mosseley H: Assessment of the optical radiation hazard from a home use intense pulsed light (IPL) source. Lasers Surg Med 2009;4:534–539.

6 Carier H, Le Pillouer-Prost A, Grognard C: Light emitting diodes (LED) (in French). Ann Dermatol Venereol 2009;136(suppl 6):S351–S358.

7 Leyden J, Stephens T: Multi-center clinical trials of home-use non-ablative fractional laser device for wrinkle reduction (abstract). Lasers Surg Med 2010;42(suppl 22):1–125.

8 Mulholland RS: Silk'n – a novel device using home pulsed light for hair removal at home. J Cosmet Laser Ther 2009;11:106–109.
9 Emerson R, Town G: Hair removal with a novel, low fluence, home-use intense pulsed light device. J Cosmet Laser Ther 2009;11:98–105.
10 Goldberg DJ: Laser and light based hair removal: an update. Expert Rev Med Devices 2007;4:253–260.
11 Leavitt M, Charles G, Heyman E, Michaels D: HairMax LaserComb laser phototherapy device in the treatment of male androgenetic alopecia: a randomized, double-blind, sham device-controlled, multicentre trial. Clin Drug Investig 2009;29:283–292.

Alia S. Brown, MD
Skin Laser & Surgery Specialists of New York & New Jersey
20 Prospect Ave 702
Hackensack, NJ 07601 (USA)
E-Mail diva71078@aol.com

Bogdan Allemann I, Goldberg DJ (eds): Basics in Dermatological Laser Applications.
Curr Probl Dermatol. Basel, Karger, 2011, vol 42, pp 166–172

Intense Pulsed Light

Nicola L. Schoenewolf · Marjam J. Barysch · Reinhard Dummer

Department of Dermatology, University Hospital Zurich, Zurich, Switzerland

Abstract

In addition to lasers, intense pulsed light (IPL) sources are widely used in medicine to treat various indications, such as vascular lesions, irregular pigmentation and hypertrichosis. In contrast to lasers, IPL systems are broadband flash lamps that emit polychromatic incoherent light ranging from visible to infrared (500–1,300 nm). Optical filters are used to tailor the polychromatic light to specific needs. As a broad range of wavelengths are delivered, treatment of multiple chromophores – including melanin, hemoglobin, water and collagen – within the same exposure is possible.

Introduction

History of Intense Pulsed Light Devices

In 1994, intense pulsed light (IPL) technology was introduced to the market as a medical device. It had gained commercial influence after Mühlbauer et al. [1] first introduced polychromatic infrared light for the treatment of vascular lesions in 1976. First-generation devices led to some side effects, such as epithelial damage, due to infrared light emission. However, second-generation devices have shown a much better reduction of the infrared portion by using water filters [2].

Light Spectrum and Chromophores

IPL systems are broadband flashlamps that emit polychromatic incoherent light ranging from visible to infrared (500–1,300 nm). In contrast to lasers, which use a flashlamp as an external energy pump to excite the laser medium, IPL technology uses the flashlamp directly. Optical cutoff filters are built in to optimize absorption of the respective target chromophores. The omission of wavelengths above 950 nm helps to reduce strong water absorption, and thus unwanted epidermal heating and potential destruction [2]. Furthermore, filters in the long-wavelength spectrum reduce the strong absorption of melanin, making the device also suitable for darker skin types and preventing adverse effects such as erythema, blistering and crusting.

Although IPL devices are not lasers, the technology is based on the principle of selective photothermolysis described by Anderson and Parrish [3].

Similar to laser systems, IPL technology is based on the principle of direct energy transfer to chromophores located in the skin. Heat induction allows target-specific treatment. Due to this target-specific absorption, maximum therapeutic versatility is achieved.

The delivery of multiple wavelengths simultaneously allows for the treatment of various chromophores – including melanin, (oxy)hemoglobin, water and collagen – even within the same exposure [4]. Melanin is the target in pigmented lesions, for hair removal and skin rejuvenation. It absorbs light wavelengths from 250 to 1,200 nm with decreasing intensity. The absorption peaks of oxyhemoglobin are 418, 542 and 577 nm, which are relevant for the treatment of vascular disorders and skin rejuvenation. Collagen fibers absorb light from 400 to 600 nm. According to the various absorption spectra and maximums of the chromophores, appropriate wavelength ranges can be selected to target them. Due to the polychromatic nature of IPL, multiple problems can be addressed in one exposure.

Intense Pulsed Light Systems

IPL systems are versatile devices that allow the individual selection of multiple parameters, such as wavelength, pulse duration, fluence and number of passes. Pulse durations available are usually between 0.2 and 100 ms. Within this millisecond domain, light can be delivered as single, double or triple pulses. Repetition rates are usually rather low (0.3–1 Hz). On most devices, pulse sequences and pulse delay times can be chosen.

The spot sizes of IPL systems usually allow the treatment of large areas. Thus, in comparison with laser systems, the treatment time can be reduced.

Safety Standards

Skin Types and Sun Exposure

As for laser systems, the skin type (according to Fitzpatrick [5]) has a large impact upon the treatment results. The darker the skin type, the more side effects have been documented [6]. With the broad range of wavelengths available, various skin types can be treated by IPL. Cutoff filters within the short-wavelength spectrum and the use of longer wavelengths allow the treatment of darker-skinned patients, as melanin absorption decreases with increasing wavelengths [7]. Sun-tanned patients need to wait for 4–6 weeks until they can undergo IPL treatment in order to prevent unwanted hypo- or hyperpigmentation. A basic rule is the avoidance of sun or tanning beds 4–8 weeks before, throughout and up to 8 weeks after IPL treatment.

Side Effects and Prevention

In the majority of cases, few severe treatment-induced side effects are observed after IPL treatment, and the recovery period is impressively short [6]. During IPL treatment, slight pain is usually reported. By strict cooling before, during and after the treatment, painful sensations can be remarkably reduced. Cooling is crucial, as it further prevents absorption of unwanted amounts of laser energy by the pigmented epidermis, and hence reduces side effects such as post-inflammatory pigment changes, blistering or even scarring. In order to avoid ice burns, pauses during the cooling process after treatment are recommended [8]. When treating vascular lesions, however, pre-cooling should be avoided.

Swelling and erythema are common side effects that persist for a short time after IPL treatment. Crusts, blisters and scars are rare and can be avoided by adequate test spot treatment and strict cooling of the epidermis.

Indications

Vascular Lesions

When treating vascular lesions with IPL, blood vessels are heated in order to coagulate and destroy them. Based on the principles of selective photothermolysis, IPL can selectively target oxyhemoglobin (absorption peaks at 540 and 580 nm), which is usually the chromophore of small and superficial clinically red vessels. Deoxyhemoglobin is typically the chromophore of deeper bluish vessels and shows absorption levels between 800 and 1,200 nm. The surrounding tissue is mostly unaffected. Heating of the blood vessels leads to thermocoagulation, necrosis and replacement by fibromatous tissue. Pulse durations in the range of milliseconds are required as larger structures need more time for adequate heat absorption.

Various vascular lesions – such as port wine stains, spider nevi, telangiectasias, angiomas and erythrosis – can be effectively approached by IPL technology. The rate of side effects is low. However, in many cases improvements of the skin lesions are poorer in one treatment session as compared to treatment by lasers such as alexandrite, Nd:YAG, KTP or flashlamp pulsed dye [9–11]. IPL treatment of vascular lesions is advantageous with respect to the lack of purpura after treatment [12].

Usually, several treatment sessions are required for most vascular lesions, with intervals of at least 4–6 weeks. As an immediate response to an effective treatment, the lesion takes on a gray-bluish color. Notably, pressure should be avoided when treating thin superficial vessels in order to avoid compression of the vessel. Erythematotelangiectatic rosacea is a frequent indication for IPL treatment. The combination of abnormal vessel ablation on the one hand and collagen remodeling on the other can be taken as an explanation for the efficient long-term IPL results in rosacea patients.

To minimize damage to the melanocytes and keratinocytes, subsequent cooling of the epidermis is required. As cooling leads to vasoconstriction, and hence to a decreased chance of thermocoagulation of the target vessels, recent IPL devices allow the user to turn off the integrated cooling system when working on superficial, diffuse and very small vessels.

Application of local anesthetizing cream prior to treatment is generally required for children if they are not narcotized or analgo-sedated. If superficial and thin vessels are targeted, local anesthetizing creams without vasoconstrictive components (such as tetracaine, Ametop®) need to be chosen, as the commonly used lidocaine and prilocaine (Emla®) lead to vasoconstriction and thus minimize the effects of IPL.

Pigmented Lesions

When treating pigmented lesions, first and foremost one has to rule out (pre)malignancy of the lesion to be treated. Furthermore, we do not advise treating melanocytic nevi with IPL or lasers. When treating benign pigmented lesions, the pigment-specific Q-switched lasers – such as the Nd:YAG, ruby or alexandrite lasers with their pulse durations within the nanosecond range – are superior and effective in targeting melanin. However, IPL with its pulse duration within the millisecond domain also shows considerable effects in the treatment of pigmented lesions, such as photo-induced dyschromia, lentigines and ephelides [2]. After treatment with IPL, a darkening of the pigmented lesions can be observed, with subsequent sloughing.

As the chromophore of pigmented lesions is melanin, sufficient absorbance and appropriate depth of penetration are achieved with wavelengths between 630 and 1,100 nm. A great advantage of novel IPL devices is the implementation of a wavelength spectrum alongside the filtering of unwanted wavelengths. Thus, treatment within the ranges of 500–670 and 870–1,400 nm (StarLux, Lux G, Palomar Medical Technologies, Burlington, USA) is able to target both, pigmented lesions and vascular lesions. Hypopigmentation is prevented in most of the cases as epidermal melanin is not affected. Depending on the depths of the lesion, different cutoff filters may be used [13]. UV protection prior to and after treatment is pivotal for minimizing the risk of post-treatment complications and for optimizing pretreatment conditions with increased effects of IPL treatment [14].

Melanocytic Lesions

Melanoma development after laser treatment of histologically confirmed melanocytic nevi has been reported [15, 16]. The relevance of IPL devices in the induction of malignant transformation is unclear and cannot be ruled out until reliable data are available. Furthermore, treatment of unrealized malignant lesions bears the risk of delayed diagnosis with a fatal outcome.

Therefore, melanocytic nevi, lentigo maligna, melanoma and other malignancies have to be ruled out by an experienced dermatologist, and should never be treated with laser or IPL devices. For this reason, application of IPL devices for the treatment of melanocytic nevi is not reported in this chapter.

Tattoos

Using IPL devices for the removal of tattoos is – despite being suggested by several manufacturers, not recommend in any situation. The treatment of tattoos requires pulse durations within the nanosecond domain as well as very high light intensities, and both criteria are only fulfilled by the Q-switched lasers. Using IPL with pulse durations in the millisecond domain leads to heat dissipation from the tattoo pigment to the surrounding tissue and subsequent side effects such as scarring [17].

Hair Removal

The wish to get rid of hairy areas, particularly in females, has a long tradition. Next to shaving, waxing and bleaching, epilation of hair has become one of the most frequently used methods. A comparison of IPL devices (with different filters), a diode laser and an alexandrite laser revealed comparable data regarding the hair removal efficiency. However, pain during treatment was most reduced under IPL treatment [14]. Hair removal nowadays is one of the key indications for IPL. Melanin serves as the main chromophore; however, the exact mechanism of action has not yet been fully elucidated. The shorter the wavelength, the better the absorption by melanin pigments. The usually large spot sizes of IPL devices allow for deeper penetration of polychromatic light. It is therefore essential to find a balance between deep hair shaft melanin and epidermal pigment absorption of the light in order to reach the target of interest in the bulbus area of the hair [14].

This conflict needs to be inter-individually resolved [8, 18]. In order to achieve long-term hair removal, it is believed that stem cell destruction is necessary [19, 20]. In repeated IPL treatment sessions, stem cells in the late anagen phase are targeted over and over again [8]. Reduced IPL treatment efficiency was reported by Grossman et al. [21] in 1996 due to waxing procedures that resulted in loss of the dermal hair shaft.

According to the principle of thermokinetic selectivity, the time of thermal relaxation needs to be longer than the epidermal relaxation but shorter than the hair follicle. Thus, the hair follicle can be optimally targeted and the epidermis prevented from unspecific heating. Histological studies have been performed in order to understand the direct on-treatment results: often extensive damage of the hair follicle was observed shortly after treatment [22, 23].

Lower fluences under 10 J/cm^2 have been shown to reduce hair growth for a short time, avoiding damage of the hair bulge. A 95% hair reduction was seen after 2 treatments in 4 weeks. This treatment was continued for further 12 weeks in a 2-week pattern. Catagen and telogen phases were effectively induced, while no permanent damage to the hair follicle was observed [24].

So far, the results are quite different with regard to the final hair reduction efficiency.

Differences in the outcome can be explained by several reasons: the skin type according to Fitzpatrick, the hair color and hair follicle geometry play a role [23, 25, 26], as well as the fluences and pulse durations chosen for treatment. Moreover, patient compliance plays another essential role, also in relation to unwanted side effects.

Minor side effects consist of crusts, erythema, scars, blisters, transient hypopigmentation or hyperpigmentation [6, 27, 28]. Severe side effects, such as paradoxical hypertrichosis as well as leukotrichia, have occasionally been reported [6, 28–30]. However, in cases of correct and responsible IPL handling, almost no severe side effects have been reported [27].

Acne

There are few and rather inhomogeneous data on the treatment of inflammatory acne with IPL, and it is far from being a standard treatment for acne lesions [2, 31]. However, positive effects, as indicated by a reduction in inflammatory acne lesions, have been shown [32–35]. However, photodynamic therapy in combination with IPL has shown better effects than IPL alone [33, 34, 36]. *Propionibacterium acnes* produces porphyrins that show absorption peaks between 400 and 665 nm. Upon irradiation with visible light of various wavelengths, bactericidal reactive oxygen species are produced that destroy the bacterium [2]. Exogenous chromophores can be applied, leading to increased target destruction upon light induction [37, 38]. A second possibility to target acne lesions by IPL is the selective photothermolysis of blood vessels. IPL has been shown to be effective in the treatment of vascular lesions. Sebaceous glands that are supplied by vessels are reduced in their secretion rate upon IPL treatment [2, 39].

Side effects vary between irritation/skin dryness, subsequent to topical or systemic acne treatment, and temporary erythema and edema, following photo treatment. Its probable that IPL treatment should best be used as an adjunct to an established systemic or topical acne therapy.

Photoaging

Photoaging and skin rejuvenation are gaining more importance for dermatologists due to the increasing exposure to UV irradiation on the one hand, and the desire for eternal youth as communicated by the media on the other hand. UV irradiation, via oxygen radicals, induces deformation of collagen and elastin fibers, resulting in decreased dermal structure and volume, e.g. lax skin, rhytides and wrinkles. Furthermore, extensive UV irradiation leads to pigmented and vascular lesions, such as solar lentigines, telangiectasias and erythrosis interfollicularis colli – the treatment of which have been discussed above. As photoaged skin presents with dyschromia and vascular changes, IPL with its broad range of delivered wavelengths is ideal to treat these changes, which can be addressed together in a single treatment.

Rejuvenation

The idea of wrinkle and rhytide reduction using IPL devices is based on secondary observations of skin texture improvement and on histological examination after IPL treatment. Contrary to the use of IPL devices for the treatment of photoinduced telangiectasias or irregular pigmentation, current data for the treatment of rhytides and wrinkles are dissonant and question the effects of IPL for this indication. However, the outcome is difficult to evaluate due to only subtle effects and interference with improvement of overall skin texture. Furthermore, study designs, evaluation of outcome and the applied IPL devices and adjustments differ, and may have led to the inhomogeneous results. However, IPL devices are favored by the patients due to high comfort levels, few side effects and reduced patient downtime [27].

The hypothesis of wrinkle and rhytide reduction by IPL devices is based on the theory of heat-induced contracture of collagen fibers performed by high-intensity light energy. Histological evaluation of IPL-treated photoaged skin proves this theory as stimulation of collagen I and III, as well as of elastin synthesis, has been found. Furthermore, hyperplasia and stimulation of fibroblasts and re-arrangement of elastin and collagen fibers within the stroma in post-treatment biopsies (compared to pretreatment biopsies) were detected [40, 41]. Higher wavelengths, particular around 1,200 nm, have been described to induce collagen I, III, and elastin stimulation due to dermal absorption by water which activates a cytokine reaction. Application of five IPL treatments (λ_{em}: 570–645 nm) has been shown histologically to lead to an epidermal thickening of 100–300 mm, formation of new rete ridges, decrease in degenerated elastic fibers, and synthesis of new dermal collagen fibers [42, 43]. The combination of IPL devices with botulinum toxin injections may bring about a more convenient collagen structure, achieved by paralyzing muscle tissue with subsequent immobilization of the dermal matrix proteins [15].

Conclusion

IPL technology allows effective and economical treatment of various skin conditions, and thus complements the spectrum covered by laser devices.

Some indications such as vascular lesions and hypertrichosis have been shown to be even more efficiently treated by means of IPL in comparison to conventional laser systems. Side effects are rare and in most cases moderate and temporary.

Further controlled studies are needed to establish treatment recommendations and regulations with regard to IPL indications and handling.

References

1 Mühlbauer W, Nath G, Kreitmair A: Treatment of capillary hemangiomas and nevi flammei with light (in German). Langenbecks Arch Chir 1976;(suppl):91–94.
2 Babilas P, Schreml S, Szeimies RM, Landthaler M: Intense pulsed light (IPL): a review. Lasers Surg Med 2010;42:93–104.
3 Anderson RR, Parrish JA: Selective photothermolysis: precise microsurgery by selective absorption of pulsed radiation. Science 1983;220:524–527.
4 Anderson RR, Parrish JA: The optics of human skin. J Invest Dermatol 1981;77:13–19.
5 Fitzpatrick TB: The validity and practicality of sun-reactive skin types I through VI. Arch Dermatol 1988;124:869–871.
6 Moreno-Arias GA, Castelo-Branco C, Ferrando J: Side-effects after IPL photodepilation. Dermatol Surg 2002;28:1131–1134.
7 Battle EF Jr, Soden CE Jr: The use of lasers in darker skin types. Semin Cutan Med Surg 2009;28:130–140.
8 Kautz G, Rick K, Sandhofer M: Photoepilation: Zur Praxis der Haarentfernung mit Licht- und Lasersystemen. Darmstadt, Steinkopff, 2004.
9 McGill DJ, MacLaren W, Mackay IR: A direct comparison of pulsed dye, alexandrite, KTP and Nd:YAG lasers and IPL in patients with previously treated capillary malformations. Lasers Surg Med 2008;40:390–398.
10 Raulin C, Weiss RA, Schonermark MP: Treatment of essential telangiectasias with an intense pulsed light source (PhotoDerm VL). Dermatol Surg 1997;23:941–945, discussion 5–6.
11 Raulin C, Werner S: Treatment of venous malformations with an intense pulsed light source (IPLS) technology: a retrospective study. Lasers Surg Med 1999;25:170–177.
12 Neuhaus IM, Zane LT, Tope WD: Comparative efficacy of nonpurpuragenic pulsed dye laser and intense pulsed light for erythematotelangiectatic rosacea. Dermatol Surg 2009;35:920–928.
13 Moreno Arias GA, Ferrando J: Intense pulsed light for melanocytic lesions. Dermatol Surg 2001;27:397–400.
14 Ciocon DH, Boker A, Goldberg DJ: Intense pulsed light: what works, what's new, what's next. Facial Plast Surg 2009;25:290–300.
15 Zipser MC, Mangana J, Oberholzer PA, French LE, Dummer R: Melanoma after laser therapy of pigmented lesions – circumstances and outcome. Eur J Dermatol 2010;20:334–338.
16 Dummer R: About moles, melanomas, and lasers: the dermatologist's schizophrenic attitude toward pigmented lesions. Arch Dermatol 2003;139:1405–1406.
17 Wenzel S, Landthaler M, Baumler W: Recurring mistakes in tattoo removal: a case series. Dermatology 2009;218:164–167.
18 Paus R, Cotsarelis G: The biology of hair follicles. N Engl J Med. 1999;341:491–497.
19 Sadick NS, Shea CR, Burchette JL Jr, Prieto VG: High-intensity flashlamp photoepilation: a clinical, histological, and mechanistic study in human skin. Arch Dermatol 1999;135:668–676.
20 Schroeter CA, Raulin C, Thurlimann W, Reineke T, De Potter C, Neumann HA: Hair removal in 40 hirsute women with an intense laser-like light source. Eur J Dermatol 1999;9:374–379.
21 Grossman MC, Dierickx C, Farinelli W, Flotte T, Anderson RR: Damage to hair follicles by normal-mode ruby laser pulses. J Am Acad Dermatol 1996;35:889–894.
22 Dierickx CC, Casparian JM, Venugopalan V, Farinelli WA, Anderson RR: Thermal relaxation of port-wine stain vessels probed in vivo: the need for 1–10-millisecond laser pulse treatment. J Invest Dermatol 1995;105:709–714.

23 Ross EV, Ladin Z, Kreindel M, Dierickx C: Theoretical considerations in laser hair removal. Dermatol Clin 1999;17:333–355, viii.
24 Roosen GF, Westgate GE, Philpott M, Berretty PJ, Nuijs TA, Bjerring P: Temporary hair removal by low fluence photoepilation: histological study on biopsies and cultured human hair follicles. Lasers Surg Med 2008;40:520–528.
25 Ort RJ, Anderson RR: Optical hair removal. Semin Cutan Med Surg 1999;18:149–158.
26 Sadick NS, Weiss RA, Shea CR, Nagel H, Nicholson J, Prieto VG: Long-term photoepilation using a broad-spectrum intense pulsed light source. Arch Dermatol 2000;136:1336–1340.
27 Fodor L, Carmi N, Fodor A, Ramon Y, Ullmann Y: Intense pulsed light for skin rejuvenation, hair removal, and vascular lesions: a patient satisfaction study and review of the literature. Ann Plast Surg 2009;62:345–349.
28 Fodor L, Menachem M, Ramon Y, Shoshani O, Rissin Y, Eldor L, Peled IJ, Ullmann Y: Hair removal using intense pulsed light (EpiLight): patient satisfaction, our experience, and literature review. Ann Plast Surg 2005;54:8–14.
29 Moreno-Arias G, Castelo-Branco C, Ferrando J: Paradoxical effect after IPL photoepilation. Dermatol Surg 2002;28:1013–1016, discussion 1016.
30 Radmanesh M, Azar-Beig M, Abtahian A, Naderi AH: Burning, paradoxical hypertrichosis, leukotrichia and folliculitis are four major complications of intense pulsed light hair removal therapy. J Dermatolog Treat 2008;19:360–363.
31 Haedersdal M, Togsverd-Bo K, Wulf HC: Evidence-based review of lasers, light sources and photodynamic therapy in the treatment of acne vulgaris. J Eur Acad Dermatol Venereol 2008;22:267–278.
32 Yeung CK, Shek SY, Bjerring P, Yu CS, Kono T, Chan HH: A comparative study of intense pulsed light alone and its combination with photodynamic therapy for the treatment of facial acne in Asian skin. Lasers Surg Med 2007;39:1–6.
33 Sami NA, Attia AT, Badawi AM: Phototherapy in the treatment of acne vulgaris. J Drugs Dermatol 2008;7:627–632.
34 Taub AF: A comparison of intense pulsed light, combination radiofrequency and intense pulsed light, and blue light in photodynamic therapy for acne vulgaris. J Drugs Dermatol 2007;6:1010–1016.
35 Chang SE, Ahn SJ, Rhee DY, Choi JH, Moon KC, Suh HS, Soyun-Cho: Treatment of facial acne papules and pustules in Korean patients using an intense pulsed light device equipped with a 530- to 750-nm filter. Dermatol Surg 2007;33:676–679.
36 Hongcharu W, Taylor CR, Chang Y, Aghassi D, Suthamjariya K, Anderson RR: Topical ALA-photodynamic therapy for the treatment of acne vulgaris. J Invest Dermatol 2000;115:183–192.
37 Degitz K: Phototherapy, photodynamic therapy and lasers in the treatment of acne. J Dtsch Dermatol Ges 2009;7:1048–1054.
38 Gold MH: Acne and PDT: new techniques with lasers and light sources. Lasers Med Sci 2007;22:67–72.
39 Lloyd JR, Mirkov M: Selective photothermolysis of the sebaceous glands for acne treatment. Lasers Surg Med 2002;31:115–120.
40 Feng Y, Zhao J, Gold MH: Skin rejuvenation in Asian skin: the analysis of clinical effects and basic mechanisms of intense pulsed light. J Drugs Dermatol 2008;7:273–279.
41 Goldberg DJ: New collagen formation after dermal remodeling with an intense pulsed light source. J Cutan Laser Ther 2000;2:59–61.
42 Hernandez-Perez E, Ibiett EV: Gross and microscopic findings in patients submitted to nonablative full-face resurfacing using intense pulsed light: a preliminary study. Dermatol Surg 2002;28:651–655.
43 Wong WR, Shyu WL, Tsai JW, Hsu KH, Pang JH: Intense pulsed light effects on the expression of extracellular matrix proteins and transforming growth factor beta-1 in skin dermal fibroblasts cultured within contracted collagen lattices. Dermatol Surg 2009;35:816–825.

Reinhard Dummer, MD
Department of Dermatology, University Hospital Zurich
Gloriastrasse 31
CH–8091 Zurich (Switzerland)
E-Mail Reinhard.Dummer@usz.ch

Bogdan Allemann I, Goldberg DJ (eds): Basics in Dermatological Laser Applications.
Curr Probl Dermatol. Basel, Karger, 2011, vol 42, pp 173–180

Light-Emitting Diode

Michael H. Gold

Tennessee Clinical Research Center, Gold Skin Care Center, and Department of Dermatology,
Vanderbilt University School of Nursing, Vanderbilt University School of Medicine, Nashville, Tenn., USA

Abstract

Light-emitting diode (LED) photomodulation has become a recognized player in the world of lasers and light sources. It is used to treat a variety of clinical entities, including photorejuvenation, erythema-induced injury following laser and other cosmetic procedures, and acne vulgaris. Its use has increased and will increase further as our understanding of LED devices deepens.

For many years, there has been a great deal of interest in utilizing low-level laser therapy (LLLT) to treat a variety of medical disorders – with some studies producing fascinating results, others questionable results. Light-emitting diodes (LEDs) have been utilized in dermatology over recent years, and although there is still debate concerning exactly where their place may be in this field, it is safe to say that LED therapy is here to stay. Therefore, the onus is on us to figure out how best to use this therapy for the benefit of our patients.

Although LEDs have been used in photomedicine to treat a variety of diseases, including infectious diseases and even hypertension, this manuscript focuses on the dermatologic application of LEDs, and where this therapy is heading in the future.

Several dermatologists have been at the forefront of LED research and it is revealing to see how their efforts have translated into medical devices that actually offer benefit to our patients – both devices used in our clinics and devices that are now available for home use.

Photorejuvenation with LED Technology

In 2002, McDaniel et al. [1] first demonstrated that when utilizing various lasers at varying wavelengths, fibroblast activity could be modulated and regulated. In their work, they were able to demonstrate that using LEDs at varying wavelengths or by using monochromatic light, different fluences could lead to upregulation of procollagen synthesis in fibroblast cultures [2]. They were able to demonstrate a clinical correlation with fibroblast culture stimulation. They measured procollagen by various microassay systems, and found that collagen synthesis was accompanied by an equal and opposite reaction which involved downregulation of matrix metalloprotease, also known as MMP-1 or collagenase. They also measured the stimulation of fibroblasts by measuring procollagen with varying fluences of a pulsed dye laser. They were able to demonstrate a broad peak of collagen stimulation at fluences that most professionals had considered too low to actually produce any significant kind of clinical

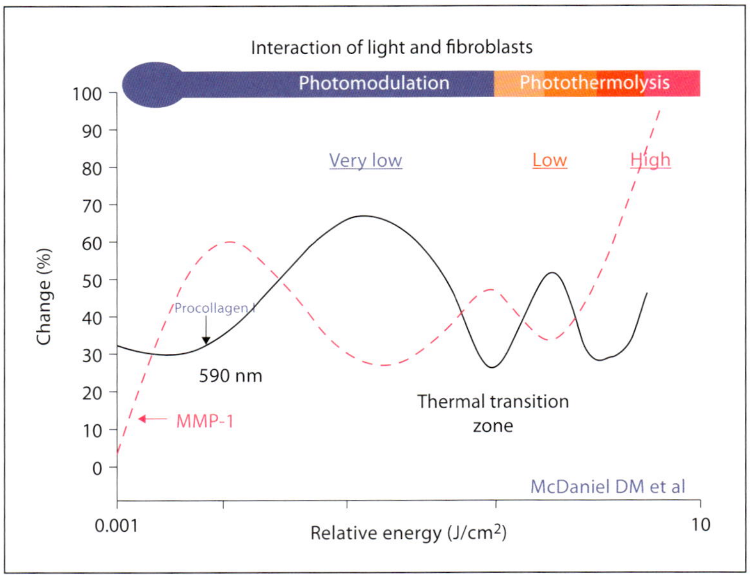

Fig. 1. The interaction of light and fibroblasts. Reprinted with permission from *Journal of Drugs in Dermatology* [19].

effect. This concept, known as photomodulation, incorporates the use of low-level narrow-band or coherent light using specific pulse sequences and pulse durations. This concept has been confirmed by many others [3, 4]. The interaction of light and fibroblasts is shown in figure 1.

LEDs in the wavelength range around 590 nm were found to have the greatest effects on stimulating cultured fibroblasts. In experiments performed on varying wavelengths of light, wavelengths in this range were found to have the most effects on fibroblast activity [1]. What was also quickly realized, and led to the term 'LED photomodulation', was that in order for this activity to occur at all, specific pulse sequences were required. Continuous application of LED light had no effect on fibroblast activity; a code was needed to turn on and off the LED activity. Thus, LED photomodulation was born as a mechanism for non-thermal stimulation of fibroblast cell growth activity.

Photomodulation occurs by stimulating mitochondrial cell organelles with the proper sequence of photons and the linked up- and downregulation of mitochondrial gene activities. As a result, upregulation of the mitochondrial cytochrome electron transport pathway and mitochondrial DNA gene modulation occurs. This has been found to correlate well with the concept that accelerated signs of aging and premature cell death are associated with genetic damage within mitochondria [5]. The ability of mitochondria to utilize light has also been demonstrated, and has been shown to occur in the cytochrome species within the mitochondrial membranes. These cytochrome membranes are synthesized by PpIX (protoporphyrin IX), which has demonstrated its peak absorption in the range of 562–600 nm [6]. Light absorption within the mitochondrial membrane causes changes in various molecules within the cell, leading to a transformation in which adenosine is converted to adenosine triphosphate, providing more energy for non-mitotic cell processes. This process is shown in figure 2.

The potential for this type of therapy soon became apparent as clinical trials looking at this effect in treating photodamaged skin began.

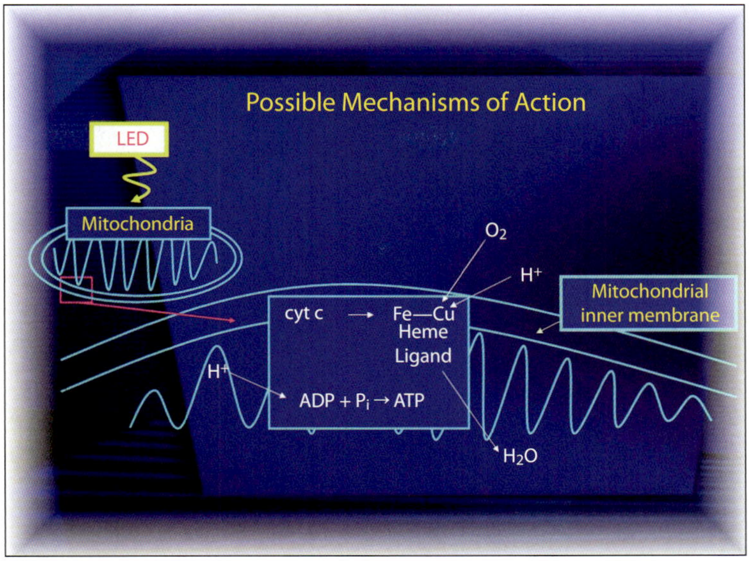

Fig. 2. Adenosine (ADP) converted to adenosine triphosphate (ATP) = non-mitotic cell processes. Reprinted with permission from *Journal of Drugs in Dermatology* [19]. All rights reserved.

A multicenter clinical trial was performed on a group of 90 individuals with photodamage. In their work, McDaniel and colleagues [7–9] looked at various parameters of photodamage and the effects of LED photomodulation on the photodamaged skin. The patients that were initially studied were re-evaluated at 1 year, looking at the effects of LED photomodulation in terms of its long-term potential [10, 11]. In their work, these investigators used an array of LEDs with a dominant wavelength of 590 nm to assess the clinical improvement in photodamaged skin using non-thermal photorejuvenation.

In their institutional review board clinical trial, Weiss et al. [9] treated 90 individuals who received a series of eight 590 nm LED arrays over a 4-week time period. The device used was known as the GentleWaves LED photomodulation unit (LightBioScience LLC, Virginia Beach, Va., USA; fig. 3). It is a device that contains 2,040 LEDs, which emit a 590 nm wavelength as their main wavelength of light. The device emits 2 pulses per cycle with 100 cycles delivered over 35 s. The preset on-time for each pulse is 250 ms and the off-time is 100 ms. The total output of energy ranges from 0.1 to 0.9 J/cm^2. In this clinical trial, the authors utilized a cumulative dose of 0.1 J/cm^2 of output for a single treatment. Patients were positioned 2 cm away from the light source with the chin fixed in position.

The patients in the clinical trial had skin types I–IV and treatments were given with a minimum of 2 days' separation between the treatments. Stereotactic standard images were taken during the course of the clinical trial and during the 1-year follow-up time period. Skin surface topography measurements were performed on 30 of the individuals using a digital optical surface profilometry. Skin biopsy specimens were taken from lateral canthus sites on 10 of the participants in the clinical trial.

Independent reviewers graded the responses to the LED therapy. Utilizing the Fitzpatrick photoaging scale, it was determined that there was a one-class improvement in the wrinkle scoring scale in the majority of the patients, which was shown to be statistically significant. The

Fig. 3. GentleWaves LED photomodulation unit (Light-BioScience LLC).

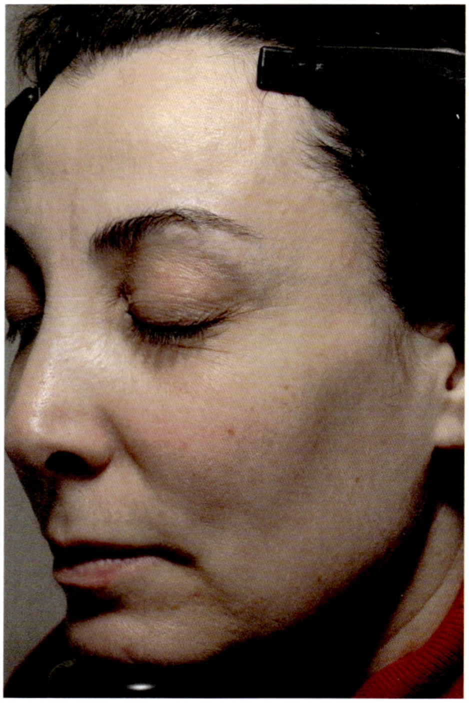

Fig. 4. Clinical example of patient before LED therapy. Courtesy of David J. Goldberg, MD.

best clinical results were noted at 4 months after therapy with over 85% of the patients showing at least a 25% improvement. In the study, elastosis and skin redness were also noted to improve. Results at 12 months were not as marked as at the 4-month time period, but still noted to be more significant than baseline evaluations; 84% of the patients noted a significant improvement in their skin texture.

Skin topography measurements showed improvement in 57% of the patients at 1 week following the LED therapy. At 1 and 3 months following the therapy, topographical improvements were noted to be 48 and 62%. Overall, the average improvement over baseline was 10%. All of the skin biopsies showed significant improvement in collagen production, with increased collagen deposition noted in the papillary dermis. Staining with collagen I antibodies demonstrated a 28% average increase while staining with MMP-1 showed an average reduction of 4%. No serious adverse events were noted during the course of the clinical trial; 2 patients noted a temporary flair of acne vulgaris while 17 patients noted improvement in their acne.

Clinical examples of patient improvements are shown in figures 4 and 5.

Another interesting caveat from this original clinical trial on LED photomodulation was the fact that there was noted an anti-inflammatory effect and erythema reduction effect in some of the participants. Observations have noted an appreciable difference in UV erythema when LED photomodulation is given within several hours after UV exposure. LED photomodulation produces significant downregulation of dermal-matrix-degrading enzymes which are stimulated by UV exposure [10]. Implications for this have also been analyzed and are showing promising results in the clinical setting following a variety of lasers and light sources, as well with other topical therapies [12].

Many find the use of LED photomodulation, as noted, to be beneficial following the use of lasers and light source therapy where the skin becomes

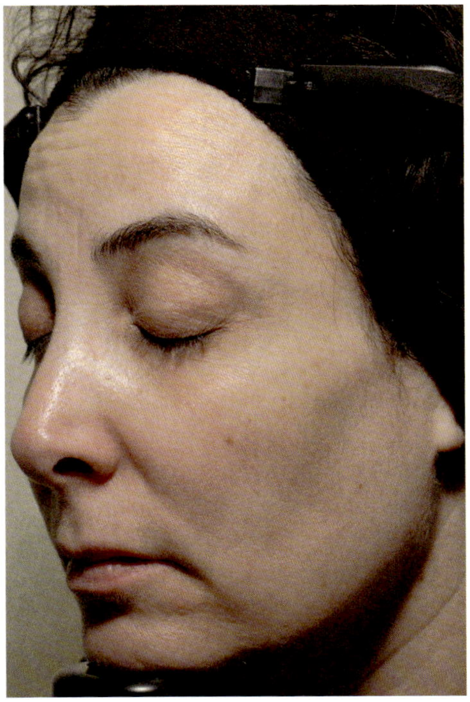

Fig. 5. Clinical example of patient with noted improvement after LED therapy. Courtesy of David J. Goldberg MD.

Fig. 6. Tanda LED blue light home-use device (Pharos Life Corporation).

inflamed and erythematous. We commonly use our LED photomodulation devices in this setting to help with the reduction of erythema following many of our more commonly used laser and light source procedures. This includes all intense pulsed light photorejuvenation therapies, all of our pulsed dye vascular treatments, and all of our non-ablative and ablative fractional resurfacing, including the newly described sublative rejuvenation therapies. Following the laser or light-based therapy, the patient has LED photomodulation therapy for 35 s as previously described and this is repeated at 48–72 h. Observational data from our clinic supports the concept that by utilizing this treatment regimen, postoperative laser course erythema can be significantly reduced and potentially enhanced patient outcomes can be achieved.

Other Uses for Light-Emitting Diode Photomodulation

LEDs have also found their way into other devices currently marketed for photorejuvenation and for the treatment of acne vulgaris. Although many of these devices are available, the two with the most clinical evidence-based data are the Tanda (Pharos Life Corporation, Cambridge, Ont., Canada) and OmniLux (Photo Therapeutics Inc, Carlsbad, Calif., USA) home devices, shown in figures 6 and 7. Blue light for the treatment of inflammatory acne vulgaris has been known for many years and the LED home blue light devices work via the same premise of utilizing LED light in the 490 nm range to effectively destroy *P. acnes* in a photodynamic therapeutic process. Published reports on the use of blue light LED support this concept [13, 14]. Utilizing red light LEDs, these same at-home devices have been shown to be effective in the treatment of photorejuvenation, when used over a period of time.

Fig. 7. OmniLux blue and red light home-use device (Photo Therapeutics Inc.).

In clinical trials with the Tanda device, an in vitro analysis showed that with blue light LED photomodulation therapy, there was a 71% reduction in *P. acnes* bacteria colonies noted at 3 min, which increased to a 86% reduction in *P. acnes* bacteria colonies at 6 min. In clinical trials with the Tanda blue LED photomodulation unit, total acne counts were found to decrease by 69% with the therapy, with a 74% reduction noted in comedonal acne lesions. Furthermore, 100% of patients noted improvement in both their inflammatory and noninflammatory acne lesions [13]. A clinical example of improvement with the Tanda home device is shown in figure 8. In a clinical study of the OmniLux home device, a combination of blue and red light LED photomodulation was given for the treatment of acne vulgaris. Again, a statistically significant reduction in the acne lesions was noted during the 4-week treatment period and during the 4-week follow-up period [14]. In this home red light LED study [14], the Tanda device was used in clinical evaluations looking at hydration of the skin and transepidermal water loss. In this study, patients noted a 75% immediate improvement in the hydration of their skin and a 67% immediate improvement in the comfort of their skin. Further, 100% of the individuals noted softer skin at the end of the clinical trial, as well as improved hydration; 92% reported an improvement in skin smoothness, and 83% noted improvement in the nourishment of the skin [Pharos Life Corp., pers. commun.]. With the OmniLux home device, 74% of the 22 subjects analyzed showed significant improvements in fine lines and wrinkles, after 4 weeks of treatments with the home device, during the 12-week follow-up after the last therapy [15].

Blue and red and combination near-infrared LED photomodulation have played in role in the home treatment of acne vulgaris and the home treatment of photorejuvenation. However, the full impact of these and other home devices in the treatment of dermatologic conditions is still unclear.

Finally, several newer LED devices have become available for clinicians to use in the treatment of other dermatologic manifestations. LEDs for the reduction of fat have recently been proposed and a device for this indication is available in the marketplace, known as Zerona (Erchonia Medical Inc, Mesa, Ariz., USA). Wands of LED light are passed over the body and circumferential reduction has been reported, as well as its potential role as an adjunct to liposuction. In 2004, Jackson et al. [16] reported on the use of this LLLT device for the enhancement of liposuction. They noted that compared to a sham device, this LLLT device enhanced the ease of performing liposuction, reduced the time in surgery, enhanced the extraction of fat, enhanced the emulsification of extracted fat, facilitated the recovery process, decreased the patient's degree of postoperative discomfort and swelling, and decreased the use of recovery medicines used for pain management in these patients. In another evaluation, Jackson et al. [17] evaluated the LLLT device (635 nm and 17.5 W) for body contouring. They noted a statistically

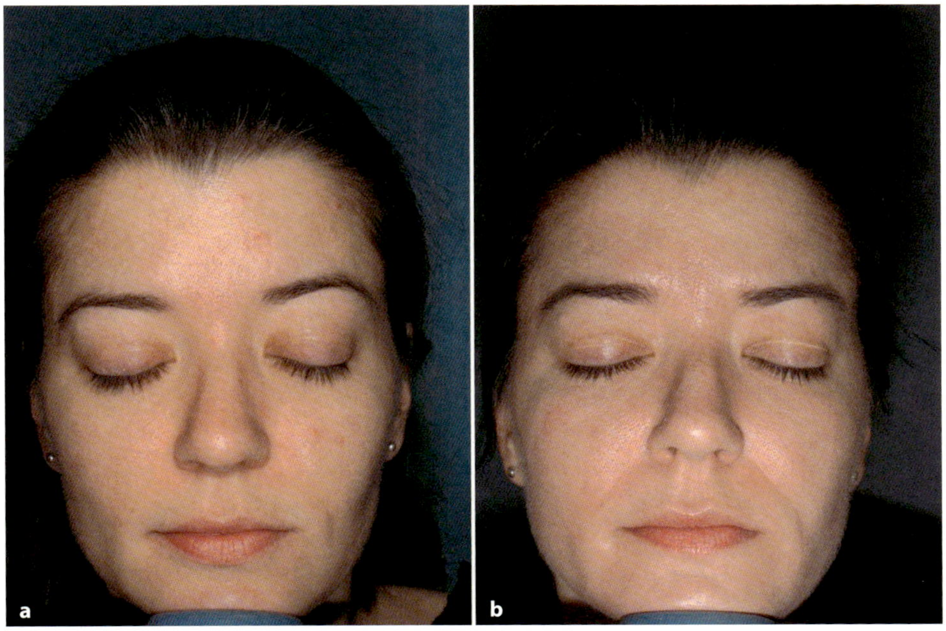

Fig. 8. Clinical example of patient with noted improvement after treatment with the Tanda home use device at baseline (**a**) and after the 8-week treatment period (**b**), with a 62% reduction in the total number of lesions. Reprinted with permission from the *Journal of Clinical and Aesthetic Dermatology* [13]. All rights reserved.

significant improvement in body contouring as a result of this therapy, performed 3 times per week for 2 weeks.

Conclusion and Personal Comments

As a personal caveat, I caution the reader to look to the medical literature and find peer-reviewed clinical and reproducible data before jumping on the band wagon that LEDs can treat all that ails us. However, there is no longer any question or argument that LLLT and LEDs can treat photorejuvenation, reduce post-laser induced erythema, and treat acne vulgaris. The incorporation of these medical devices into your clinical practice will enhance patient outcomes.

References

1 McDaniel DH, Weiss R, Geronemus R, Ginn L, Newman J: Light-tissue interactions I: photothermolysis vs. photomodulation laboratory findings. Lasers Surg Med 2002;14:25.

2 McDaniel DH, Weiss R, Geronemus R, Ginn L, Newman J: Light-tissue interactions II: photothermolysis vs. photomodulation clinical applications. Lasers Surg Med 2002;14:25.

3 Whelan HT, Buchmann EV, Dhokalia A, Kane MP, Whelan NT, Wong-Riley MT, Eells JT, Gould LJ, Hammamieh R, Das R, Jett M: Effect of NASA light-emitting diode irradiation on molecular changes for wound healing on diabetic mice. J Clinic Laser Med Surg 2003;21:67–74.

4 Whelan HT, Connelly JF, Hodgson BD, Barbeau L, Post AC, Bullard G, Buchmann EV, Kane M, Whelan NT, Warwick A, Margolis D: NASA light-emitting diodes for the prevention of oral mucositis in pediatric bone marrow transplant patients. J Clin Laser Med Surg 2002;20:319–324.
5 Trifunovic A, Wredenberg A, Falkenberg M, Spelbrink JN, Rovio AT, Bruder CE, Bohlooly Y-M, Gidlöf S, Oldfors A, Wibom R, Törnell J, Jacobs HT, Larsson NG: Premature aging in mice expressing defective mitochondrial DNA polymerase. Nature 2004;429:417–423.
6 Voet D, Voet JG: Biochemistry, ed 2. New York, John Wiley & Sons, 1995.
7 McDaniel DH, Newman J, Geronemus R, Weiss RA, Weiss MA: Non-ablative non-thermal LED photomodulation – a multicenter clinical photoaging trial. Lasers Surg Med 2003;15:22.
8 Geronemus R, Weiss RA, Weiss MA, McDaniel DH, Newman J: Non-ablative LED photomodulation – light activated fibroblast stimulation clinical trial. Lasers Surg Med 2003;25:22.
9 Weiss RA, McDaniel DH, Geronemus RG, Weiss MA: Clinical trial of a novel non-thermal LED array for reversal of photoaging: clinical, histologic, and surface prolfilometric results. Lasers Surg Med 2005;36:85–91.
10 McDaniel DH, Geronemus R, Weiss RA, Weiss MA, Newman J: LED photomodulation reverses acute UV induced skin damage. Lasers Surg Med 2004;16:30.
11 Weiss RA, McDaniel DH, Geronemus R, Weiss MA, Newman J: Non-ablative, non-thermal light emitting diode (LED) phototherapy of photoaged skin. Lasers Surg Med 2004;16:31.
12 Zitzlsperger K, Counters J, Tasi S, Zelickson B: Photomodulation using Gentlewaves LED with and without rejuvenating masque products. Lasers Surg Med 2004;16:31.
13 Gold MH, Biron JA, Andriessen A: Clinical efficacy of self-applied blue light therapy for mild to moderate facial acne. J Clin Aesthet Derm 2009;2:44–50.
14 Sadick NS: Handheld LED array device in the treatment of acne vulgaris. J Drugs Dermatol 2008;7:347–350.
15 Sadick NS: A study to determine the efficacy of a novel handheld light-emitting diode device in the treatment of aging skin. J Cosmet Dermatol 2008;7:263–267.
16 Jackson RF, Roche G, Butterwick KJ, Dedo DD, Slattery KT: Low-level laser-assisted liposuction: a 2004 clinical study of its effectiveness for enhancing ease of liposuction procedures and facilitating the recovery process for patients undergoing thigh, hip, and stomach contouring. Am J Cosmet Surg 2004;21:191–198.
17 Jackson RF, Dedo DD, Roche GC, Turok DI, Maloney RJ: Low-level laser therapy as a non-invasive approach for body contouring: a randomized, controlled study. Lasers Surg Med 2009;41:799–809.
18 Weiss RA, Weiss MA, Geronemus RG, McDaniel DH: A novel non-thermal non-ablative full panel LED photomodulation device for reversal of photoaging: digital microscopic and clinical results in various skin types. J Drugs Dermatol 2004;3:605–610.

Michael H. Gold, MD
2000 Richard Jones Road, Suite 220
Nashville, TN 37215 (USA)
Tel. +1 615 383 2400, E-Mail goldskin@goldskincare.com

Bogdan Allemann I, Goldberg DJ (eds): Basics in Dermatological Laser Applications.
Curr Probl Dermatol. Basel, Karger, 2011, vol 42, pp 181–192

Photodynamic Therapy

Michael H. Gold

Tennessee Clinical Research Center, Gold Skin Care Center, and Department of Dermatology,
Vanderbilt University School of Nursing, Vanderbilt University School of Medicine, Nashville, Tenn., USA

Abstract

Photodynamic therapy (PDT) utilizes a photosensitizer, a light source, and molecular oxygen to selectively destroy cells. In dermatology, the photosensitizers accumulate in non-melanoma skin cancer cells, actinically damaged skin cells and actinic keratoses (AK), and in the pilosebaceous units. This has forged great interest and clinical research towards utilizing PDT for the treatment of non-melanoma skin cancers, AK and associated photorejuvenation, and inflammatory acne vulgaris. Here, the evidence-based medicine supporting the use of PDT in the USA will be presented, and how we utilize PDT in our clinic will also be discussed. It is hoped that with this knowledge, more and more clinicians will feel comfortable using this therapy, thereby benefitting their patients to an even greater extent.

Photodynamic therapy (PDT) has found its place in dermatology with more and more clinicians utilizing this form of therapy to treat a myriad of clinical conditions routinely encountered in dermatology practices around the world. Dermatologic conditions treated with PDT are shown in table 1.

PDT, in its simplest form, is a therapy which utilizes a photosensitizer and a light source and – in the presence of molecular oxygen – selectively destroys cells. Concerning dermatologic uses for PDT, evidence has shown that the currently available photosensitizers (described below) are selectively absorbed into non-melanoma skin cancer cells, actinically damaged skin cells and actinic keratoses (AK), and the pilosebaceous units.

Photosensitizers Available in the USA

The most commonly utilized photosensitizer in dermatology for PDT is called 5-aminolevulinic acid (ALA). ALA is known as a prodrug. Kennedy et al. [1] first described this photosensitizer in 1990, in particular its ability to penetrate the stratum corneum and to be converted from its prodrug form into its active form, protoporphyrin IX (PpIX). PpIX, a natural part of the heme pathway in the body, has its own unique absorption spectrum for light, and this is shown in ure 1. This figure demonstrates several peaks important for light to be successful in activating PpIX, including blue light (Soret band) as well as several smaller bands in which a variety of lasers and light sources can be used to activate PpIX, including red light. The heme biosynthetic pathway is under a closed feedback loop, not allowing the build-up of exogenous PDT, and thus minimizing the potential

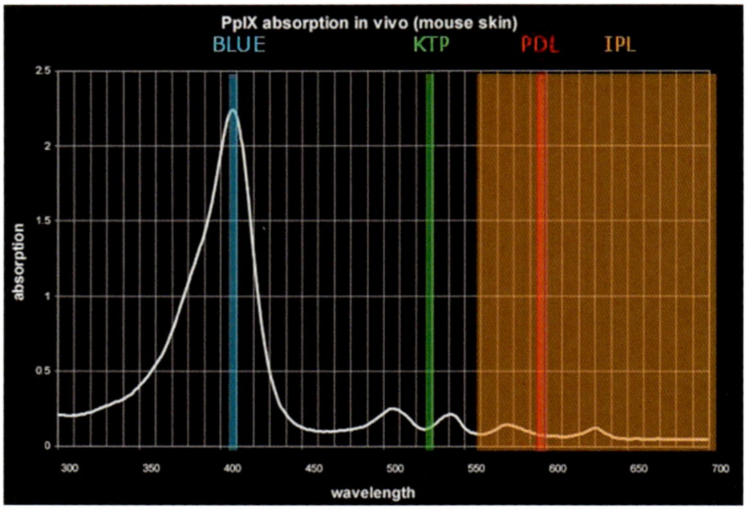

Fig. 1. PpIX absorption spectrum. Reprinted with permission from G Ital Dermatol Venereol 2010;145:525–541.

Table 1. Dermatologic uses of 5-aminolevulinic acid PDT

Actinic keratoses
Photodamage and associated actinic keratoses[1]
Bowen's disease
Superficial basal cell carcinoma
Superficial squamous cell carcinoma
Cutaneous T cell lymphoma
Kaposi's sarcoma
Malignant melanoma
Actinic cheilitis
Keratoacanthoma
Psoriasis vulgaris
Human papillomavirus
Molluscum contagiosum
Alopecia areata
Hirsutism
Acne vulgaris[1]
Sebaceous gland hyperplasia[1]
Hidradenitis suppurativa[1]

[1] Common indications for 5-aminolevulinic acid PDT in the USA.

for phototoxicity, the main adverse event noted in PDT treatments [2].

Two photosensitizers are currently available for PDT in the USA. The first one is a 20% 5-ALA solution, commonly known as Levulan® Kerastick® (DUSA Pharmaceuticals, Wilmington, Mass., USA). The second photosensitizer, commonly known as Metvixia™ (Metvix® in Europe), has been available in the USA since 2009 (Galderma, Fort Worth, Tex., USA, and PhotoCure ASA, Oslo, Norway). Metvix has been available in Europe and elsewhere for the past 10 years. These photosensitizers have each undergone a great deal of clinical evaluation, which has shown their usefulness in the treatment of AK, photorejuvenation, skin cancers (for Metvix), and inflammatory acne vulgaris. The remainder of this chapter will focus on the evidence-based research investigating these photosensitizers.

Levulan Kerastick

The first photosensitizer to be described is the 20% 5-ALA solution, known commercially as Levulan Kerastick (fig. 2). The only approved FDA indication for this medicine is in the treatment of non-hyperkeratotic AK of the face and scalp following

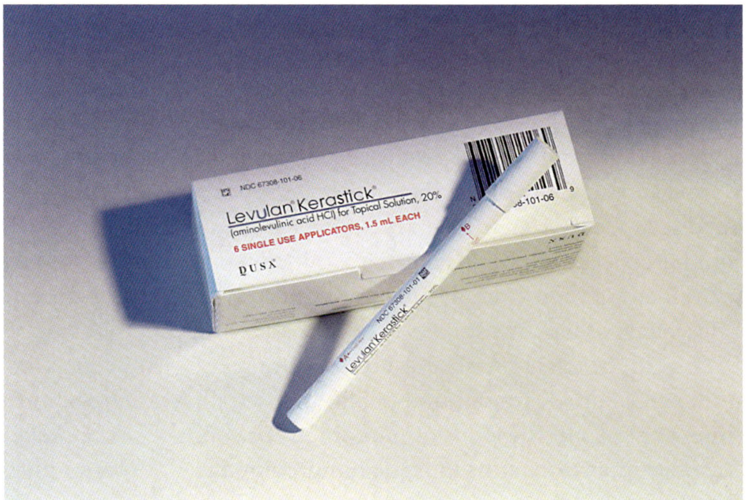

Fig. 2. Levulan Kerastick (DUSA Pharmaceuticals).

a 14- to 18-hour drug incubation time period. Drug activation is then commenced with a blue light source for a treatment time of 16 min 40 s. The most common of the blue light devices available in the USA is known as the Blu-U® (DUSA Pharmaceuticals) and is shown in figure 3. All other uses for Levulan are considered off-label in terms of FDA compliance. The FDA clinical trials showed that in phase II, 39 patients were treated by applying Levulan to individual AK on the face and scalp. After the 14- to 18-hour drug incubation period, these individuals had blue light therapy, as described above. Eight weeks following these treatments, patients showed a 66% resolution of AK [3]. Those lesions which did not clear received another treatment, and 8 weeks following the therapy, 85% of the treated AK had complete resolution. A phase III clinical trial followed – a large multicenter placebo-controlled trial – in which 243 patients were enrolled and treated, as noted above. A 70% complete clearance rate of AK was noted at 12 weeks; those AK not cleared were treated again and at 24 weeks; 88% of the individuals in the clinical trial had a greater than or equal to 75% clearance of their AK as compared to 20%

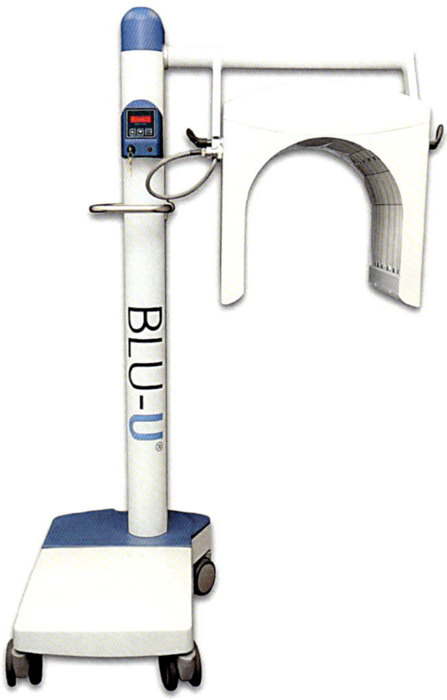

Fig. 3. Blu-U (DUSA Pharmaceuticals).

in the placebo arm. Adverse events in both of the clinical trials were noted. These included pain associated with the therapy, erythema, and crusting, which lasted upwards of 1 week. As a secondary endpoint in the phase III clinical trials, patients were asked to comment on their cosmetic appearance as a result of their PDT treatments. Ninety-four percent of the patients noted a significant improvement in their overall skin appearance [4].

Several other investigators began studying PDT with 20% 5-ALA. These investigations not only looked at AK resolution but also focused on improvements in the signs of photorejuvenation. In addition, investigators looked at improving the therapy protocols in general, as it is difficult in the 'real world' to bring patients into the office to have an application of a medicine and then to have them return some 14–18 h later for a second visit and treatment. The clinical trials have been reported elsewhere and can be summarized by noting that utilizing a 1-, 2-, or 3-hour drug incubation time was as effective in treating AK as the 14- to 18-hour drug incubation time period. Furthermore, this shorter drug incubation, when applied to the full face, also resulted in improvements in the signs of facial photodamage. These clinical trials also allowed us to identify that in addition to blue light, other light sources including intense pulsed light (IPL) and pulsed dye laser (PDL) were equally as effective in treatments for AK and photorejuvenation utilizing the 20% 5-ALA solution [5–10]. Next, several important split-face clinical trials were performed looking primarily at IPL and PDL devices so as to verify the work of the previous investigations. Five US-based split-face trials [11–15] all demonstrated that utilizing ALA-IPL or ALA-PDL was more effective than using IPL or PDL alone in treating AK and the signs of photodamage. In fact, one study by Marmur et al. [13] demonstrated ultrastructually that when using ALA-IPL, there is a greater amount of type I and III collagen produced as compared to IPL alone, thus confirming some of the clinical findings of improved skin texture in many of the treated patients. By using this new short-contact full-face therapy, most of the associated pain with the therapy as well as the downtime associated with the treatments has been minimized. Recent work by Orringer et al. [16] confirmed these findings.

The currently accepted way to apply and utilize Levulan will be described at the end of this chapter.

Metvixia and Metvix

Metvixia in the USA took on a different path to FDA approval. The European product, known as Metvix, is the methylated form of 20% 5-ALA. The FDA insisted on a different name; hence, Metvix in Europe, Metvixia in the USA. It is more lipophilic than the 20% 5-ALA solution, so as to increase the penetration depth of the cream product as compared to the solution. Early work in Europe focused primarily on non-melanoma skin cancers, AK, and Bowen's disease. Numerous clinical trials showed the effectiveness of Metvix as compared to traditional treatments, including cryotherapy and surgery. Complete clinical cure rates showed consistently high results for the treatment of non-melanoma skin cancers, AK, and Bowen's disease. These clinical studies have been published in other manuals and can be referenced as needed [17, 18]. In the USA, Metvixia has FDA approval for the treatment of non-hyperkeratotic AK of the face and scalp, utilizing the 20% 5-aminolevulinic acid methyl ester cream applied to the lesions after removing the keratotic material with a gentle curettage. The cream is occluded with an occlusive dressing for 3 h and then incubated with a red light source at 630 nm. Metvix is shown in figure 4 and the most common red light source, known as Aktilite® (PhotoCure ASA and Galderma) is shown in figure 5. The clinical results from two parallel phase III clinical trials showed the effectiveness of Metvix in the treatment of AK. Clearing of AK in these clinical trials was shown to be 83–86% at 12 weeks after the last treatment. Pain associated with the treatments was also noted and there was an

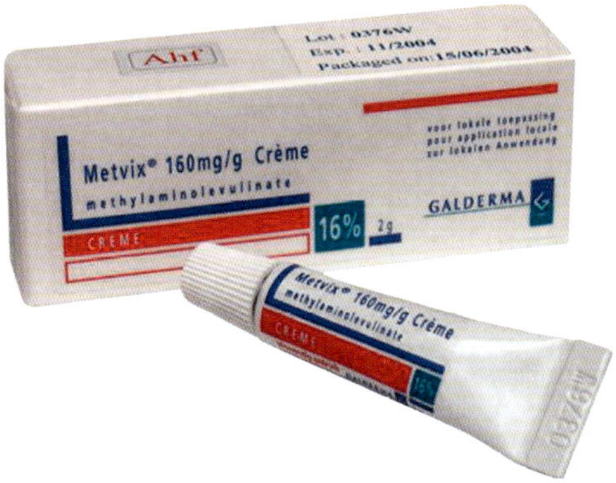

Fig. 4. Metvix (PhotoCure ASA and Galderma).

associated downtime, similar to the 20% 5-ALA solution clinical trials [19, 20]. It is interesting to note that Metvixia is not FDA approved for the treatment of non-melanoma skin cancers. Several very important clinical guidelines for Metvix therapy have been published in the literature [21, 22].

Further investigations with Metvix have shown its usefulness in the treatment of photorejuvenation. Several clinical studies have looked at the signs of photodamage when utilizing Metvix therapy and noted that improvements can be made in the majority of the parameters of photodamage with Metvix and red light. Further clinical investigations have also explored the possibilities of using different drug incubation times, whether occlusion is necessary or not, and how to make the overall experience more palatable for all of our patients [23–25]. How I personally use Metvixia in clinical practice will be summarized shortly.

Application in Inflammatory Acne Vulgaris

Both Levulan Kerastick and Metvixia can be used in the treatment of inflammatory acne vulgaris, as the photosensitizers have an affinity for the pilosebaceous units in the skin. Many clinical investigations with Levulan have shown its effectiveness in

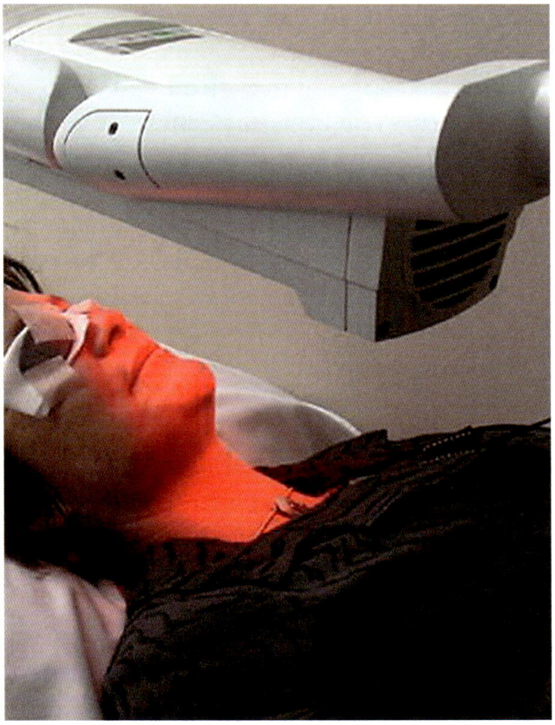

Fig. 5. Red light source: Aktilite® (PhotoCure ASA).

treating acne vulgaris with a variety of lasers and light sources; these include the blue light, the IPL, the PDL, and the KTP laser systems [26–32]. Three published split-face clinical studies utilizing ALA-IPL versus IPL alone demonstrated superiority of the ALA-IPL-treated over the IPL-treated side [33–35] and one split-face PDL-ALA study showed the superiority of PDL-ALA over PDL alone [36]. Unfortunately, two phase II clinical evaluations utilizing ALA and blue light failed to show statistical significance between the combination versus blue light alone, which already has FDA clearance for the treatment of acne vulgaris. Most firmly believe there is a benefit to using 20% 5-ALA in the treatment of moderate-to-severe inflammatory acne vulgaris, with many different light sources; however, these benefits are currently not amenable to US insurance company reimbursements at this time. For those of us who utilize this therapy for our severe acne patients, once again, short-contact full-face therapy seems to work the best.

Metvix has also been studied for its effectiveness in the treatment of inflammatory acne vulgaris. In the original work, significant clearing was noted but unfortunately with the drug occlusion, pustulation (noted 24 h after the therapy) was common. Since then, various investigators have evaluated different protocols using Metvix with clearing noted in the case reports in the literature [37, 38]. It is interesting to note that phase II clinical investigations with a lower percent formulation of methyl-ALA also failed to show significance when compared to red light therapy alone; thus, making it unclear whether further methyl-ALA clinical trials for acne will be conducted in the future.

Key Points for Utilization in Clinical Practice

So what is the best way to utilize these medicines? There are no true 'absolute' standards that state empirically that one way or another is the best way or the only way clinicians should approach the utilization of PDT in the USA. Each clinician will develop their own procedures which will optimize its use in their office. A US consensus paper [39] outlines steps which one may follow when introducing ALA-PDT into the clinical setting. The following description shows how my own clinic utilizes ALA-PDT, and is similar to how I have used ALA-PDT for many years now.

The first discussion is a thorough review of the risks and benefits of the procedure. You must explain every detail and thoroughly document this has been completed. A signed written informed consent will verify that you have reviewed the necessary information and potential sequelae with the patient. Remember that PDT utilized for photorejuvenation is an off-label use of PDT according to the FDA, and this too must be explained. The use of PDT for AK is an approved use for the use of PDT.

The face will be used as the example for the procedure. It must be thoroughly cleansed with a mild cleanser. A degreasing procedure is needed to enhance the skin's absorption of the photosensitizer. The two main degreasing procedures for PDT are microdermabrasion or an acetone scrub. We use an acetone scrub as it is a less expensive procedure, but there is growing evidence that microdermabrasion enhances the penetration of ALA [40]. If a microdermabrasion is to be done, there are options here including standard microdermabrasion with crystals, crystal-less microdermabrasion, paddle-enhanced microdermabrasion, and microdermabrasion with iontophoresis to enhance the photosensitizers' penetration into the skin. We utilize crystal-less microdermabrasion systems (the paddle method) or the iontophoresis system when we choose to use microdermabrasion in our patient population. Recently, the use of a supersonic jet technology has been shown to be useful for this purpose and is gaining favor amongst many clinicians.

Next, if using the 20% 5-ALA solution, the Levulan® Kerastick™ is mixed. The Levulan Kerastick is a 20% weight/volume ALA solution with 48% alcohol. It has a special roll-on dermatologic applicator at one of the ends to allow easy and accurate application of the medicine to the area(s) being treated. This applicator tip is applied to a flexible glass tubing which contains two glass vials. One of the vials contains the ALA in a powder form; the other contains the ethanol. Light manual pressure upon the glass vials in the Kerastick will break the vials allowing for mixture of the two by gentle rotation in a back and forth direction. Three minutes of mixing is recommended prior to the application of the 20% 5-ALA.

Once the Kerastick is prepared, it will be 'painted' onto the skin in a uniform manner covering the forehead, cheeks, chin, and nose. If there are AK present, a second coat of medicine is routinely applied. The drug is then allowed to incubate on the skin. In our clinic, we incubate the patient for 1 h for photorejuvenation and AK, 30–45 min for inflammatory acne vulgaris, and 1 h for sebaceous gland hyperplasia. For each successive treatment, drug incubation is increased by approximately 15–30 min.

The next step is the drug activation by a light source. If an IPL light source is going to be used, it is recommended that the skin be once again thoroughly cleansed prior to the use of the light source. One will find that the coupling medium, or gel, will not remain on the skin surface after the application of the ALA. If a blue light source or PDL is to be used, most would still recommend that the skin be cleansed once again, although it is not mandatory. All of our treatments are carried out with the aid of cold air cooling, which the patient holds and uses as needed. The treatment settings for a variety of devices are shown in table 2.

After the therapy is completed, any remaining ALA needs to be thoroughly removed from the skin. Many clinicians are finding that if they perform an IPL or PDL treatment, a subsequent short 5-min blue light 'quenching' will, for all practical purposes, remove any excess ALA. It may also serve to give some additional benefit to the patient, although clinical trials with regard to this concept have not been performed. Next, we employ the use of ice onto the treated areas which helps with any patient discomfort or skin burning. During this time, we begin discussions of proper skin care and sun protection which are crucial for the patient to follow to avoid phototoxicity. Phototoxicity can be minimized or even eliminated by following several very simple rules. The patient must remain out of the sun for the first 24–48 h following the procedure. Sunscreen use, with a minimum SPF of 30, is required, and in our office this is applied prior to the patient leaving our office. There are many new creams and lotions which are being evaluated to help reduce any erythema which may be associated with lasers and light sources without the use of ALA; these are also being investigated for their potential with ALA as well. These include Neocutis® Bio-Restorative Skin Cream [41], Biafine® [42], and Avene® Gel Deau [43]. Patients may note some desquamation of the skin after several days, but proper use of moisturizers or 'spray' waters will minimize this effect.

Patients must also have a thorough understanding of how many treatments with ALA-PDT they will require. The answer to this question remains elusive, and further clinical studies will help provide guidelines for all of us utilizing PDT in our offices. However, clinical trials always do not correlate into real world experiences, so it is not possible to give a precise answer when asked how many treatments one will require to achieve a certain result. Each patient must be treated as an individual and clinical trial results should be used as guidelines for a given patient's needs. Also, skin care is a must with patients who undergo an ALA-PDT procedure, and maintenance therapy will be required, although no clinical trials have demonstrated when or how many.

Clinical examples of the use of Levulan® PDT for AK are shown in figures 6 and 7.

Metvixia is also used in our clinic on a regular basis – 'off-label' more often than not. We apply it to the entire area being treated and use a drug incubation time of between 30 min and 1 h routinely for our patients treated for AK. We do not occlude the methyl-ALA and we utilize the red light source as previously indicated. Clinical examples of methyl-ALA-PDT are shown in figures 8 and 9. Post-operative care and aftercare for Metvixia are the same as for Levulan.

Conclusions

PDT remains a good therapy and is growing every year as more and more clinicians are utilizing this therapy in their daily dermatologic practices. As we continue to evolve further in PDT research, perhaps more photosensitizers will be developed and even newer and better light sources to deliver the product to our patients. The future for PDT remains bright.

Table 2. Parameters for lasers/light sources useful in PDT

Lumenis One[1] (Lumenis Ltd., Yokneam, Israel)	– 560 nm filter – double-pulse (4.0/4.0 ms with 20-ms delay) – single pass with no overlap – 15–25 J/cm^2
IPL Quantum SR[1] (Lumenis Ltd., Yokneam, Israel)	– 560 nm filter – double-pulse (2.4/4.0 ms with 10-ms delay) – single pass with no overlap – 25–35 J/cm^2
EsteLux Pulsed Light System[1] (Palomar Medical Technologies, Burlington, Mass., USA)	– 20 ms – 19–30 J/cm^2 – single pass with no overlap
Cynergy[1] (Cynosure Inc., Chelmsford, Mass., USA)	– 585 or 595 nm wavelength PDL – 10-mm spot size – 40-ms pulse width – 7.5 J/cm^2 – 2 passes with 50% overlap
V Beam Perfecta (Candela Corp., Wayland, Mass., USA)	– 595 nm PDL – 10-mm spot size – 6-ms pulse width – 7.5 J/cm^2 – 2 passes with 50% overlap
ClearLight[1] (Lumenis Ltd., Yokneam, Israel)	– 405 to 420 nm blue light – 8–10 min under light
BluU[1] (Dusa Pharmaceuticals, Wilmington, Mass., USA)	– 417 nm (+/– 5 nm) blue light – 8–15 min under light
SkinStation[1] (Radiancy, Orangeburg, N.Y., USA)	– 500 to 1,200 nm pulsed light – 2 passes – 45 J/cm^2
eMax (Syneron Medical Ltd., Yokneam, Israel)	– 580–980 nm – optical energy of 16–22 J/cm^2; RF energy of 16-22 J/cm^2 – single pass
Photosilk (DEKA Lasers, Carlsbad, Calif., USA)	– 550 nm filter – 8 J/cm^2 – 15–20 ms – single pass
Sciton BBL[1] (Sciton Inc., Palo Alto, Calif., USA)	– 560 nm broadband light filter – 14 J/cm^2 – 12 ms – single pass

[1] Devices used by author (other settings from recognized experts in the field).

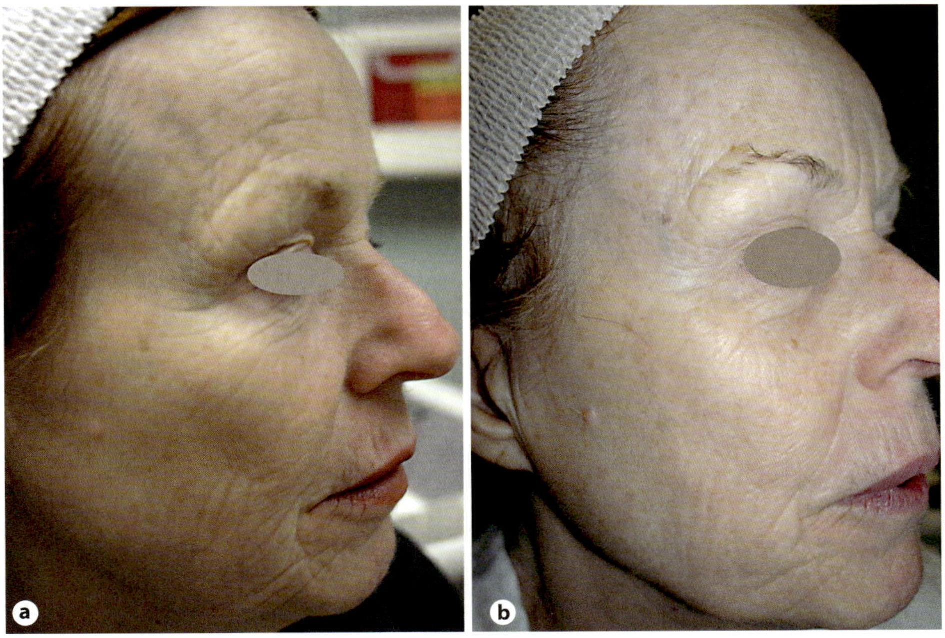

Fig. 6. Levulan PDT for AK. Female patient presented with AK before treatment (**a**) and after two ALA-PDT treatments (**b**). Reprinted with permission from G Ital Dermatol Venereol 2010;145:525–541.

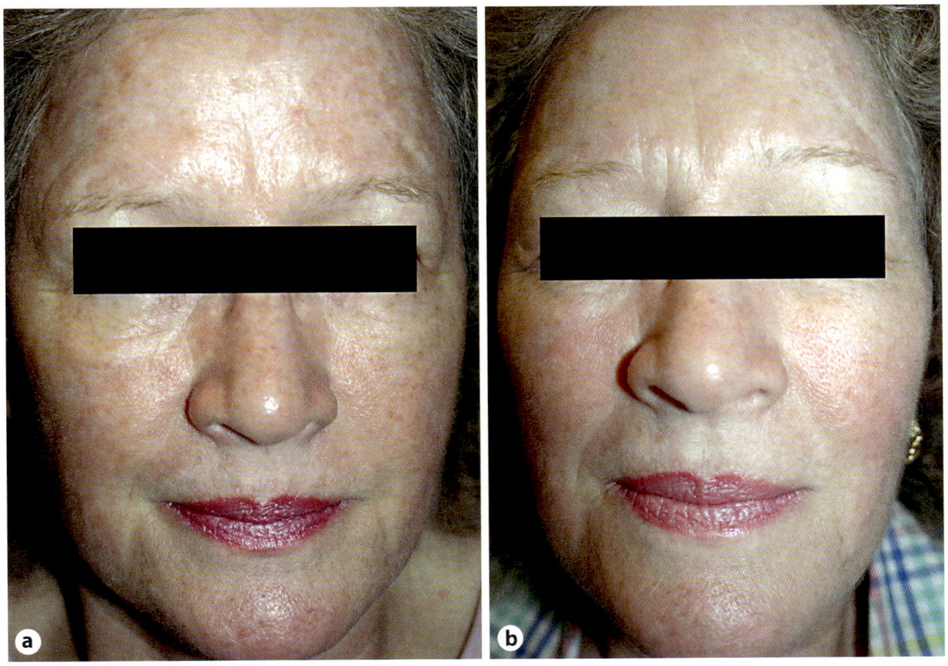

Fig. 7. Levulan PDT for AK. Female patient presented with AK before treatment (**a**) and after ALA-PDT treatment (**b**). Reprinted with permission from G Ital Dermatol Venereol 2010;145:525–541 .

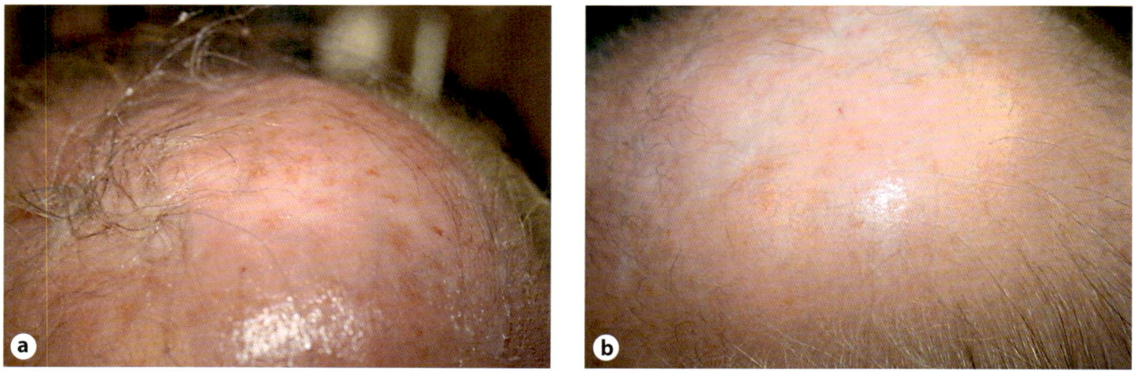

Fig. 8. MAL-PDT for AK. Male patient presented with AK before treatment (**a**) and after two MAL-PDT treatments (**b**). Reprinted with permission from G Ital Dermatol Venereol 2010;145:525–541.

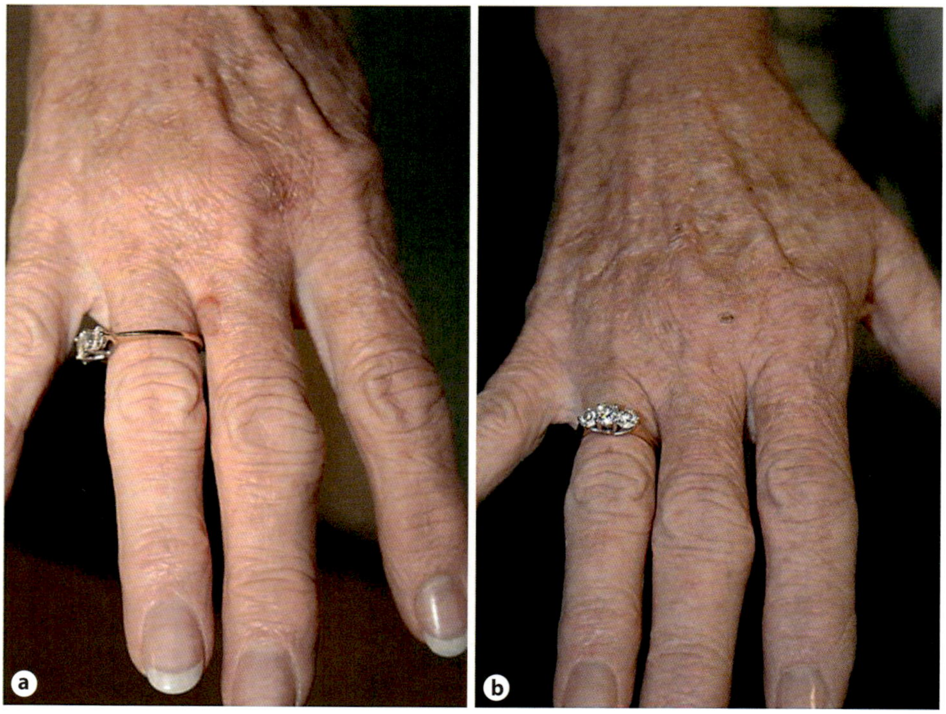

Fig. 9. MAL-PDT for AK. Female patient presented with AK before treatment (**a**) and after two MAL-PDT treatments (**b**). Reprinted with permission from G Ital Dermatol Venereol 2010;145:525–541.

References

1 Kennedy JC, Pottier RH, Pross DC: Photodynamic therapy with endogenous protoporphyrin IX: basic principles and present clinical experiences. J Photochem Photobiol B 1990;6:143–148.

2 Gold MH, Goldman MP: 5-Aminolevulinic acid photodynamic therapy: where we have been and where we are going. Dermatol Surg 2004;30:1077–1084.

3 Jeffes EW, McCullough JL, Weinstein GD, Kaplan R, Glazer SD, Taylor JR: Photodynamic therapy of actinic keratoses with topical aminolevulinic acid hydrochloride and fluorescent blue light. J Am Acad Dermatol 2001;45:96–104.

4 Jeffes EW: Levulan: the first approved topical photosensitizer for the treatment of actinic keratosis. J Dermatolog Treat 2002;13:S19–S23.

5 Touma D, Yaar M, Whitehead S, Konnikov N, Gilchrest BA: A trial of short incubation, broad-area photodynamic therapy for facial actinic keratoses and diffuse photodamage. Arch Dermatol 2004;140:33–40.

6 Ruiz-Rodriguez R, Sanz-Sanchez T, Cordoba S: Photodynamic photorejuvenation. Dermatol Surg 2002;28:742–744.

7 Gold MH, Bradshaw VL, Boring MM, Bridges TM, Biron JA, Lewis TL: Treatment of sebaceous gland hyperplasia by photodynamic therapy with 5-aminolevulinic acid and a blue light source or intense pulsed light source. J Drugs Dermatol 2004;3:S6–S9.

8 Goldman MP, Atkin D, Kincad S: PDT/ALA in the treatment of actinic damage: real world experience. J Lasers Surg Med 2002;14(suppl):24.

9 Avram DK, Goldman MP: Effectiveness and safety of ALA-IPL in treating actinic keratoses and photodamage. J Drugs Dermatol 2004;3(suppl): S36–S39.

10 Alexiades-Armenakas MR, Geronemus RG: Laser-mediated photodynamic therapy of actinic chelitis. J Drugs Dermatol 2004;3:548–551.

11 Alster Ts, Tanzi EL, Welch EC: Photorejuvenation of facial skin with topical 20% 5-aminolevulinic acid and intense pulsed light treatment: a split-face comparison study. J Drugs Dermatol 2005;4:35–38.

12 Key DJ: Aminolevulinic acid-pulsed dye laser photodynamic therapy for the treatment of photoaging. Cosmet Derm 2005;18:31–36.

13 Marmur ES, Phelps R, Goldberg DJ: Ultrastructural changes seen after ALA-IPL photorejuvenation: a pilot study. J Cosmet Laser Ther 2005;7:21–24.

14 Dover JS, Bhatia AC, Stewart B, Arndt KA: Topical 5-aminolevulinic acid combined with intense pulsed light in the treatment of photoaging. Arch Dermatol 2005;141:1247–1252.

15 Gold MH, Bradshaw VL, Boring MM, Bridges TM, Biron JA: Split-face comparison of photodynamic therapy with 5-aminolevulinic acid and intense pulsed light versus intense pulsed light alone for photodamage. Dermatol Surg 2006;32:795–801.

16 Orringer JS, Sachs DL, Bailey E, Kang S, Hamilton T, Voorhees JJ: Photodynamic therapy for acne vulgaris: a randomized, controlled, split-face clinical trial of topical aminolevulinic acid and pulsed dye laser therapy. J Cosmet Dermatol 2010;9:28–34.

17 Gold MH: Aminolevulinic acid photodynamic therapy for actinic keratoses and photorejuvenation. Expert Rev Dermatol 2007;2:391–402.

18 Gold MH: Aminolevulinic acid photodynamic therapy for cosmetic uses on the skin. G Ital Dermatol Venereol 2007;142:311.

19 Pariser D, Loss R, Jarratt M, Abramovits W, Spencer J, Geronemus R, Bailin P, Bruce S: Topical methyl-aminolevulinate photodynamic therapy using red light-emitting diode light for treatment of multiple actinic keratoses: a randomized, double-blind, placebo-controlled study. J Am Acad Dermatol 2008;59:569–576.

20 Szeimies RM, Matheson RT, Davis SA, Bhatia AC, et al: Topical methyl aminolevulinate photodynamic therapy using red light-emitting diode light for multiple actinic keratoses: a randomized study. Dermatol Surg 2009;35:586–592.

21 Gollnick H, Cunliffe W, Berson D, Dreno B, Finlay A, Leyden JJ, Shalita AR, Thiboutot D, Global Alliance to Improve Outcomes in Acne: Management of acne: a report from a Global Alliance to Improve Outcomes in Acne. J Am Acad Dermatol 2003;49:S1–S37.

22 Morton C, Campbell S, Gupta G, et al: Intraindividual, right-left comparison of topical methyl aminolaevulinate-photodynamic therapy and cryotherapy in subjects with actinic keratoses: a multicentre, randomized controlled study. Br J Dermatol 2006;155:1029–1036.

23 Zane C, Capezzera R, Sala R, Venturini M, Calzavara-Pinton P: Clinical and echographic analysis of photodynamic therapy using methylaminolevulinate as sensitizer in the treatment of photodamaged facial skin. Lasers Surg Med 2007;39:203–209.

24 Ruiz-Rodríguez R, López L, Candelas D, Pedraz J: Photorejuvenation using topical 5-methyl aminolevulinate and red light. J Drugs Dermatol 2008;7:633–637.

25 Issa MC, Piñeiro-Maceira J, Vieira MT, Olej B, Mandarim-de-Lacerda CA, Luiz RR, Manela-Azulay M: Photorejuvenation with topical methyl aminolevulinate and red light: a randomized, prospective, clinical, histopathologic, and morphometric study. Dermatol Surg 2010;36:39–48.

26 Goldman MP, Boyce S: A single-center study of aminolevulinic acid and 417 nm photodynamic therapy in the treatment of moderate to severe acne vulgaris. J Drugs Dermatol 2003;2:393–396.

27 Gold MH: The use of a novel intense pulsed light and heat source and ALA-PDT in the treatment of moderate to severe inflammatory acne vulgaris. J Drugs Dermatol 2004:3:S14–S18.

28 Gold MH, Rao J, Goldman MP, Bridges TM, Bradshaw VL, Boring MM, Guider AN: A multicenter clinical evaluation of the treatment of mild to moderate inflammatory acne vulgaris of the face with visible blue light in comparison to topical 1% clindamycin antibiotic solution. J Drugs Dermatol 2005;4:64–70.

29 Gold MH: The use of a novel intense pulsed light and heat source and ALA-PDT in the treatment of moderate to severe inflammatory acne vulgaris. J Drugs Dermatol 2004:3:S14–S18.

30 Taub AF: PDT for the treatment of acne: a pilot study. J Drugs Dermatol 2004;3:S10–S14.

31 Alexiades-Armenakas M: Long-pulsed dye laser-mediated photodynamic therapy combined with topical therapy for mild to severe comedonal, inflammatory, or cystic acne. J Drugs Dermatol 2006;5:45–55.
32 Miller A, Van Camp A: Treatment of cane vulgaris with photodynamic therapy: the use of aminolevulinic acid and green light. Cosm Derm 2006;10:624–627.
33 Santos AV, Belo VG, Santos G: Effectiveness of photodynamic therapy with topical 5-aminolevulinic acid and intense pulsed light versus intense pulsed light along in the treatment of acne vulgaris: comparative study. Dermatol Surg 2005;31(part 1):910–915.
34 Rojanamatin J, Choawawanich P: Treatment of inflammatory facial acne vulgaris with intense pulsed light and short contact of topical 5-aminolevulinic acid: a pilot study. Dermatol Surg 2006;32:991–996, discussion 996–997.
35 Oh SH, Ryu DJ, Han EC, Lee KH, Lee JH: A comparative study of topical 5-aminolevulinic acid incubation times in photodynamic therapy with intense pulsed light for the treatment of inflammatory acne. Dermatol Surg 2009;35:1918–1926.
36 Orringer JS, Kang S, Hamilton T, Schumacher W, Cho S, Hammerberg C, Fisher GJ, Karimipour DJ, Johnson TM, Voorhees JJ: Treatment of acne vulgaris with a pulsed dye laser: a randomized controlled trial. J Am Med Assoc 2004;291:2834–2839.
37 Wiegall SR, Wulf HC: Photodynamic therapy of acne vulgaris using methyl aminolevulinate: a blinded, randomized, controlled trial. Br J Dermatol 2006;154:969–976.
38 Horfelt C, Funk J, Frohm-Nilsson M: Topical methyl aminolevulinate photodynamic therapy for treatment of facial acne vulgaris: results of a randomized, controlled study. Br J Dermatol 2006;155:608–613.
39 Nestor MS, Gold MH, Kauvar AN, Taub AF, Geronemus RG, Ritvo EC, Goldman MP, Gilbert DJ, Richey DF, Alster TS, Anderson RR, Bank DE, Carruthers A, Carruthers J, Goldberg DJ, Hanke CW, Lowe NJ, Pariser DM, Rigel DS, Robins P, Spencer JM, Zelickson BD: The use of photodynamic therapy in dermatology: results of a consensus conference. J Drugs Dermatol 2006;5:140–154.
40 Fang JY, Lee WR, Shen SC, Fang YP, Hu CH: Enhancement of topical 5-aminolaevulinic acid delivery by erbium:YAG laser and microdermabrasion: a comparison with iontophoresis and electroporation. Br J Dermatol 2004;151:132–140.
41 Gold MH, Biron JA: A novel skin cream containing a mixture of human growth factors and cytokines for the treatment of adverse events associated with photodynamic therapy. J Drugs Dermatol 2006;5:796–798.
42 Gold MH: Single-center, split-face, double-blind, randomized clinical trial of the efficacy of a trolamine-containing topical emulsion and a topical petrolatum-based product in reducing redness and peeling after treatment with a MicroLaser peel. Poster, Am Acad Dermatol Ann Meet, San Francisco, 2006.
43 Ribet V, Gral N, Emmenegge V, Rostain G, Gold MH, Goldberg DJ, Gerwen DV, Adatto M, Drosner M: Calming efficacy and tolerance of Avene Spring Water Gel after epilatory laser treatment combined to healing cream in adults. Poster, Am Acad Dermatol Ann Meet, San Francisco, 2006.

Michael H. Gold, MD
2000 Richard Jones Road, Suite 220
Nashville, TN 37215 (USA)
Tel. +1 615 383 2400, E-Mail goldskin@goldskincare.com

Author Index

Subject Index